The Complete Baofeng & GMRS Radio Field Manual

Beginner Guide to Ham Radio, GMRS Licensing, CHIRP Programming, Repeaters, Privacy Tones, and Off-Grid Emergency Comms

Baofeng Radios - A Bootcamp for Beginners

The Baofeng Radio Revolution

GMRS Radio Field Manual

The Baofeng Radio Family Lifeline

MorseCode Publishing

For permissions, bulk sales, or special inquiries, contact:
Jared@MorseCodePublishing.com

Contents

Introduction: Four Guides. One Field Manual. A Radio You Can Actually Use.

You probably did not buy a Baofeng or GMRS radio because you wanted another complicated gadget sitting in a drawer.

You bought it because some part of you wanted to be ready.

Maybe you wanted a backup plan for storms, blackouts, road trips, camping, hunting, overlanding, or a family emergency. Maybe you wanted to get into ham radio but felt overwhelmed by the language.

Maybe you bought a Baofeng UV-5R because everyone online said it was cheap, powerful, and useful, then you turned it on and realized the radio was beeping, flashing numbers, and speaking in menu codes that made absolutely no sense.

If that sounds familiar, you are in the right place.

This book is not written for electrical engineers. It is not written for the guy at the radio club who has been building antennas since 1978 and thinks every beginner should already know what an offset is. And it is definitely not written to impress other experts with complicated theory.

This book is for the person who looks at a handheld radio and thinks, "I just want this thing to work."

That is the whole point of *The Complete Baofeng & GMRS Radio Field Manual*. This is a practical, beginner-first, four-books-in-one guide designed to move you from confused radio owner to confident operator. Instead of sending you down a dozen YouTube rabbit holes, forum threads, product manuals, and half-finished checklists, this field manual puts the entire learning path in one place.

Inside this complete edition, you are getting four connected guides working together as one system.

First, you will learn how to take a Baofeng from the box to real use. We will walk through the buttons, the screen, VFO mode, Memory mode, squelch, volume, weather channels, repeaters, and the basic settings that cause most beginners to quit.

Second, you will expand that foundation into real-world Baofeng operating confidence. You will learn how to use your radio for simplex communication, repeaters, scanning, accessories, range improvement, radio etiquette, troubleshooting, and practical preparedness scenarios.

Third, you will move into GMRS and learn how it fits into the bigger radio world. We will cover GMRS licensing, family communication, privacy tones, repeaters, plain-language radio use, and how to build a simple radio plan that normal people in your household can actually follow.

Fourth, you will bring everything together into a practical family emergency communication system. That means CHIRP programming, channel planning, repeater offsets, household protocols, off-grid check-ins, go-bag setup, battery planning, weekly drills, and the simple scripts that help you speak clearly when your adrenaline is high.

Because here is the truth: owning a radio is not the same thing as having communication.

A Baofeng in the glovebox is not a plan. A GMRS handheld in a kitchen drawer is not a network. A programming cable still sitting in its plastic bag is not preparedness. And a radio that only one person in the family understands is not a family lifeline.

The difference between gear and readiness is training.

And that training does not have to be complicated.

You do not need to understand every circuit inside your radio. You do not need to become a full-time radio hobbyist. You do not need to speak in cryptic codes or memorize every possible frequency. What you need is a clear path: what to press, what to program, what to listen for, what to say, what not to touch, and what to do when something stops working.

That is what this field manual is built to give you.

We will start where beginners actually start: confusion. The beeps, the two lines on the screen, the random numbers, the static, the silence, the fear of pressing the wrong button, and the nagging worry that you might accidentally transmit somewhere you should not. Then we will slow everything down and turn the radio into something understandable.

You will learn how to listen before you transmit. You will learn why a repeater can sound loud and clear but still not hear you. You will learn why privacy tones are not actually private. You will learn why range claims on the box are often wildly optimistic. You will learn why the wrong squelch setting can make a perfectly good radio seem broken. You will learn how to troubleshoot the most common problems without panicking or factory-resetting everything in frustration.

And just as important, you will learn how to bring other people into the plan.

That matters because emergency communication is rarely just about one person. It is about your spouse getting home safely when cell towers are overloaded. It is about your teenager knowing which channel to use if phones fail. It is about a road trip convoy staying together in a dead zone. It is about checking in during a storm, getting weather information when the power is out, or coordinating with neighbors when normal communication gets fragile.

That is why this book keeps coming back to simple systems: a short channel plan, a laminated quick card, a repeatable check-in schedule, clear roles, plain-English scripts, and drills that take ten minutes instead of an entire Saturday.

There is also an important legal reality we will be clear about from the beginning. Radios are powerful tools, and different radio services have different rules. Ham radio, GMRS, FRS, MURS, and emergency use are not all the same thing. Licensing matters. Equipment rules matter. Responsible operation matters. This book will help you understand those differences in plain English so you can listen, learn, practice, and operate responsibly.

But this book is not here to scare you away from using your radio.

It is here to help you finally use it with confidence.

By the end, your radio should no longer feel like a mysterious plastic brick. It should feel like a tool you understand. You should know how to turn it on, find useful signals, program channels, use repeaters, understand tones, build a family communication plan, troubleshoot common problems, and prepare a simple kit that is ready before you need it.

The goal is not perfection.

The goal is capability.

The goal is to be the calm person who knows what to do when everyone else is staring at a phone with no signal.

So if your Baofeng has been sitting in a drawer, if your GMRS radios have never made it past the factory settings, if CHIRP programming sounds intimidating, if repeaters feel confusing, or if you simply want a practical off-grid communication plan for your family, this complete field manual was built for you.

Four guides. One system. One clear path from confusion to confidence.

Let's turn the radio on and make it useful.

Jared Johnson

Ham Callsign: KF0RTU

GMRS Callsign: WSLC230

Email: Jared@MorseCodePublishing.com

Structure of a UV-5R Radio

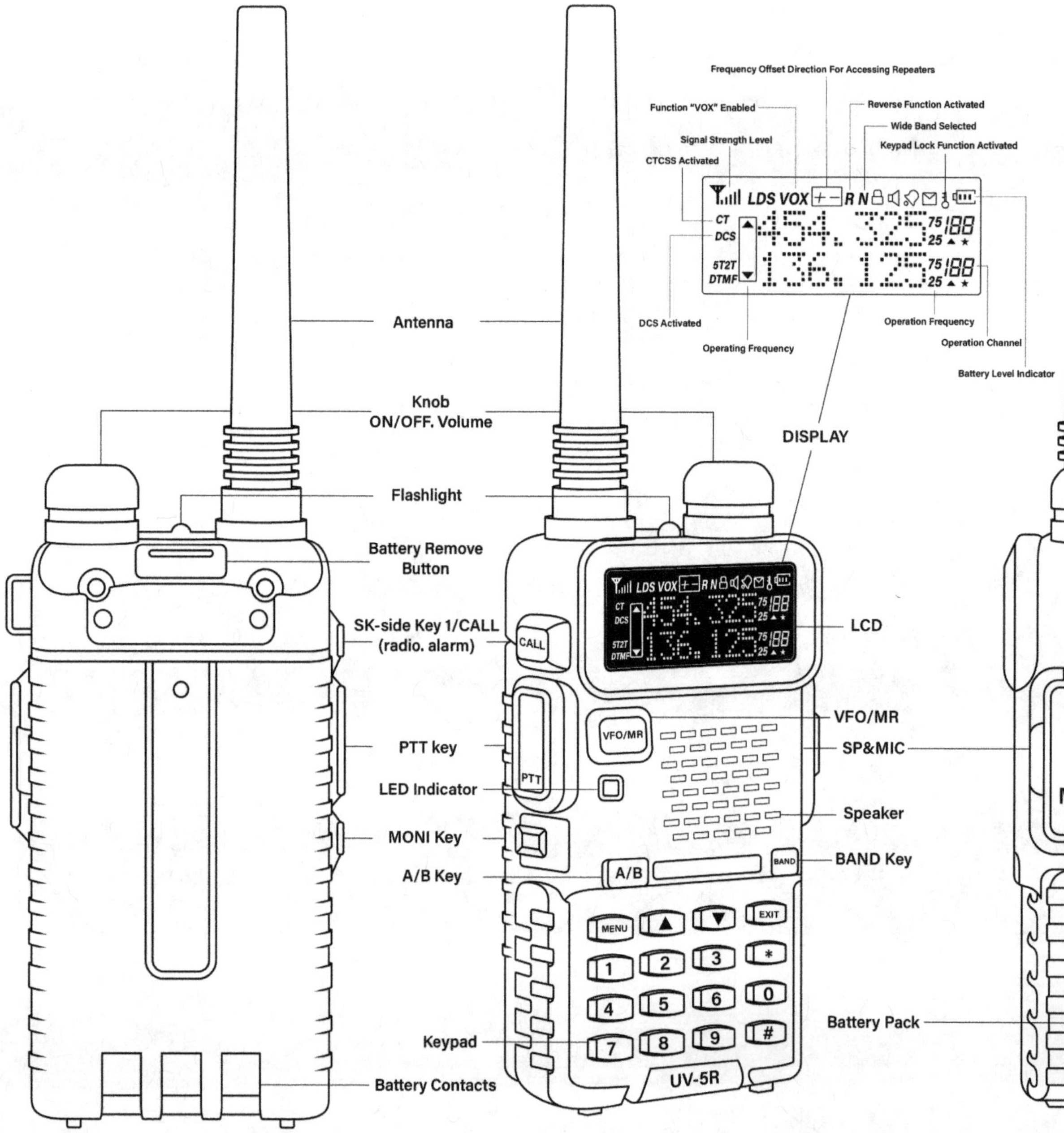

Baofeng Radios - A Bootcamp for Beginners

Unbox to First Contact in 60 Minutes, Menus Made Simple, Repeater + Channel Programming, First-Call Script, and Troubleshooting Trees for Emergency Comms

MorseCode Publishing

Disclaimer: Safety, Legality, and Common Sense

Let's get the serious stuff out of the way so we can get to work.

The Baofeng UV-5R (and its many clones) is an incredible tool, but it is not a walkie-talkie from a toy store. It is a powerful radio capable of transmitting on frequencies reserved for amateur radio operators, emergency services, and commercial businesses.

Here is the golden rule: You can *listen* to almost anything legally. But you cannot *transmit* (press the push-to-talk button) on amateur radio bands without an amateur radio (Ham) license issued by your local government (like the FCC in the United States).

This book is a practical training guide, not legal advice.

- Do not use this book as an excuse to transmit illegally.
- We will not provide "secret frequency lists" for rule-breaking.
- We highly encourage you to get your basic amateur radio license, it's easier than you think, and it unlocks the full power of your radio.

For anything related to local laws, frequencies, or emergency use exemptions, consult official sources and your local radio clubs. Now, let's learn how to use this thing.

How to Use This Book

You bought this radio for peace of mind, preparedness, or as your first step into a new hobby. But right now, you're staring at a little screen with confusing numbers, and the manual that came in the box reads like a poorly translated physics textbook.

Take a breath. **You can do this.** You don't need to be an engineer to use a Baofeng. You just need a good coach. We have designed this book to deliver on one specific promise: *You can go from unopened box to hearing traffic, saving channels, and making a clean first transmission (legally) in exactly one hour.*

To make this easy, this book is split into three specific paths. You are in control of the journey.

Path 1: The First Hour (Start Here)

If you just took the radio out of the box, start at Chapter 1. We will ignore 90% of the useless menus and focus only on the buttons that get you hearing voices and making your first transmission within 60 minutes.

Path 2: Fix It Fast

Already messed up your settings? Radio making a weird noise? Can't hear anything? Jump to **Part 2: Troubleshooting**. It is a symptom-based triage tree. You look up your problem, and we give you the 60-second fix. No theory, just solutions.

Path 3: The Family Plan

If you bought a multi-pack of radios to keep your family connected during a blackout, cell outage, or camping trip, jump to **Part 3**. We'll show you how to set up a stress-proof communication plan that anyone in your family can use, even if they've never read this book.

Gear Checklist

You do not need a lot of expensive gear to be successful. Let's look at what you actually need right now.

The Minimum (Start With This):

- Your Baofeng UV-5R radio.
- A fully charged battery.
- The stock "rubber duck" antenna that came in the box.

Strongly Recommended:

- **A spare battery and charger.** Radios always seem to die right when you figure them out. Having a backup keeps the momentum going.

Optional (Nice to Have Later):

- **A programming cable and CHIRP software.** This lets you program the radio from your computer. It's amazing, but we won't need it for your first hour. We are going to learn how to drive the radio manually first.

[FIELD NOTE] Avoid wasting money on... Giant, 15-inch "tactical" whip antennas. Yes, they look cool, and yes, they give you a bit more range. But right now, they just make the radio top-heavy and clumsy to hold while you are trying to learn the buttons. Stick to the stock antenna for your first week.

Your Quick-Start "Mission Map"

Keep this page bookmarked. This is your itinerary for the next 60 minutes. We are going to build your skills layer by layer.

Minute 0–5: Power, Volume, and Silence We'll turn it on, set a comfortable volume, and adjust the "squelch" so the radio stops making that horrible static hissing noise.

Minute 5–15: Find a Lighthouse Before we try to talk to anyone, we need to prove the radio works. We will find a local, reliable signal (like a weather broadcast) to listen to.

Minute 15–35: The Repeater Puzzle We'll demystify "repeaters", the towers that give your little radio massive range. We'll translate confusing terms like *offsets* and *tones* into plain English.

Minute 35–55: Program and Save This is where most people quit. You won't. We will manually program a repeater into your radio and save it to a memory channel so you never have to type it in again.

Minute 55–60: Your First Transmission You'll press the button. We'll give you a simple, embarrassment-proof script to use for your very first legal transmission.

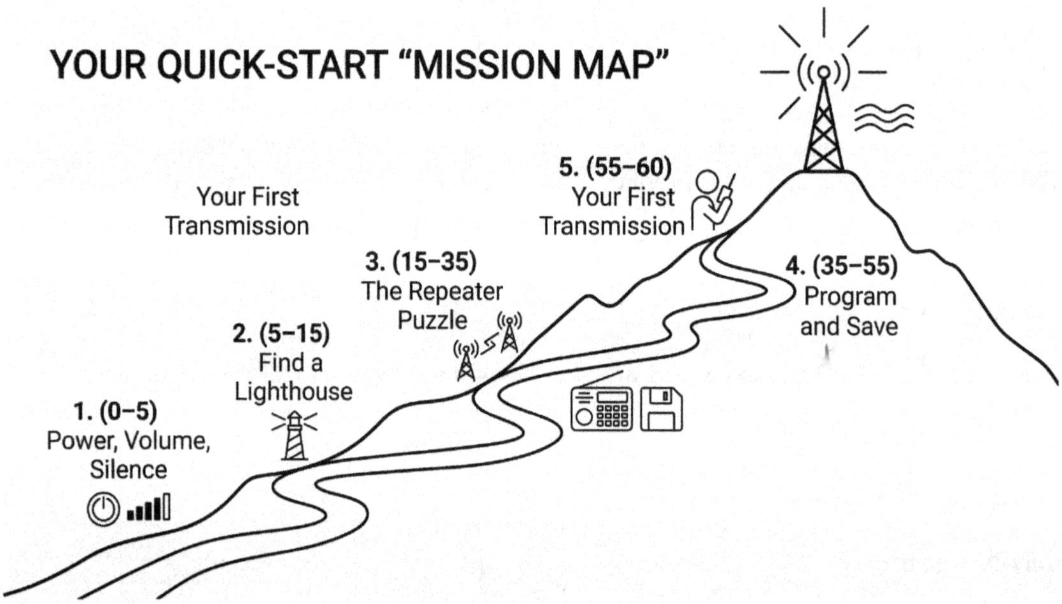

Ready? Turn the page, and let's start the clock on your First Hour.

- Jared Johnson (KF0RTU) – Jared@MorseCodePublishing.com

Part 1:

The First Hour (Get Working Fast)

Chapter 1

Your First 60 Minutes: What "Success" Looks Like

You bought the radio. Now it's just beeping at you. The screen is flashing randomly, a robotic voice prompt is yelling numbers every time you brush a key, and you're afraid to press a button in case you accidentally call in a rescue helicopter or break a federal law.

Relax. Take your hand off the radio, take a deep breath, and let's get oriented. You are exactly where every beginner starts.

What you'll accomplish:

- Define the 3 wins so you know exactly what we are trying to achieve today (and what we can safely ignore).
- Understand VFO vs. Memory (this is the single most important concept in this entire book).
- Prevent the 5 most common beginner mistakes that cause people to throw their radios in a drawer and quit.

What you need:

- Your Baofeng UV-5R (keep it powered off for now so it doesn't distract you).
- A comfortable place to sit.

The 3 Wins (What Success Feels Like)

When you look at a radio with 21 buttons and a screen full of tiny icons, it's incredibly easy to get overwhelmed. We are going to ignore almost all of them today. For your first 60 minutes, we only care about achieving three specific, measurable wins:

1. **Hear:** We need to prove the radio works out of the box by catching a strong, local signal. Hearing traffic validates your equipment and builds your confidence before you ever try to transmit.
2. **Store:** We are going to find a local frequency and save it to the radio's permanent memory. You are going to build a "rolodex" of safe channels so you never have to blindly type numbers in an emergency.
3. **Talk:** You are going to make a short, clean, legal transmission (using your ham radio license) without freezing up or feeling stage fright.

That's it. If you can hear, store, and talk, you have mastered 90% of what this radio was built to do. Everything else in the manual is just optional bells and whistles.

The Two Modes (VFO vs. Memory)

If you only remember one thing from this chapter, let it be this: **Your radio has a split personality.** It exists in one of two distinct modes, and understanding the difference is the secret key to unlocking the Baofeng.

FIG 2

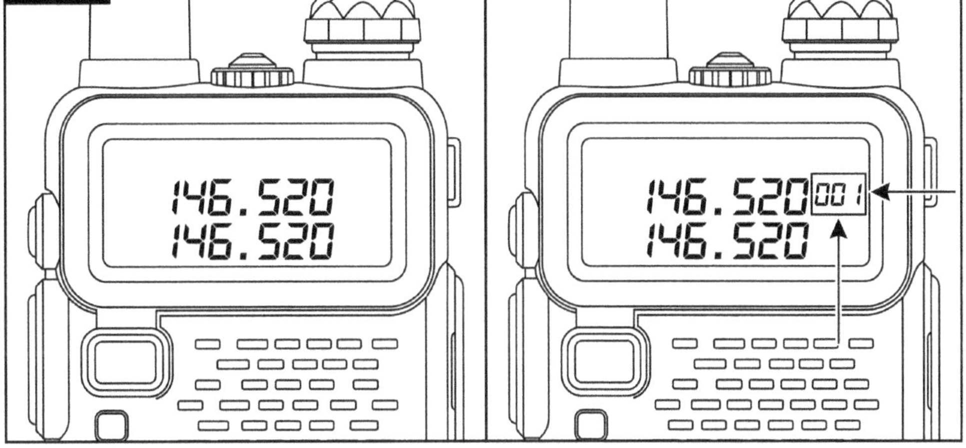

VFO MODE (Frequency) MEMORY MODE (Channel)

Mode 1: VFO Mode (Frequency Mode) Think of VFO mode like the manual tuning dial on a classic car radio. You are turning the knob, searching through the static to see what is out there. In this mode, you use the keypad to type in exact, raw frequencies (like 1 6 2 . 5 5 0 for a local National Weather Service broadcast). You are freely exploring the airwaves. *Crucially, you must be in VFO mode to program new channels into the radio.*

Mode 2: Memory Mode (Channel Mode) Think of Memory mode like the preset push-buttons on your car radio. Once you find a frequency you like in VFO mode, you save it as a "Channel" (like Channel 1, Channel 2, etc.). When you are in Memory mode, the danger of accidentally typing the wrong number disappears. You just flip safely through your pre-saved, trusted channels.

[FIELD NOTE] How to switch between them: On the UV-5R, there is an orange button labeled **VFO/MR**. Pressing this button toggles the radio back and forth between manual tuning (VFO) and your saved presets (Memory/MR).

Why You Feel Lost (The Baofeng Reality)

If you feel confused by the menu system, it's not your fault. These radios were originally designed by engineers, for other engineers. The menus are not alphabetical, the abbreviations on the screen don't always make obvious sense, and the factory manual is famously unhelpful. The radio provides zero context clues to tell you what a button actually does.

Don't let this intimidate you. Put the factory manual in a drawer right now, you don't need to memorize the menus to be highly effective. In this book, we will give you the exact, step-by-step keypad combos you need. Treat the radio like a calculator: you don't need to know how the circuit board works. Just punch in the exact formula we give you, and you will get the right result every single time.

The 5 Time-Wasters

Thousands of beginners have walked this exact path before you. We know exactly where they stumble. Keep an eye out for these 5 common traps, we will solve all of them together in the coming chapters:

1. **Tone Confusion:** If you grew up using cheap walkie-talkies from a sporting goods store, you probably learned about "privacy codes." On a Baofeng, trying to use these tones (CTCSS/DCS) when you don't actually need them results

in nobody being able to hear you.

2. **Wrong Mode:** The frustration of typing a raw frequency while stuck in Memory mode and watching the radio ignore you, or trying to save a channel while in Memory mode.

3. **Wrong Step Size:** Typing in a frequency and watching the radio magically change the last digit on you, feeling like the machine is fighting your inputs.

4. **Saving Order:** The Baofeng requires you to save the listening frequency and the talking frequency in a very specific, quirky order. It isn't intuitive. If you mess up the order, the channel simply won't work.

5. **Squelch Too High:** The "squelch" is the radio's static filter. Setting this invisible gate so high that the radio goes completely deaf is the number one reason beginners think their new radio is broken.

[STOP] Don't push buttons just to see what happens. If you dive into the menu and start changing random settings to see what they do, you will eventually lock yourself out of transmitting or receiving. Stick to the guide.

Quick Orientation: The Big Four

Before we actually turn the radio on in the next chapter, physically locate these four things on the face of your device:

- **The Keypad:** This feels like an old flip phone. It is used for typing frequencies directly and selecting numbered menu options.
- **The MENU Button:** Located in the center of the up/down arrows. This acts as your "Enter," "Select," or "Confirm" key.
- **The A/B Display:** Your screen has a top line (A) and a bottom line (B). The radio can monitor two frequencies at once, but you can only actively control one line at a time. The blue **A/B button** flips your control between the top and bottom. You will know which line is active by the small black triangle pointing to it on the left side of the screen.
- **The PTT Button:** This is the large, tactile rubber button on the left side of the radio housing. PTT stands for "Push To Talk." *Do not press this today until we explicitly tell you to.*

Quick Win Checklist

- [ ] I understand the critical difference between VFO mode (manual typing) and Memory mode (saved presets).
- [ ] I know that I use the orange VFO/MR button to switch back and forth between the two modes.
- [ ] I have located the MENU button (which acts like the "Enter" key).
- [ ] I have located the A/B button (which switches my control between the top and bottom lines of the screen).
- [ ] I have located the big rubber PTT button on the side, but I will strictly avoid pressing it for now.

If you're stuck, go to...

- *Chapter 10: I Can't Hear Anything* (if you are playing with the radio early and it's dead silent).
- *Chapter 12: My Channels Won't Save* (if you forgot the rule about VFO vs. Memory).

Chapter 2

Five-Minute Setup

Power, Volume, Squelch, and a Clean Receive

YOU CAN'T FIX WHAT you can't hear. If your radio is completely silent, or if it is constantly blasting you with a wall of ear-piercing static, you are never going to catch a signal. Before we can even think about programming channels, we need to get the physical hardware running smoothly so you have a clean, reliable baseline.

What you'll accomplish:

- Set your volume and "squelch" correctly (finding the perfect Goldilocks zone for listening).
- Use the physical Monitor button to prove your speaker actually works.
- Learn how to lock the keypad so you don't accidentally ruin your progress.

What you need:

- Your Baofeng radio.
- The stock antenna (screwed firmly into the top).
- A charged battery.

Step 1: The Hardware Check

Before we turn the knob, we need to assemble the hardware correctly.

1. **Seat the battery:** Slide the battery onto the back of the radio, pressing upward until you hear a firm, satisfying *click*. Don't be afraid to use a little force with your thumbs, brand new radios often have very stiff plastic rails. If the battery is loose, the radio will randomly shut off right in the middle of a transmission.
2. **Attach the antenna:** Screw the stock rubber antenna onto the top brass connector. Thread it until it is snug, but do not crank it down with superhero strength.

[STOP] Never press the Push-To-Talk (PTT) button if the antenna is unscrewed. Transmitting without an antenna creates a backlog of power that has nowhere to go. It will reflect right back into the radio's circuit board and literally fry the internal components. Always have an antenna attached.

Step 2: Power and Volume

Turn the volume knob on the top of the radio clockwise. You will feel a mechanical *click* as the radio powers on.

The screen will light up, flashing an orange or purple backlight, and a loud robotic voice will likely announce "FREQUENCY MODE" or "CHANNEL MODE."

Turn the knob about halfway up. Do not crank it to maximum volume just yet, when a signal does come through, the Baofeng speaker is surprisingly loud and can easily startle you. Right now, you probably won't hear anything else, which leads us to the most important testing button on the side of the radio.

Step 3: The Monitor Button (Proving the Speaker Works)

Right below the big PTT button on the left side of the radio, there is a smaller button labeled **MONI** (Monitor).

Press and hold the **MONI** button right now.

You should immediately hear a loud, rushing hiss of white noise. This is the raw sound of empty radio waves. While holding the button, adjust your top volume knob until the static is at a comfortable listening level.

Let go of the MONI button. The static should instantly stop. *Congratulations, you just proved your speaker and receive path are working perfectly.*

Step 4: Setting the Squelch (The Invisible Gate)

Why did the static stop when you let go of the MONI button? Because of the **Squelch**.

The squelch is like an invisible dam that holds back the constant river of background static in the atmosphere. When the squelch is working correctly, your radio stays dead silent. But when a real, powerful signal (like someone talking) hits your antenna, it "breaks the squelch", spilling over the dam and opening the speaker so you can hear them.

If your squelch is set too low (Level 0), the dam is gone, and you hear non-stop hissing. If your squelch is set too high (Level 9), the dam is too tall. Even strong, local signals won't be able to break through, and your radio will seem totally deaf.

Let's set your squelch to the perfect beginner level (Level 1 or 2):

1. Press the **MENU** button.
2. Type the numbers **0** then **0** on the keypad. Your radio has 40 different numbered menus, and Menu 0 is always Squelch. The screen should now say SQL (Squelch).
3. Press the **MENU** button again. You will see a small arrow drop down to the bottom line of the screen. You are now unlocked and editing the setting.
4. Type the number **2**.
5. Press the **MENU** button one more time to save it. You will hear the robotic voice say "Confirm."
6. Press the **EXIT** button to leave the menu.

You just navigated your first Baofeng menu! You used the exact "formula" you will use for the rest of this book: Menu, find the setting, Menu to unlock, change it, Menu to save, Exit.

Step 5: Lock it Down

Once you have your radio set up exactly the way you like it, it is incredibly easy to accidentally bump a keypad button with the palm of your hand, or have it rub against the inside of your jacket pocket, completely changing your frequency without you noticing.

Look at the bottom right key on your keypad. It has a # symbol and a tiny picture of a padlock. Press and hold the **# / Padlock key** for about two seconds. You will hear a beep, and a tiny padlock icon will appear at the top center of your screen. Your keypad is now locked. You can still use the volume knob, and you can still press the PTT button to talk, but you cannot accidentally change channels or dive into menus. To unlock it, just press and hold the # key for two seconds again until the padlock icon disappears.

[GOOD ENOUGH] If your battery is seated, your antenna is on, your squelch is set to 2, and you've locked your keypad so you don't accidentally ruin things, you are in perfect shape. Your radio is primed and ready to catch its first real signal.

Quick Win Checklist

- [ ] My antenna is firmly attached to the radio.
- [ ] I understand that I should never press the PTT button if the antenna is removed.
- [ ] I have used the side MONI button to hear raw static and adjust my volume knob to a comfortable level.
- [ ] I understand that "Squelch" acts as an invisible gate to block background static.
- [ ] I have successfully navigated the menu and set my Squelch (SQL) to Level 2.
- [ ] I know how to lock and unlock my keypad using the # key.

If you're stuck, go to...

- *Chapter 10: I Can't Hear Anything* (if you pressed the MONI button and heard absolutely zero sound).

Narrative Interlude: The Beeping Brick

Character: Mark Jensen (Midwest Suburbs)

Skills reinforced:

- VFO vs. Memory Mode
- Volume and Squelch setup
- Keypad Lock
- Setting "Beginner-Friendly" Defaults

Where to learn the steps: Chapters 1 & 2

The box sat on the center of the kitchen island like a tactical alien artifact. Mark Jensen, age forty-two and officially a licensed Amateur Radio Technician as of three days ago, stared at it with a mixture of pride and growing dread. To the uninitiated, it was a thirty-dollar piece of electronics; to Mark, it was a gateway to the global airwaves, a piece of gear that signaled he was a man with a plan for the next blizzard.

"Is it supposed to look that... small?" his wife, Jen, asked, leaning over a bowl of salad. She poked the stubby antenna through the plastic wrapping.

"It's compact, Jen. Ruggedized," Mark corrected, though he was secretly thinking the same thing. In the online photos, it looked like something a paratrooper would carry; in his kitchen, next to a bowl of cherry tomatoes, it looked like a high-tech garage door opener. "It's a masterpiece of Chinese engineering. Industrial. Utilitarian. It's a Baofeng UV-5R."

"It looks like a walkie-talkie I had in 1994," his son, Owen, chimed in, finally looking up from his phone with a glimmer of teenage skepticism. "Does it have Snake on it? Or at least a flashlight?"

"It has the power to communicate across the Twin Cities during a grid-down scenario, Owen," Mark said, trying for a tone of steady, military-grade competence. "And yes, it actually does have a flashlight." He reached for the box, carefully extracting the radio from its crinkly plastic cocoon.

The unboxing started well. The battery clicked onto the rails with a satisfying mechanical snap that sounded like a magazine being loaded into a rifle. The antenna screwed on smoothly. Mark felt his confidence rising, until he twisted the top knob.

"BEEP. FREQUENCY MODE."

The radio didn't just talk; it shouted. A loud, female, digitized voice echoed off the granite countertops, sounding like a very stern flight attendant.

"Whoa!" Owen jumped. "Is that the emergency? Is the world ending right now?"

"No," Mark muttered, his face reddening. He tried to turn the volume down, but his fingers were slightly sweaty, and he accidentally grazed the keypad.

"BEEP. SIX. BEEP. SEVEN. BEEP. MENU."

The radio was now screaming random integers at the ceiling. Mark frantically pushed a button, he thought it was the 'Exit' button, but instead, the radio emitted a long, sustained, screeching siren tone that sounded like a smoke alarm having a nervous breakdown in the middle of a library.

"Mark, honey," Jen said, her voice strained as she covered her ears. "Make it stop. Please."

Mark mashed the orange button. Then the blue one. Then he just started clicking the top knob back and forth like he was trying to win a game of arcade whack-a-mole. Finally, he managed to twist the power off. The silence that followed was heavy and judgmental.

"Okay," Mark said, clearing his throat and adjusting his glasses. "That was just... a diagnostic check. Calibrating the audio levels. It's an advanced feature."

"You looked like you were trying to disarm a bomb and forgot which wire to cut," Owen said, returning his gaze to his phone. "I give it ten minutes before the FBI knocks on the door to see why you're broadcasting a dial-up modem sound to the neighbors."

Mark ignored him and turned the radio back on, moving with the agonizing caution of a man handling nitroglycerin. He kept the volume at a whisper. But now, as he stared at the dual-line display, there was a new problem. The radio was dead silent. He could see two lines of numbers, 144.000 on top and 440.000 on the bottom, but even when he turned the volume knob all the way to the right, nothing happened. No static, no voices, just a cold, digital stare.

"I think it's broken," Mark whispered. "It was beeping a second ago, and now it's just... dead. I must have fried the circuit board when I was button-mashing."

He felt that familiar beginner's spiral. He'd had the radio for exactly four minutes and he'd already turned it into a tactical paperweight. He tried typing in the National Weather Service frequency he'd seen in the book, 162.550, but every time he hit the '1', the radio just let out a sharp, dismissive beep and stayed stubbornly on 144.000.

"It won't even let me type," he groaned. "It's locked. Or haunted. I've bought a haunted radio."

Jen walked over, wiping her hands on her apron. She picked up the field manual Mark had left on the counter, the one with the bright yellow 'Mission Map' on the front.

"Did you actually do the five-minute setup?" she asked, pointing to a section labeled 'The Invisible Gate.'

"Jen, I'm a licensed ham. I have a callsign. I studied for weeks. I don't need to, "

"Mark," she interrupted gently, tapping a finger on the page. "Read Step 2."

Mark sighed and looked at the page. *The Squelch Wall.* He looked back at his radio. He realized that in his earlier panic, he'd dived into the menu and changed 'SQL' to 9, thinking higher numbers meant more 'power' or better 'quality.' Instead, he'd built a digital wall so thick that not even a lightning bolt from the gods could break through to the speaker.

"Wait," Mark said, his curiosity finally overriding his embarrassment. He followed the instructions. *Menu -> 0 -> Menu.* He dropped the squelch from 9 down to a sensible 2. *Menu -> Exit.*

Nothing. Still silence.

"It's still dead, Jen. The wall is down, but nobody's home."

"Try the side button," she suggested, pointing at the checklist again. "The Monitor one. It says it bypasses the gate entirely."

Mark pressed the small, rubberized button labeled **MONI** located just below the large Push-To-Talk button.

SHHHHHHHHHHHHHHHHHH!

A glorious, rushing waterfall of white noise exploded from the speaker. He jumped, then laughed, a huge grin breaking across his face. "Okay! The speaker isn't fried. It's alive! That's the sound of the universe right there."

"Now look at the right side of the screen," Jen said, reading ahead. "Does it have tiny numbers next to the frequencies?"

Mark squinted at the display. "Yeah. A little 001 and 002. What does that mean?"

"The book says that's Memory Mode. You're in the 'locked display case.' You can only use what's already saved there. To type new numbers, you have to push the orange button to get to the 'VFO Canvas.'"

Mark pressed the orange **VFO/MR** button. The tiny 001 vanished. The radio felt different, unlocked. He tentatively typed the digits: 1 6 2 5 5 0.

"...mostly cloudy with a sixty percent chance of rain tonight. Lows in the mid-forties, winds from the North-Northwest at ten miles per hour..."

The robotic, rhythmic voice of the National Weather Service filled the kitchen. It was clear, steady, and incredibly satisfying. Mark stood a little taller, adjusted his posture, and looked at his son.

"Look at me," he said, gesturing to the radio. "I'm basically NASA. Mission Control is officially online."

"You're a guy listening to a weather report in his kitchen, Dad," Owen said, though he actually looked a bit impressed that the beeping had been replaced by actual information. "But hey, at least it's not screaming at us anymore. That's a win."

Mark realized he'd spent the last ten minutes fighting the radio because he was trying to run before he could walk. He'd assumed silence meant hardware failure, when it really just meant he'd closed the gate too tight and was standing in the wrong room.

He held down the # key for two seconds. A tiny padlock icon appeared on the screen with a cheerful *chirp*.

"There," Mark said, setting the radio down. "Keypad lock. Now nobody, not even me, can accidentally mess up the settings while I'm making coffee."

"I'm not the one who turned it into a smoke alarm," Owen reminded him.

Mark just smiled, listening to the weather bot. He had his first win. He'd navigated the menus, conquered the squelch, and proven the receiver worked. The beeping brick was now a functioning tool.

"Okay," Mark said, looking at the next chapter in the guide. "I'm hearing the lighthouse. Now I just need to figure out where the other ships are."

Takeaway

Silence usually isn't a broken radio, it's just a setting out of place. Most beginner frustration comes from being in the wrong Mode (Memory vs. VFO) or having the Squelch gate set too high. Use the Monitor button to prove your hardware is okay, then check your mode.

Bridge: Mark's radio is receiving loud and clear. But to talk to someone miles away, he needs to understand the towers that make it possible. It's time to look for a "Lighthouse."

Chapter 3

Find Something to Listen To (Legally)

BEFORE YOU EVEN THINK about pressing that Push-To-Talk button to broadcast your voice to the world, we need to prove your radio can actually hear the world around it. Right now, you are holding a piece of plastic and metal that feels full of potential, but if it stays completely silent, that potential quickly turns into frustration. We need to find a 'known-good,' continuous signal to validate that your hardware is actually working. We are going hunting for a lighthouse.

What you'll accomplish:

- Find and tune into "known-good" signals that broadcast continuously.
- Confirm your receive path works in the real world (not just listening to raw static).
- Build massive confidence in using VFO mode to explore the airwaves without the pressure of transmitting.

What you need:

- Your Baofeng radio, powered on with the antenna attached.
- Your squelch set to Level 2 (from Chapter 2).
- A piece of paper and a pen.

The Lighthouse Concept

When a sailor is navigating a dark, unfamiliar sea, they look for a lighthouse, a fixed, reliable, constant beam that proves their position.

When you are a beginner with a new radio, you need a lighthouse signal. If you just type in a random frequency and hear nothing, you have no idea if the frequency is dead, if you are out of range, or if your radio is broken. We need to find signals that we *know* are broadcasting right now.

Target 1: The National Weather Service (NOAA)

For users in North America, the absolute best lighthouse is the NOAA Weather Radio system. These stations broadcast automated weather reports 24 hours a day, 7 days a week. It is a continuous, non-stop loop of spoken audio. If you can tune into this, you immediately prove your antenna, receiver, and speaker are perfectly functional.

NOAA broadcasts on seven specific frequencies. You will use VFO mode to check them one by one until you hear the robotic weather voice.

Step 1: Look at your screen. Make sure you are in **VFO Mode** (you should see just frequency numbers, not small channel numbers on the right). If you see channel numbers, press the orange **VFO/MR** button once.

Step 2: Use the blue **A/B button** to make sure the tiny black arrow is pointing to the top line.

Step 3: Use the keypad to type in the first frequency: 1 6 2 . 4 0 0.

FIG 3

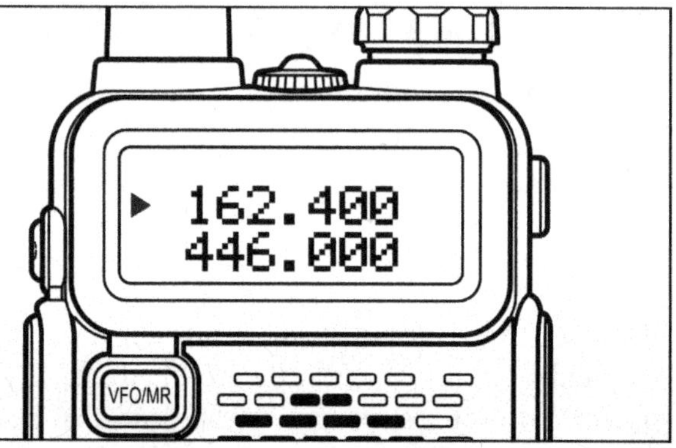

Note: As you type the numbers, the radio will speak them out loud. When you hit the 6th number, it will load the frequency.

Did the squelch open? Do you hear a voice reading a weather forecast?

- If **YES**, stop! You found your lighthouse. The voice will likely sound highly synthesized, repetitive, and robotic, this is completely normal.
- If **NO**, the station might be too far away. Type in the next frequency on the list:
    - 162.425
    - 162.450
    - 162.475
    - 162.500
    - 162.525
    - 162.550

YOU'RE WINNING

If you hear the weather broadcast, take a moment to celebrate. Your Baofeng is officially receiving real-world radio traffic! You can use the volume knob to dial in the perfect sound level.

Target 2: Local Ham Radio Repeaters

If you aren't in North America, or if you live in a deep valley where weather radio doesn't reach, your next best lighthouse is a local amateur radio repeater.

Repeaters are large, powerful radio towers placed on tall buildings or mountains by local radio clubs. They listen for weak signals from handheld radios (like yours) and blast them back out at high power so everyone in the city can hear.

- **How to find them:** You will need to use your smartphone or computer. Go to a website like **RepeaterBook.com** or download their free mobile app. Search for your local city or county.
- **What to look for:** Look for repeaters operating in the "2 Meter" (144-148 MHz) or "70 Centimeter" (420-450

MHz) bands, as these are what your Baofeng can hear.

- **The Catch:** Unlike weather radio, repeaters are not continuous. You will only hear something if two people are actively having a conversation. You might have to type in the frequency (e.g., 1 4 6 . 5 2 0) and just let the radio sit on your desk for twenty or thirty minutes while you do other things. Don't stare at it. Just wait for the squelch to suddenly break and surprise you with a conversation.

What Good Receive Sounds Like (vs. Bad Receive)

As you start exploring, you will hear a variety of different audio qualities. It takes a little practice to train your ear. Here is what to listen for:

- **Full Quieting (Excellent):** The person's voice comes through crystal clear, loud, and rich. There is almost zero background hiss. It sounds like they are standing in the room with you. This means you have a perfectly strong signal.

- **Scratchy/Picket-Fencing (Marginal):** You can hear the voice, but it is buried under a layer of static. If the person is driving a car while talking, the signal might rapidly chop in and out (this rhythmic chopping is called 'picket-fencing,' because it sounds exactly like listening to a car drive by while you stand behind a wooden picket fence).

- **Static Crash (Lost Signal):** You hear a sudden, loud burst of pure static, and then the squelch snaps the speaker shut again. The signal was strong enough to break the squelch gate for a split second, but too weak to hold it open.

FIELD NOTE

Your body is part of the antenna system. If you are listening to a scratchy, marginal signal, try standing up, walking near a window, or simply holding the radio higher in the air. UHF/VHF radio waves are strictly "line-of-sight." Even moving two feet to the left can completely clear up a bad signal.

Exercise: The Receive Log

Grab your pen and paper. Today, we are just listening. Spend the next ten minutes typing in different frequencies from your local RepeaterBook search or the weather radio list.

Create a quick log with three columns:

1. **Frequency:** (e.g., 162.550)

2. **What I Heard:** (e.g., *Automated weather voice / Two older guys talking about antennas*)

3. **Signal Quality:** (e.g., *Crystal clear / Very scratchy*)

Writing this down does two things: it builds your muscle memory for typing on the VFO keypad, and it gives you a list of "known-good" frequencies that we are going to permanently save to your radio in Chapter 5.

Quick Win Checklist

- [ ] I have successfully switched my radio to VFO mode (manual typing).
- [ ] I understand why we hunt for a continuous "lighthouse" signal first.
- [ ] I have successfully tuned into a NOAA Weather Radio broadcast or a local repeater.
- [ ] I understand the difference between a "Full Quieting" signal and a scratchy signal.
- [ ] I have started a simple paper "Receive Log" and noted at least one working frequency.

If you're stuck, go to...

- *Chapter 10: I Can't Hear Anything* (if you have tried all the weather frequencies and still hear only silence).

Narrative Interlude: The Lighthouse Test

Character: Nate Brooks (Southeast)

Skills reinforced:

- Known-good receive
- Confidence check
- Receive log
- "Listen-first" mindset

Where to learn the steps: Chapter 3

The humidity in the garage was a physical weight, the kind that signaled hurricane season was no longer a date on the calendar, but a looming reality. Nate Brooks stood at his workbench, surrounded by the orderly chaos of preparedness. He accounted for his mission-essential inventory: four-stage water filtration, primary and secondary illumination, and a trauma kit stocked for 72 hours of field care.

In the center of the bench sat the new Baofeng UV-5R.

Nate didn't have his ham license yet, that was a target for next month, but he was a man who believed in redundant systems. If the cell towers went down during a Category 3 surge, he needed a way to hear the outside world. He'd read the warnings about these radios being complicated, but he figured his years in the field and his disciplined approach to gear would see him through.

He clicked the radio on.

"BEEP. FREQUENCY MODE."

He kept the volume low. He didn't need his wife, Kara, coming out to the garage to hear the radio shouting at him. He'd promised her this wouldn't be another "project" that required a three-hour lecture to use. He needed it to be a tool: reliable, simple, and proven.

Nate began tapping the up-arrow key, scanning through the 144.000 MHz range. He heard a burst of static. Then a snippet of a voice, "...roger that, 73s..." then silence. He hopped to another frequency. More static. He tried scanning, but the radio seemed to skip over everything or stop on nothing but the rhythmic *chug-chug-chug* of local electrical interference.

Ten minutes in, Nate felt a flicker of uncharacteristic frustration.

"Is this it?" he muttered. He looked at the stock antenna. It was small, flexible, and felt... cheap. Maybe the reviews were right.

Maybe he'd bought a thirty-dollar paperweight. If he couldn't hear a clear sentence from three miles away, how was he going to hear emergency coordinates when the sky turned purple and the power lines started snapping?

He almost reached for his phone to order a larger, "tactical" antenna, the kind that looked like it belonged on a tank. He had momentarily fallen into the gear-trap, assuming a hardware upgrade would compensate for a lack of verified procedure. He was searching for a single point of failure in the antenna when the real failure was in his diagnostic process. But then, he stopped. He looked at the "Ghostwriter Guide" pinned to the corkboard above his bench. He'd underlined a specific sentence in Chapter 3: *Before you can trust a tool, you need a repeatable test.*

He realized he was "frequency hopping" like a kid looking for a radio station on a road trip. He hadn't established a baseline. He hadn't found his lighthouse.

Nate sat down and pulled a fresh notebook toward him. At the top of the page, he wrote: **RECEIVE LOG.**

He checked the guide for his local "known-good" signals. The first was the National Weather Service. He knew the tower was located about twelve miles to the northwest. It was a massive, high-power broadcast. If the radio could hear anything, it would be that.

He typed the digits slowly: 1 6 2 4 0 0.

Silence.

His heart sank. "Nothing," he whispered.

Then he remembered the lesson from the previous chapter. He initiated a manual override of the squelch gate, holding down the **MONI** button to force the audio channel open and conduct a perimeter check of the signal threshold.

SHHHHHHHHH,

And underneath the hiss, a faint, rhythmic pulse.

Nate stood up and walked to the garage door. He hit the opener, and as the heavy metal door rumbled upward, the signal transformed. The static retreated, and a clear, robotic voice filled the garage.

"...Highs in the lower nineties. Heat index values up to one hundred and five. A slight chance of thunderstorms in the evening..."

"Target acquired," Nate said. He felt a surge of relief, but he didn't stop there. He went back to the bench and made his first entry in the log.

- **Freq:** 162.400 (NWS)
- **Location:** Garage (Door Open)
- **Quality:** 5/5 (Clear)
- **Note:** Signal drops inside the garage when the door is closed.

He moved to the next "lighthouse", the primary local ham repeater. He'd found the frequency on a local club's website: 146.760. He typed it in and waited.

For three minutes, there was nothing. Nate didn't touch the buttons. He didn't scan. He simply listened. He practiced the "listen-first" mindset. He knew the repeater was like a lighthouse; it didn't shine all the time, only when someone was using it.

Finally, the radio chirped.

"K4XYZ monitoring," a voice said. It was crisp, loud, and sounded like the person was standing right in the driveway.

Nate checked the signal bars on the screen. Full strength. He didn't transmit. He couldn't, he wasn't licensed, and he respected the discipline of the airwaves. But he'd proven the most important part of the chain: **The Receive.**

He logged the second entry:

- **Freq:** 146.760 (Charleston Repeater)
- **Location:** Garage Workbench
- **Quality:** 5/5
- **Note:** High-traffic tower. Reliable for local info.

By the time Kara walked out to the garage twenty minutes later, Nate wasn't button-mashing or looking frustrated. He was calmly writing in his notebook.

"How's the new toy?" she asked, leaning against the doorframe.

"It's not a toy, Kara. It's a verified system," Nate said, turning the volume down. "I've established a receive baseline. I know exactly where to go for weather and exactly where to go for neighborhood status. I've documented the dead zones in the house."

Kara looked at the notebook, then at the radio. "You actually made it work? No lectures?"

"No lectures," Nate promised. "Just data. I know the radio works because I found the lighthouses."

Nate felt a new level of confidence as he closed the garage door. He didn't need a bigger antenna yet. He didn't need more gadgets. He had a receive chain he could trust because he'd bothered to test it before he needed it.

He looked at the radio sitting on the bench. It wasn't a "paperweight" anymore. It was a window into the local world.

"Now," Nate said to himself, looking at the next section of the guide. "If I can hear them, I need to understand why they might not be able to hear me. It's time to solve the offset puzzle."

Takeaway

A radio is a tool of perception before it is a tool of transmission. Finding a "lighthouse signal" (like the NWS or a high-power local repeater) proves your hardware is functional and your settings are correct. Logging these successes creates a "map of certainty" you can rely on when a real crisis strikes.

Bridge: Nate has confirmed his radio is a capable listener. But a listener is only half of a conversation. To reach out and touch those distant towers, he has to master the math of the Repeater Puzzle.

Chapter 4

The Repeater Puzzle Made Simple (Offset + Tone in Plain English)

You are tuning around in VFO mode, and suddenly, you hear a booming, crystal-clear voice. You've found a local repeater! Excited, you press the Push-To-Talk button to say hello, but the person keeps talking as if you don't even exist. You try again. Nothing. You pull away from the radio and wonder: *"I can hear the repeater... why can't it hear me?"*

What you'll accomplish:

- Understand "Offsets" with a simple, visual mental model (no physics degree required).
- Understand what a "Tone" is, what it isn't, and why it acts like a bouncer at a club.
- Use a dead-simple, 60-second decision chart to know exactly what to type in.

What you need:

- Your Baofeng radio.
- A listing for a local repeater (from an app like RepeaterBook or a local club website).

The Repeater in One Diagram

To understand why the repeater can't hear you, you have to understand how repeaters actually work.

Imagine two people standing on opposite sides of a massive mountain. They have small handheld radios, but the radio waves smash directly into the rock. They cannot talk to each other.

Now, imagine we build a giant tower on the very top of the mountain. This tower has two things: a massive **microphone** pointing down at the city, and a massive **speaker** blasting audio out across the entire valley.

If the tower's microphone and speaker were on the *exact same frequency* at the *exact same time*, it would instantly create a deafening feedback loop, like holding a microphone too close to a concert speaker. To prevent this, the repeater uses **two different lanes on the highway.**

- It **listens** for your voice on one frequency (the uplink).
- It **transmits** your voice out to the city on a different frequency (the downlink).

FIG 4: THE REPEATER PUZZLE (HOW IT WORKS)

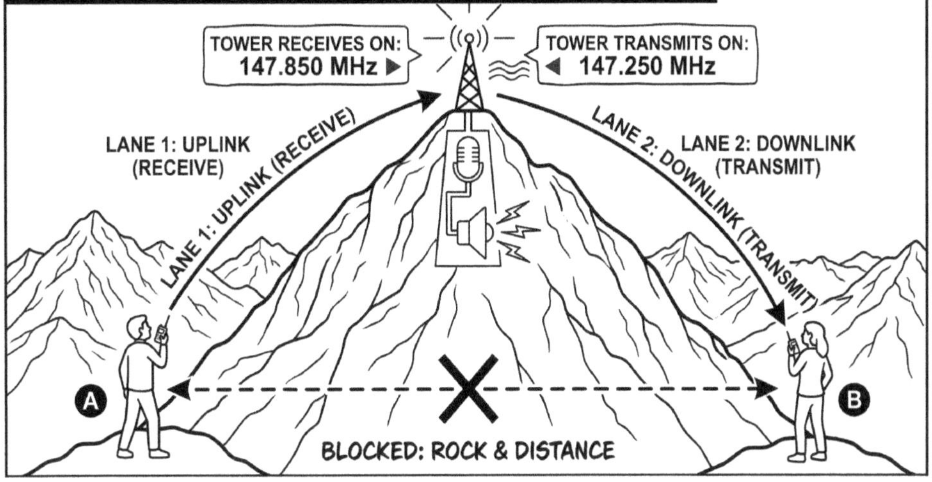

When you typed a frequency into your radio in Chapter 3, you typed in the repeater's *speaker* frequency. That's why you can hear it! But when you press the button to talk, your radio is transmitting on that same frequency. The repeater's microphone is tuned to a completely different channel, so it never hears you.

Offset: The Math Between the Lanes

To fix this, we have to tell your Baofeng to automatically jump to the repeater's listening lane whenever you press the PTT button. This jump is called the **Offset**.

There are two parts to an Offset:

1. **The Direction (Duplex):** Does the radio need to jump *Up* (+) or *Down* (-) from the listening frequency?
2. **The Amount:** How far does it need to jump? (Usually 0.600 MHz for the 2-meter band, or 5.000 MHz for the 70-centimeter band).

Example: A repeater is listed at 146.700 with a Negative (-) offset of 0.600. When you listen, your radio stays on 146.700. When you press PTT, your radio does the math instantly and jumps down to 146.100 to talk to the tower. When you let go, it jumps right back to listen. *Don't panic, you do not have to do this math in your head. You just program the 0.600 once, and the radio handles the jumping automatically.*

Tones: The VIP Password

Let's say you perfectly programmed the offset. You press the button, and the repeater *still* ignores you. What gives? You are missing the VIP password.

In busy cities, there is a lot of random static, interference, and overlapping radio traffic. Repeaters don't want to accidentally broadcast garbage to the whole city. So, they put an invisible "bouncer" at the door.

This bouncer demands a password. In the radio world, this password is a sub-audible tone (a low-pitched hum that your ear can't hear, but the computer can).

- These are officially called **CTCSS** (Continuous Tone-Coded Squelch System) or **DCS** (Digital Coded Squelch).
  - Many older hams just call them 'PL Tones' (which stands for Private Line, an old Motorola trademark that unfortunately makes beginners think it's a private, encrypted channel).

If your radio transmits your voice *without* adding this invisible tone to the background, the repeater's bouncer refuses to open the door, and your transmission hits a brick wall.

FIELD NOTE

Tones are NOT encryption. They do not hide your voice or make your transmission private. Anyone tuning to that frequency can still hear every word you say. They are purely a password to wake up the repeater tower. In the next chapter, we will show you exactly which keypad buttons to press to hand the bouncer your password.

The 60-Second Decision Chart

When you look at a repeater listing online, it can look like a spreadsheet of confusing numbers. Use this simple chart to strip away the noise and focus only on what matters:

1. What is the Receive (RX) Frequency?

- *Action:* Write this down. This is your baseline number.

2. Does the listing show an Offset (+ or -) and an amount?

- *Action:* Write down the direction (+ or -) and the number (like 0.600).

3. Does the listing show an "Uplink Tone" (sometimes just called Tone or CTCSS)?

- *If YES:* Write down the number (e.g., 100.0). You will need to program this into the T-CTCS menu.
- *If NO:* **Leave your tone off.**

STOP

The Tone-Squelch Trap. Do not touch the menu settings labeled R-CTCS (Receive Tone) or Tone Squelch. If you set a receive tone, you are putting a bouncer on *your own radio.* You will block out the repeater entirely and hear nothing but silence. You only ever want to set a **Transmit Tone** (T-CTCS) so you can open the repeater's door. Keep your own door wide open.

Quick Win Checklist

- [ ] I understand that repeaters listen on one frequency and transmit on another to avoid feedback loops.
- [ ] I understand that the "Offset" is simply the math my radio does to jump between those two lanes.
- [ ] I know that a CTCSS/PL Tone is just an invisible password required to wake up the repeater tower.
- [ ] I know that I should only ever program a Transmit Tone, never a Receive Tone (unless explicitly needed).
- [ ] I have looked up one local repeater and written down its Receive Frequency, Offset, and Tone.

If you're stuck, go to...

- *Chapter 9: I Can Hear Them, But They Can't Hear Me* (if you suspect you have the tone wrong).
- *Chapter 11: I Can Hit One Repeater but Not Another* (if you are confused by DCS vs CTCSS tones).

Chapter 5

Program Your First Repeater (Keypad Method)

You understand how the repeater works. You have the frequency, the offset, and the tone written down on a piece of paper in front of you. Now, you have to actually put them into the radio so they stay there.

This is the exact moment most people quit. The Baofeng menu system was designed by engineers, for engineers. It is incredibly unforgiving, and the specific sequence in which you press the buttons matters deeply. If you mess up the order, the radio won't give you an error message, it will silently pretend to save it, leaving you with a broken, frustrating channel.

Don't quit. We have the exact, step-by-step formula. Do not skip a single step.

What you'll accomplish:

- Enter the repeater receive frequency manually.
- Program the offset and tone correctly using the menu.
- Save the entire package to a permanent memory channel so it actually sticks.

What you need:

- Your Baofeng radio.
- Your piece of paper from Chapter 4 with the Repeater Frequency, Offset (+ or -), Offset Amount, and Transmit Tone.

Step 1: Prepare the Canvas (VFO Mode)

You cannot program a channel while you are in Memory mode. You must be in the "manual typing" mode.

1. Press the orange **VFO/MR** button until your screen shows raw frequency numbers without any small channel numbers on the right edge.
2. Press the blue **A/B** button until the little black triangle arrow is pointing to the **TOP line** of the screen. (Always program on the top line to avoid glitching the radio).

Step 2: Delete the Target Memory Channel

The Baofeng is stubborn. If you try to save a new frequency into Channel 001, but Channel 001 already has something in it from the factory, the radio will silently reject your save. You *must* sweep the floor before you build the house. Let's delete Channel 001 to make room.

1. Press **MENU**.
2. Type **2 8**. The screen will say DEL-CH (Delete Channel).
3. Press **MENU** to unlock it. The arrow drops down.
4. Use the Up/Down arrow keys (or type 0 0 1) to select Channel 001.
5. Press **MENU** to confirm the deletion. The voice will say "Confirm."
6. Press **EXIT**. The channel is now empty and ready for our data.

Step 3: Type the Receive Frequency

On the keypad, type in the main listening frequency of your repeater. *(Example: 1 4 6 . 7 0 0)*

Step 4: Set the Offset Direction

We need to tell the radio which way to jump when you press the PTT button.

1. Press **MENU**.
2. Type **2 5**. The screen will say SFT-D (Shift Direction).
3. Press **MENU** to unlock.
4. Use the arrow keys to select + (Plus), - (Minus), or **OFF** (based on your notes).
5. Press **MENU** to confirm.

Step 5: Set the Offset Amount

Now we tell it *how far* to jump.

1. While still in the menu, press the **Up Arrow** once to go to Menu **2 6**. The screen will say OFFSET.
2. Press **MENU** to unlock.
3. Type the amount using exactly six digits. The Baofeng is not a smart device; if you just type '6 0 0', it will get confused. You must feed it all the leading zeros. *(Example: For a 0.600 MHz shift, type 0 0 0 6 0 0. For a 5.000 MHz shift, type 0 0 5 0 0 0).*
4. Press **MENU** to confirm.

Step 6: Set the Transmit Tone (The Password)

If your repeater requires a tone to open the door, we set it here. If your repeater does *not* require a tone, you can skip this step entirely!

1. Type **1 3** while in the menu. The screen will say T-CTCS (Transmit CTCSS).
2. Press **MENU** to unlock.
3. Use the arrow keys to find your exact tone number *(Example: 1 0 0 . 0).*
4. Press **MENU** to confirm.
5. Press **EXIT** to leave the menu system.

Step 7: The Final Save (Crucial Sequence)

You have successfully built the profile in VFO mode. Now, we have to lock it into the permanent memory slot (Channel 001) that we cleared earlier.

FIG 5

1. Press **MENU**.
2. Type **2 7**. The screen will say MEM-CH (Memory Channel).
3. Press **MENU** to unlock.
4. Use the arrow keys (or type 0 0 1) to select Channel 001.
5. Press **MENU**. The robotic voice will say **"Receiving Memory."**

STOP

Do not press Exit yet! The Baofeng just saved the *listening* frequency, but it hasn't locked in the *talking* frequency and offset math yet. We have to save it a second time.

1. Press **MENU** again. The arrow drops down.
2. Make sure Channel 001 is still selected.
3. Press **MENU** one more time. The voice will say **"Transmitting Memory."**
4. Now, press **EXIT**.

YOU'RE WINNING

You did it! Press the orange **VFO/MR** button to switch over to Memory Mode. Navigate to Channel 001. You should see your repeater frequency. When you press the PTT button, you should see the frequency on the screen physically jump to the new talking lane, and jump back when you release it. You have successfully programmed a Baofeng manually!

Quick Win Checklist

- [ ] I always clear/delete a memory channel before trying to save new data into it.
- [ ] I successfully typed my repeater frequency into VFO mode.
- [ ] I set my Offset Direction (Menu 25) and Offset Amount (Menu 26).
- [ ] I set my Transmit Tone (Menu 13) if the repeater requires one.
- [ ] I followed the double-save sequence in Menu 27, hearing "Receiving Memory" and then "Transmitting Memory."
- [ ] I have switched back to Memory Mode and tested the channel recall.

If you're stuck, go to...

- *Chapter 12: My Channels Won't Save* (if you switch to Memory Mode and the channel is blank or behaves strangely).

Narrative Interlude: Why the Repeater Hears You, Or Doesn't

Offset for Humans

Character: Mark Jensen (Midwest Suburbs)

Skills reinforced:

- Offset / Duplex understanding
- Transmit Tone vs. Tone-Squelch
- Keypad Programming
- The "Double Save" Order

Where to learn the steps: Chapters 4 & 5

Mark Jensen stood at the end of his driveway, bathing in the crisp Minnesota evening air, standing as tall as a suburban sentinel. He was convinced he was exactly one minute away from greatness.

In his right hand, he held the Baofeng UV-5R. He was officially a licensed ham. He had conquered the squelch wall in his kitchen two days ago, and now, he was ready to graduate from merely eavesdropping on the automated weather robot to talking to actual human beings on the primary Twin Cities repeater.

He had his cheat sheet in his left hand. *Receive Frequency: 146.700. Minus offset. Tone 114.8.*

"Okay. This is... a lot of numbers," Mark muttered to himself. "But I am a man of science and precision."

He punched 1 4 6 7 0 0 into the keypad. Instantly, the radio crackled to life. Two guys were discussing the best type of coax cable to run through a dry-walled garage ceiling. Mark grinned. He was at the door of the club. Now he just needed to walk in.

He pressed the Push-To-Talk button. He took a breath, ready to announce his callsign to the world.

Nothing happened. The two guys just kept talking right over him.

Mark frowned. He tried again, holding the radio closer to his face like a stubborn slice of pizza. Still nothing. The radio didn't even make that polite, secondary *beep* the book said repeaters make when they hear you.

"Right, the password," Mark said out loud. He dove into the menus. He remembered seeing something about 'Tones'. He aggressively pressed the menu button, scrolled until he found R-CTCS, and proudly typed in 114.8. *Receive Tone. Obviously,*

Mark thought. *I want to receive them, don't I? Perfect.* Then he remembered the offset thing. He went to OFFSET and typed 000600.

He hit Menu 27 to save it to Channel 001, heard the robotic voice say *"Receiving Memory,"* and excitedly pressed Exit.

He switched to Memory Mode, pulled the radio to his mouth, and pressed PTT.

Silence. Not only was the repeater ignoring him, but the two guys talking about coax cable had completely vanished. The radio was as dead as a brick.

"I touched one button and now I'm in another universe," Mark groaned, letting his arms flop to his sides. "I've broken it. I've been banned from the airwaves before I even said hello."

"Having a fight with the tactical walkie-talkie?"

Mark jumped, feeling like a teenager caught playing with a toy lightsaber. His neighbor, Ron, was leaning over the property line fence, holding a pair of garden clippers. Ron was in his late sixties, a retired electrical engineer, and possessed an aura of relentless, practical calm. More importantly, Ron was an Extra Class ham radio operator. He had an antenna on his roof that looked like a metallic spider web.

"Ron. Hey," Mark said, trying to casually hide the Baofeng behind his leg. "Just... testing propagation. Atmospheric stuff."

"Right," Ron chuckled, walking over to the edge of Mark's driveway. "Let me guess. You can hear the repeater, but it acts like you don't exist. So you went into the menus, changed a bunch of things, and now you can't hear anything at all."

Mark blinked. "Are you psychic?"

"I'm experienced," Ron smiled. "I heard the Baofeng error beeps from my tomato garden. Hand it here."

Mark handed over the radio. Ron adjusted his glasses, glanced at the screen for half a second, and immediately shook his head. "Well, first off, you put a bouncer on your own door, Mark."

"A what?"

"Menu 11. R-CTCS. You set a receive tone. That tells your radio to ignore everything in the world unless the incoming signal has a secret password. The repeater doesn't send passwords back to you, so your radio just locked itself inside a soundproof vault." Ron pressed a few buttons with blazing speed. "Turned that off. Keep your ears open. Now, let's look at your transmit side."

"I did the offset," Mark protested defensively. "Six hundred. And I saved it."

"Let's gather the four numbers first," Ron said patiently, holding up four fingers. "No guessing. What's the receive frequency, the offset direction, the offset amount, and the transmit tone?"

Mark read them off his wrinkled cheat sheet.

"Good. Now, here is the secret of the repeater," Ron explained, gesturing with his garden clippers. "It's a two-lane highway. You listen on the southbound lane, but you have to talk on the northbound lane. If you just shout into the southbound lane, you're going the wrong way, and the tower's microphone never hears you."

Mark nodded slowly. "So the offset is just... the radio changing lanes?"

"Exactly," Ron said. "But the Baofeng is a very stubborn student. You have to teach it exactly how to change lanes, and you have to save it in a very specific order."

Ron handed the radio back to Mark. "Do it with me. Let's sweep the floor first. Menu 28. Delete Channel 001. Get rid of the corrupted file you just made."

Mark followed the steps. *Delete Channel. Confirm.*

"Now," Ron instructed. "Top line. VFO mode. Type the receive frequency."

Mark typed 1 4 6 7 0 0.

"Menu 25. Shift Direction. Make it a minus." Mark did so. "Menu 26. Offset. Type 000600." Mark confirmed it. "Menu 13. Transmit Tone, or T-CTCS. This is your VIP password for the tower. Set it to 114.8."

"Done," Mark said, his thumbs starting to sweat.

"Now, the trapdoor," Ron said, stepping back and folding his arms. "Menu 27. You have to do the double-save. If you just save it once and exit, the radio only remembers the listening lane. You have to save the talking lane, too."

Mark went to Menu 27. He selected Channel 001. He pressed Menu. *"Receiving Memory,"* the robot voice chimed. "Don't touch Exit," Ron warned. "Press Menu again." Mark pressed it. The arrow dropped. He pressed it a third time. *"Transmitting Memory."*

"Boom," Ron said. "You just saved both lanes. Exit out. Switch to Memory mode. Try it."

Mark pressed the orange VFO/MR button. The tiny 001 appeared on the right. He pulled the radio up to his mouth. He pressed the PTT button, but before he could speak, he looked at the screen. The numbers had instantly jumped from 146.700 down to 146.100.

He let go of the button. The numbers jumped right back to 146.700. Then, the radio let out a crisp, polite *beep* from the speaker, followed by a tiny burst of static.

"The repeater tail," Ron grinned. "The tower heard you, opened the door, and closed it again."

"Look at that," Mark breathed, staring at the plastic device with newfound reverence. "It actually worked. The math actually worked."

"Repeaters aren't hard, Mark," Ron said, tapping the Baofeng's antenna lightly with his clippers before turning back toward his yard. "Guessing is hard. Stop guessing. Get your four numbers, check your bouncers, and follow the exact saving order. You'll be fine."

Mark stood in the driveway alone again. The sun was setting, casting a golden hour glow over the suburbs. He had done it. He had a fully functioning, programmed repeater saved to Channel 1. He wasn't locked out anymore.

He looked down at the PTT button. The technical hurdle was gone. The only thing standing between him and the rest of the amateur radio world was his own voice.

"Okay," Mark said aloud to the empty driveway. "This is the part where I pretend I'm not terrified of pressing the button."

Takeaway

Repeaters aren't hard, guessing is hard. Never set a Receive Tone (R-CTCS) unless you absolutely have to, or you will lock yourself out of hearing traffic. Gather your four pieces of data (RX Freq, Offset Direction, Offset Amount, TX Tone) and always finish with the "Double-Save" to ensure both the listening and talking lanes are locked into memory.

Bridge: Mark's radio is programmed perfectly. Now comes the scary part: pressing PTT and speaking to the world for the first time.

Chapter 6

Your First On-Air Moment (What to Say + What NOT to Do)

You've programmed the repeater perfectly. You switch over to Memory Mode, select Channel 1, and the screen looks exactly right. Now, you're staring at the Push-To-Talk (PTT) button like it's the launch button for a nuclear missile.

"Mike fright" is incredibly common. The thought of broadcasting your voice to a whole city can cause your brain to freeze. You worry about sounding foolish, breaking a rule, or using the wrong jargon. Take a deep breath. Every single expert on the radio today was terrified the first time they pressed that button. Think of this like using an intercom at a hardware store to ask a quick question, not giving a presidential address. Nobody expects you to sound like a professional broadcaster. We are going to keep it short, polite, and completely painless.

What you'll accomplish:

- Learn the absolute simplest, stress-free script for your first legal transmission.
- Understand the 5 golden rules of radio etiquette that prevent awkwardness.
- Understand basic identification rules so you stay on the right side of the law.

What you need:

- Your Baofeng radio, tuned to your newly saved repeater channel.
- An Amateur Radio License (Required before you press the PTT button!).
- Your official Call Sign.

The 5 Golden Rules of Etiquette

Before we talk, we need to know how to behave. Ham radio is a community, and like any community, there are unwritten rules of politeness.

1. **Listen First:** Never turn on a radio and immediately start talking. The channel might be in the middle of a serious conversation, an emergency, or a weekly club meeting. Wait at least 30 to 60 seconds to ensure the airwaves are clear.

2. **Press, Pause, Speak:** When you press the PTT button, do not start speaking immediately. It takes about a half-second for the repeater tower's computer to hear your invisible tone, verify your password, and physically open its gates. Think of it like walking up to the automatic doors at a grocery store, you have to pause for a beat to let them open before you walk through. If you speak instantly, your first word will bounce off the closed door and be cut off. Press the button, take a quick breath, and *then* speak.

3. **No "CQ" on Repeaters:** You might have seen movies where radio operators yell "CQ, CQ, CQ" (calling any station). This is for long-distance, static-filled global radio. On a crisp, local city repeater, yelling CQ sounds bizarre. We just use plain English.

4. **Don't Overstay Your Welcome:** Repeaters have automatic timers built-in. If you hold the button down and talk non-stop for 3 minutes, the repeater will automatically shut off (time-out) to prevent overheating. Keep your transmissions to 30 seconds or less. Talk, then listen.

5. **No Profanity:** The airwaves are public, monitored by the FCC (or local equivalents), and families are often listening. Keep it PG-rated.

The Simplest Script: "Monitoring"

You do not need a reason to talk on a repeater. You don't need a specific topic. The most common transmission in ham radio is simply letting people know you are alive, your radio works, and you are listening if anyone wants to chat.

Here is the exact script. Just replace the bracketed text with your official call sign (e.g., K-1-A-B-C).

(Press PTT... wait half a second...) "This is *Your Call Sign*, monitoring." *(Release PTT)*

That's it. Two words. You have officially made your first transmission. If the repeater heard you, you will usually hear a short burst of static, or a polite "beep" from the tower acknowledging you, when you release the button.

If someone is out there driving in their car, they might reply: "Hey *Your Call Sign*, this is *Their Call Sign*. Good morning, how are you doing today?" If they do, just reply normally! "Doing great, just testing out a new Baofeng radio. Thanks for coming back to me."

Alternative Script: The Radio Check

If you specifically want to know how your audio sounds, you can ask for a "signal report." Hams love giving signal reports; it makes them feel helpful.

(Press PTT... wait half a second...) "This is *Your Call Sign*, testing a new radio. Can anyone give me a quick signal report?"

(Release PTT)

Someone will likely come back and say something like, *"You are loud and clear, full quieting,"* or *"You are a little scratchy, maybe step outside."* Thank them, give your call sign one more time to close out the conversation, and you are done.

Practice: The Mirror Drill

Before you press the button for real, practice out loud in an empty room.

FIELD NOTE

Phonetic Alphabet: Say your call sign phonetically if you can (e.g., Kilo-One-Alpha-Bravo-Charlie). Phonetics are used because letters like 'B', 'C', 'D', and 'E' all sound identical over static. However, plain letters are perfectly fine for a beginner.

Hold the radio about two inches from your mouth. Speak *across* the face plate of the microphone rather than pointing it directly into your lips like a rockstar, this prevents your breath from 'popping' loudly on the audio. Speak in your normal, calm, indoor voice. Do not yell. Say the script out loud three times. Muscle memory defeats anxiety every single time.

Quick Win Checklist

- [ ] I understand that I must possess a valid amateur radio license to legally perform this step.
- [ ] I have listened to my chosen channel for at least 60 seconds to ensure it is clear.
- [ ] I understand the "Press, Pause, Speak" rule to prevent my first words from being cut off.
- [ ] I have practiced my short script ("This is *Call Sign*, monitoring") out loud in an empty room.
- [ ] I know not to yell "CQ" on a local FM repeater.
- [ ] I have successfully pressed the PTT button and made my first on-air transmission!

If you're stuck, go to...

- *Chapter 9: I Can Hear Them, But They Can't Hear Me* (if you made your call, but nobody replied, and you didn't hear the polite "beep" from the repeater).

Chapter 7

Save a Clean "Home Channel Plan" (5 Channels You'll Actually Use)

You've successfully programmed one repeater and made your very first transmission. Take a moment to appreciate that. You are lightyears ahead of most people who buy this radio, turn it on, hear static, and throw it into a closet in frustration.

But having just one working channel isn't quite enough for real preparedness, and falling into the trap of having 128 random, unnamed frequencies you downloaded from the internet is overwhelming. A radio full of mystery channels isn't a communication tool; it's just a noisemaker.

If an actual emergency happens, a severe winter storm knocks out regional power, or cellular networks go down due to overload, you do not want to be fumbling with VFO mode, offset math, and confusing menu settings in the dark. You need a small, highly organized, hyper-reliable set of channels you trust with your life. You need to build a "Home Profile."

What you'll accomplish:

- Understand the psychology behind the 5-channel "Home Profile" and why less is actually more.
- Organize your memory channels logically so you can find exactly what you need in the dark, under stress.
- Create a physical, weatherproof paper backup so your family can use the radio even if you aren't around to help them.

What you need:

- Your Baofeng radio.
- A small index card or heavy piece of paper.
- A thick permanent marker.
- Clear packing tape (to laminate and waterproof your card).

The "Top 5" Philosophy: Defeating Cognitive Overload

Your Baofeng holds up to 128 distinct channels in its memory bank. When new owners discover this, the immediate temptation is to plug the radio into a computer and fill every single slot with police scanners, fire dispatch, ambulance channels, and distant repeaters from three counties away.

Don't do it. Treat your radio like a neatly organized toolbox, not a junk drawer.

When human beings are under severe stress, like during a natural disaster or a medical emergency, our brains drop complex tasks. Fine motor skills degrade, your heart rate spikes, and tunnel vision sets in. If you have to desperately scroll through 90 channels of chaotic static and irrelevant police dispatch to find the *one* frequency your family uses to talk, the radio has become a liability, not an asset.

We highly recommend saving your 5 most important, mission-critical frequencies into Memory Channels 001 through 005. Think of these exactly like the emergency speed-dial favorites on your smartphone. You want to be able to turn the radio on and click up exactly three times, purely by muscle memory, knowing exactly who will be on the other end.

The 5 Channels You Actually Need

Here is the perfect, stress-proof template for a beginner's Home Profile:

Channel 001: The Primary Local Repeater This is the reliable "Lighthouse" we found and programmed in Chapters 3 and 5. This is your daily driver. It is your most powerful tool for reaching out across the city to get critical information, ask for assistance, or make contact with local emergency nets.

Channel 002: The Backup Repeater Radio towers lose power, too. Backup generators run out of diesel, and heavy winds can snap antennas. You need redundancy. Jump back on RepeaterBook and find a second repeater, preferably one located on a completely different mountain or the opposite side of town, and program it exactly like you did the first one. If Channel 1 goes dead, you immediately click over to Channel 2.

Channel 003: Simplex 1 (Primary Family Channel) "Simplex" means pure, radio-to-radio direct communication without using a repeater tower as a middleman. It is strictly short-range (usually 1 to 3 miles in a suburban neighborhood), but it is the ultimate backup because it relies on zero external infrastructure. It works even if the entire city grid collapses. Pick a standard 2-meter simplex frequency (like 146.520, the national calling frequency, or 146.415) and save it. No offsets, no tones. Just your radio talking directly to their radio.

Channel 004: Simplex 2 (Backup Family Channel) If Channel 3 is suddenly busy with other people talking, you need a pre-planned alternate lane to switch your family over to. Pick a frequency in the 70-centimeter band (like 446.000) and save it. Having one 2-meter backup and one 70-centimeter backup gives you options if one frequency band is acting up due to local interference.

Channel 005: Local Weather Radio (Receive Only) The automated NOAA weather broadcast you found in Chapter 3. Save this to Channel 5. *Important note: The Baofeng physically cannot transmit on weather frequencies, but having it in your top 5 quick-dial means you can instantly get emergency weather updates with a single click, without having to type numbers in VFO mode.*

Memory Layout Strategy (The A/B Defensive Setup)

Because the Baofeng has a top line (A) and a bottom line (B) on the display screen, you have the unique ability to actively monitor two of these vital channels at the exact same time.

A fantastic defensive posture for a storm or regional power outage is to set the Top Line (A) to **Channel 001** (your primary city repeater to listen for wide-area news and emergency nets) and set the Bottom Line (B) to **Channel 003** (your family's short-range simplex channel).

The radio will rapidly scan both simultaneously. If the city tower speaks, you hear it. But if your spouse calls you from down the street on the simplex channel to say they are walking home safely, the radio will instantly switch down to the bottom line to let you answer them.

The Physical Backup (The "Hit By A Truck" Rule)

Electronics fail. Batteries die in the cold. Memory banks occasionally get accidentally wiped by a clumsy thumb pressing the wrong menu button in a dark tent. If your radio resets to the factory defaults during a crisis, all your hard work is instantly gone. Furthermore, if you are injured or away from the house, your family members need to know how to use the radio without reading this entire book.

Take a small, heavy index card. Write your 5-Channel Home Profile on it in dark, bold permanent marker so it is highly visible.

- **CH 1:** Primary RPT (146.700 / - / Tone 100.0)
- **CH 2:** Backup RPT (444.100 / + / Tone 123.0)
- **CH 3:** Family Direct 1 (146.415 / No Tone)
- **CH 4:** Family Direct 2 (446.000 / No Tone)
- **CH 5:** Weather Alert (162.550)

Cover both sides of this index card entirely in clear packing tape to make it waterproof and tear-resistant. Tape this card directly to the back of the radio (right over the battery block), or fold it up and place it prominently inside the padded case you keep the radio in. If someone else needs to pick up your radio, they can look at this card, look at the screen, and instantly understand exactly how to reach you.

Quick Win Checklist

- [ ] I understand why a small, trusted "Home Profile" is vastly superior to a cluttered 128-channel list during a crisis.
- [ ] I have successfully programmed my Primary and Backup repeaters into Channels 1 and 2.
- [ ] I have successfully programmed two direct "Simplex" channels into Channels 3 and 4 for pure, short-range family use.
- [ ] I have successfully programmed a receive-only Weather frequency into Channel 5.
- [ ] I understand how to use the A/B dual-display to monitor both the city repeater and my family channel at the same time.
- [ ] I have written my 5-channel plan down on a physical index card, waterproofed it, and attached it to my radio kit.
- [ ] I have completed Part 1! I can officially operate a Baofeng radio with confidence!

If you're stuck, go to...

- *Chapter 15: Reset and Recover* (if you accidentally filled your memory channels with junk, got overwhelmed, and want to completely wipe the radio clean to start fresh).

Narrative Interlude: Thumb Off the Trigger

Character: Nate Brooks (Southeast)

Skills reinforced:

- First-call script (practice)
- Etiquette (listen-first mindset)
- The 5-channel Home Profile
- Creating a paper backup

Where to learn the steps: Chapters 6 & 7

The thick South Carolina heat had finally broken, leaving behind a muggy, cricket-filled evening that smelled faintly of distant thunderstorms and wet asphalt. Nate Brooks sat at his garage workbench, the overhead LED lights casting a stark, surgical glow over his neatly arranged gear. Everything had its place: the water filtration kits on the bottom shelf, the medical go-bags on the middle, and his communications array taking center stage. In front of him, the Baofeng UV-5R stood upright in its charging cradle base, its green screen illuminated as it quietly scanned the local frequencies.

He had spent the last hour meticulously following the programming guide. He had successfully locked the primary Charleston repeater into Channel 1, complete with the correct minus offset and the exact DCS tone required to digitally open the tower's squelch gate. He had done the math. He had swept the memory floor clean of junk frequencies and executed the double-save precisely as instructed. The infrastructure was built.

Suddenly, the radio broke the silence with a crisp burst of static, followed by a calm, authoritative voice that filled the garage.

"This is W4ABC, Net Control for the Tri-County Skywarn training net. We are now taking check-ins from mobile and portable stations. Please call now."

Nate leaned forward, resting his forearms on the workbench. He could hear a slight pause, then another voice chimed in, clear and professional.

"W4ABC, this is Kilo-Four-Echo-Mike. Portable station, Mount Pleasant. Sky is overcast, no precipitation. Over."

"Acknowledged, K4EM. Thank you for the check-in," the Net Control operator replied warmly.

It was an orderly, disciplined roll call of people proving their equipment worked. There was no talking over one another, no CB-radio slang, and no panic. Just a methodical network of citizens creating a web of situational awareness across the county. It was exactly the kind of network Nate wanted to be a part of.

Nate picked up the Baofeng. The plastic casing felt solid and utilitarian in his palm. His thumb naturally found its way to the large, textured rubber Push-To-Talk (PTT) button on the left side of the chassis.

Just one click, a voice in his head rationalized. *Just give a fake name and ask for a quick radio check. How else are you going to know if the offset math actually worked? You need to verify the transmit chain before hurricane season officially starts. It'll take five seconds.*

He applied a fraction of an ounce of pressure to the PTT button. He could feel the mechanical switch beneath the rubber beginning to yield, right on the edge of breaking the squelch gate and broadcasting his voice to three counties.

Then, he stopped. He looked at the radio, then at the storm-prep bins stacked neatly against the garage wall. Each bin represented a commitment to doing things the right way, no shortcuts, no half-measures. He pulled his thumb completely off the trigger and set the radio back down on the workbench with a soft clack.

"Discipline," Nate muttered quietly to himself, staring at the green screen.

He wasn't licensed yet. He didn't have a call sign. Pressing that button right now wouldn't be a valid test of his gear; it would be a failure of his own operational discipline. The radio was just a machine, it would blindly do whatever he commanded. The real vulnerability in his preparedness plan wasn't the antenna, the battery, or the microchips. It was the operator. It was his own impatience. If he couldn't follow the basic, established legal rules of the airwaves on a perfectly calm Tuesday night, how could he trust himself to stay grounded and methodical when a Category 4 hurricane was churning off the coast and the cell networks were dead?

Nate took a deep breath, letting the tension leave his shoulders. The radio worked perfectly. He had proven the receive side beyond a shadow of a doubt. The transmit side would simply have to wait until he had earned the legal right to use it. But that didn't mean he couldn't finalize the architecture of his system tonight.

He opened his logbook notebook to a blank page. If an emergency happened while he was at work, he wouldn't be the only one needing to operate this radio. His wife, Kara, had made her boundaries incredibly clear when he first brought the box home: *One page. No menus. No lectures.* She was pragmatic. She wasn't going to memorize frequency offsets, tone squelch codes, or standard operating procedures. She needed a tool, not a new hobby.

Nate grabbed a bright yellow index card, chosen specifically for high visibility by flashlight, and a thick black permanent marker. He began to draft his "Home Profile." He needed a small, locked-down ecosystem of channels that required zero guesswork for anyone who picked it up.

In crisp, block letters, he wrote:

CH 1: PRIMARY (Charleston City Repeater - City Info) CH 2: SECONDARY (North County Backup Repeater) CH 3: DIRECT 1 (Neighborhood Simplex - 1 Mile) CH 4: DIRECT 2 (Backup Simplex) CH 5: WEATHER (Receive Only - NWS)

He reviewed the list, nodding in satisfaction. It was clean. It was redundant. It provided two avenues for long-range city intelligence, two avenues for short-range neighborhood security and family coordination, and an automated weather beacon. Five channels were all they realistically needed. Anything more would just become mental clutter during a crisis.

Nate pulled a roll of clear, heavy-duty packing tape from his tool drawer. He carefully laid strips of tape over the front and back of the yellow index card, pressing out the air bubbles with his thumb until it was completely sealed, waterproof, and tear-resistant. He punched a small hole in the top corner and used a short loop of high-visibility orange paracord to tie the card directly to the zipper of the padded radio pouch.

"Okay," Nate said, looking at the finished product. "The system is built and documented. Now we drill."

He picked up the radio again. He deliberately kept his thumb away from the PTT button this time. Instead, he looked at the blank garage wall, visualizing a real grid-down scenario. He practiced the etiquette of the "listen-first" mindset, counting a full *one-Mississippi, two-Mississippi* in his head to wait for the imaginary repeater tail to drop. Muscle memory, he knew, wasn't just for your hands; it was for your vocal cords, too.

He held the radio two inches from his mouth, angled slightly past his cheek to avoid popping his P's into the microphone, and spoke aloud in the empty garage.

"This is... well, pending call sign. Monitoring." He paused, feeling slightly foolish talking to a wall, but knowing the absolute necessity of vocal repetition under zero stress. "Let's try the check-in format. Who, Where, What, Next."

He cleared his throat and tried again, speaking with a steady, flat cadence designed to cut through static. "Kara, this is Nate. I am secure at the neighborhood entrance. Roads are clear of debris. I will check back on this channel at the top of the hour. Over."

"Loud and clear," a voice said from the doorway.

Nate turned. Kara was standing there, the condensation dripping down the sides of two glasses of iced tea she was holding. She walked over, set one down on the workbench, and let her eyes fall on the bright yellow, laminated index card tied to the radio pouch.

She picked it up, her eyes scanning the five channels.

"Channel Three to talk to you directly, Channel One to hear what the city is doing," she read aloud, tracing the heavy marker lines with her finger. She looked at him, raising an eyebrow in genuine surprise. "You actually distilled it down to five lines? No digging into menus? No three-hour lecture on radio wave physics?"

"No menus," Nate confirmed, taking a sip of the iced tea. "You turn the top knob on. You use the arrows to go to Channel 3. You press the side button and talk like a normal person. That's the entire system."

Kara nodded slowly, clearly approving of the brutal, elegant simplicity. "I can do that. It makes sense. Does it actually reach the city tower from here?"

"The receive does, I've confirmed that," Nate said. "I can't verify the transmit yet. Not legally, anyway."

He reached past the radio, opened his laptop on the corner of the bench, and quickly navigated to the local amateur radio club's website. He clicked on the testing schedule, found the session listed for that upcoming Saturday morning at the local community center, and typed in his credit card information to pay the fifteen-dollar exam fee.

"But I will be able to verify it by Saturday afternoon," Nate added, closing the laptop with a definitive snap.

He looked back at the radio. It was no longer a mystery box of confusing buttons or a temptation to break the rules. It was a fully prepped, documented, and organized communications node. It was ready for the exact moment he earned his call sign. There was no panic, and there was no guessing. He had built the plan, established the discipline, and locked in the procedure before he ever needed to press the button.

Takeaway

Discipline beats impulse. You can build the system now, even before you transmit. By establishing a clean, 5-channel Home Profile and creating a laminated, physical backup card, you ensure your family can use the radio safely without needing a technical lecture. Practice your scripts out loud so muscle memory takes over when it counts.

Bridge: But if something breaks, I need a triage plan, not a spiral.

Part 2:

Fix It Fast (Troubleshooting Tree)

Chapter 8

Troubleshooting: How to Diagnose in 60 Seconds

You press the Push-To-Talk button, expect to hear a friendly reply, and instead... nothing happens. Or worse, you turn the radio on and it immediately screams a wall of deafening static at you. When your communication equipment fails, especially if you are outdoors, in the dark, or in the middle of a stressful situation, your heart rate naturally goes up.

Your first instinct is to assume you bought a defective piece of junk or that you broke it beyond repair. You don't want to read a twenty-page chapter on radio wave propagation theory or dig through a badly translated factory manual right now. You just want to know exactly which button to press to make the machine work again.

Welcome to the triage tent. Take a deep breath. In this section, we abandon deep theory entirely and focus strictly on rapid, symptom-based problem solving to get you back on the air in 60 seconds or less.

What you'll accomplish:

- Shift your mindset from "my radio is permanently broken" to "my radio just has a setting out of place."
- Master the "Big Three" checkpoints that solve 95% of all Baofeng problems in under a minute.
- Build massive confidence in your ability to fast-debug your own gear in the field without relying on the internet.

What you need:

- Your misbehaving Baofeng radio.
- A calm, methodical mindset.

The Triage Mindset

When a car won't start, a good mechanic doesn't immediately assume the engine has exploded and the car is totaled. They check to see if there is gas in the tank, they check the battery connections, and they verify the key is actually turned.

Your Baofeng is a tough little industrial computer. It can survive being dropped in the dirt, left in a cold truck, or bumped around in a hiking backpack. It is highly unlikely that the internal circuit board is physically broken. Almost every single "failure" a beginner faces is simply a software setting that got bumped by a stray palm, changed by accident, or saved incorrectly in the confusing menu system.

If you can shift your mindset from "this radio is broken" to "a setting is out of place," you will save yourself hours of frustration. If we systematically check the core settings, we will find the cure every single time.

The "Big Three" Checkpoints

Before you dive into the specific symptom chapters ahead, run your radio through this quick 60-second diagnostic flow. These three checkpoints solve the vast majority of user errors.

Checkpoint 1: Are you in the right mode? The Baofeng's split personality (VFO vs. Memory) is the number one cause of user lockouts. If you are trying to type a raw frequency manually using the keypad, but you see tiny channel numbers (like 001 or 015) on the far right edge of the screen, the radio is going to actively fight you. It will beep angrily and ignore your typing.

- *The Fix:* Press the orange **VFO/MR** button until the small channel numbers on the right side of the screen completely disappear. You are now safely in VFO mode. Remember: Memory mode is for securely listening to what you've already built; VFO mode is your blank canvas for making changes.

Checkpoint 2: Is the black arrow pointing at the right line? The Baofeng has a dual-watch display featuring a top line (A) and a bottom line (B). It is incredibly common to meticulously program a perfect frequency on the top line, setting your offsets and tones exactly right, but accidentally leave the active control arrow pointing to the bottom line. When you press Push-To-Talk, you are broadcasting on completely the wrong frequency. You feel like the radio is ignoring you, but it's actually doing exactly what the arrow told it to do.

- *The Fix:* Always look for the tiny black triangle on the left side of the screen. Before you press the PTT button, press the blue **A/B** button to physically move that little black arrow to the line you actually intend to use. Making this a habit prevents 50% of all transmission errors.

Checkpoint 3: Is the battery actually seated? If the radio is acting truly possessed, turning off randomly in your hand, the screen flickering or dimming when you press the PTT button to talk, or the radio completely restarting itself in the middle of a sentence, you do not have a software problem. You have a physical power delivery problem. Brand new Baofengs have famously stiff plastic battery rails, and it's easy to think the battery is locked when it's actually loose. A microscopic disconnect of the metal contacts will cause the radio's computer to instantly reboot and reset your active screen.

- *The Fix:* Remove the battery entirely by pressing the release latch. Wipe the metal contacts on both the battery and the back of the radio with a clean shirt to remove any factory grease or dust. Slide the battery back onto the rails, and press upward firmly with your thumbs until you hear a loud, undeniable *click*.

YOU'RE WINNING

If you can memorize these three checkpoints, you are officially the designated "radio tech" for your family or friend group. You will look like an absolute wizard to other beginners just by pressing the VFO/MR or A/B button to magically "fix" their broken radios.

Quick Win Checklist

- [ ] I understand that 95% of radio problems are just bumped settings, not broken hardware.
- [ ] I know to check VFO vs. Memory mode first if the keypad is ignoring me.
- [ ] I know to check where the A/B active arrow is pointing before I transmit.
- [ ] I know to check my physical battery connection if the radio acts erratic or restarts.
- [ ] I am ready to use the symptom chapters ahead to fix my specific problem.

If you're stuck, go to...

- *Chapter 9* (if your symptom is: you can hear them, but they can't hear you).
- *Chapter 10* (if your symptom is: the radio is completely silent or blasting static).
- *Chapter 12* (if your symptom is: your saved channels keep disappearing).

Chapter 9

I Can Hear Them, But They Can't Hear Me

The Symptom: You are tuned perfectly to a local repeater. You hear the automated weather bot, or you hear two older guys talking about their antennas with crystal-clear audio. Taking a deep breath, you press the Push-To-Talk (PTT) button, proudly announce your call sign, and release the button.

Dead silence. Crickets. Nobody replies. You don't even hear the polite, robotic "beep" from the repeater tower confirming it heard your transmission.

This is undeniably the most frustrating beginner symptom in the entire hobby because it makes you feel completely invisible, or worse, like you've broken your brand new gear. But don't worry, this is a rite of passage. Almost every new radio owner experiences this exact moment. The fix is almost always a simple software setting related to the invisible "bouncer" at the repeater's door.

What you'll accomplish:

- Diagnose and fix incorrect Offset (Duplex) settings by visually watching the screen.
- Diagnose and fix incorrect CTCSS/Tone settings (the VIP password).
- Verify your physical power output and understand your environmental range limitations.

What you need:

- Your Baofeng radio.
- The original RepeaterBook listing (Frequency, Offset, and Tone) for the specific tower you are trying to hit.

Fix Tree Step 1: Confirm the Duplex (Offset)

Remember the "two lanes of the highway" concept from Chapter 4? Repeaters listen on one lane and talk on another to avoid a deafening feedback loop. If you don't automatically jump over to the repeater's listening lane the exact moment you press PTT, you are just shouting at a brick wall.

1. Go to the channel or frequency you are having trouble with.
2. Press the **PTT** button briefly while keeping your eyes glued to your Baofeng's screen.
3. *Did the frequency numbers physically change while you were holding the button down?*

- **If NO:** Your offset is missing. The numbers stayed stubbornly frozen, meaning you are trying to talk to the tower's speaker instead of its microphone. You need to go into **Menu 25 (SFT-D)** and set it to + or - according to your notes. Then, go to **Menu 26 (OFFSET)** and ensure the amount (usually 000600 or 005000) is typed in.

- **If YES:** You literally saw the numbers drop or jump (e.g., from 146.700 down to 146.100). The math is working perfectly. Move to Step 2.

Fix Tree Step 2: Confirm the Transmit Tone

If the radio is jumping to the correct frequency lane, but the tower is still completely ignoring you, the tower's automated "bouncer" is rejecting your signal. You are standing at the right door, but you aren't whispering the secret VIP password.

1. Check your repeater notes. Does the tower require an "Uplink Tone" (like 100.0)?

2. Press **MENU** and go to **Menu 13 (T-CTCS)**.

3. Is it set to OFF, or is it set to the wrong number?

4. Unlock the menu, use the arrows to carefully select the exact tone listed in your notes, and press **MENU** to confirm. Remember, this tone is sub-audible, you will never actually hear it with your human ears, but the tower's computer will.

STOP

The Receive Tone Trap: Check **Menu 11 (R-CTCS)** right now. Is there a number in there? If so, turn it **OFF**. You should almost never have a Receive Tone turned on as a beginner. Setting a receive tone is like putting heavy earplugs in your own ears. You are telling your Baofeng, *"Ignore everyone and keep my speaker muted unless they transmit a secret password to me."* Because repeaters rarely send passwords back to you, your radio will stay totally silent, even when people are trying to answer your call.

Fix Tree Step 3: Check Your Power Level

If your offset is right and your tone is perfect, but the tower still can't hear you, you might just be whispering too quietly for the distant microphone to pick you up.

Think of transmit power like the volume of your own voice. Your Baofeng can transmit at High Power (usually 5 to 8 watts) or Low Power (1 watt). Low Power is like an indoor conversational voice, great for talking to a buddy in the same campsite and saving battery life, but terrible for shouting across town.

1. Look at your screen. Do you see a tiny **"L"** on the left side? If so, you are in Low Power mode.

2. *The Fix:* Tap the **# / Padlock key** once quickly (do not hold it down). The "L" should immediately disappear, putting you back into High Power mode (the equivalent of shouting). Try your transmission again.

Fix Tree Step 4: Move Your Body (The Line-of-Sight Rule)

VHF and UHF radio waves act exactly like a beam of light from a flashlight. They do not bend around the earth, and they absolutely hate thick concrete, steel framing, and dense rocky hills.

If a flashlight beam cannot pass through the wall of your house, neither can your radio signal. If you are sitting on the couch in your basement trying to hit a repeater 15 miles away, the laws of physics are working against you.

- *The Fix:* Change your environment. Walk outside into the yard. Stand on an elevated back deck. If you are sitting in your car, remember that the vehicle is essentially a giant metal cage that bounces your signal right back at you, roll the window down or physically step out of the vehicle. Sometimes, just holding the radio three feet higher in the air or taking ten steps to the left can be the difference between a perfectly clear signal and being completely ignored by the tower.

Quick Win Checklist

- [ ] I watched my screen while pressing PTT to confirm my frequency is actually shifting (Offset).
- [ ] I checked Menu 13 (T-CTCS) to ensure my Transmit Tone password is set correctly.
- [ ] I checked Menu 11 (R-CTCS) to ensure I didn't accidentally lock my own radio by setting a receive tone.
- [ ] I verified my radio is in High Power mode (no tiny "L" on the screen).
- [ ] I physically moved outside or higher up to improve my line-of-sight to the tower.

If you're stuck, go to...

- *Chapter 11: I Can Hit One Repeater but Not Another* (if you applied all these fixes, and it works for Tower A, but Tower B is still stubbornly ignoring you).

Chapter 10

I Can't Hear Anything (Or It's Random and Noisy)

The Symptom: Your buddy is standing fifty feet away from you, pressing his PTT button and talking directly to you. But your radio is dead silent. Or, alternatively, your radio sits quietly on the desk and then randomly blasts you with five seconds of terrifying, ear-piercing static, then goes silent again.

If you can't hear incoming traffic, your radio is essentially a paperweight. Let's get your ears open.

What you'll accomplish:

- Fix the most common squelch and volume errors.
- Diagnose "Step Size" issues that push you off the correct frequency.
- Tame a radio that randomly blasts static for no reason.

What you need:

- Your Baofeng radio.

Fix Tree Step 1: The Squelch Wall

The absolute most common reason a radio goes deaf is that the invisible static gate (the Squelch) has been built too high.

1. Press the **MONI** button on the side of the radio. Do you hear rushing static?
   - *If YES:* Your speaker works. The problem is your settings.
   - *If NO:* Your volume knob is turned all the way down, or your battery is dead.
2. Press **MENU** and go to **Menu 0 (SQL)**.
3. If this number is set to 5, 8, or 9, your squelch is violently high. It is blocking out legitimate signals.
4. *The Fix:* Change the Squelch down to Level **1** or **2**.

Fix Tree Step 2: The Receive Tone Trap

We mentioned this in the last chapter, but it bears repeating because it ruins so many beginners' days.

If you set a Receive Tone (CTCSS/DCS), you are telling your Baofeng: *"Keep the speaker shut off unless the incoming signal has a secret VIP password."* If your buddy is transmitting to you on a standard simplex channel, he is not sending a password. Therefore, your radio ignores him.

- *The Fix:* Press **MENU** and go to **Menu 11 (R-CTCS)**. Ensure it is set to **OFF**. Do the same for **Menu 10 (R-DCS)**. Ensure it is also set to **OFF**. Keep your receiving doors wide open.

Fix Tree Step 3: The "Step Size" Rounding Error

The Symptom: You try to type 146.520 into the keypad. But when you hit the last zero, the radio magically changes it to 146.525. It feels like the machine is mocking you.

Your radio tunes through frequencies in specific "steps," like walking up a staircase. If your step size is too large (like 25.0kHz), the radio's computer physically cannot step on 146.520. It rounds up to the nearest large step.

- *The Fix:* Press **MENU** and go to **Menu 1 (STEP)**. Change this value to **2.5K**. This is a tiny, precise staircase that will allow you to type in almost any exact frequency without the radio fighting your inputs.

Fix Tree Step 4: Random Blasts of Static

The Symptom: The radio is quiet, then suddenly blasts raw static for a few seconds, then stops. Nobody is talking.

This usually means your Squelch is set *too low* (Level 0 or 1), or you are sitting near a massive source of electromagnetic interference. Cheap LED lightbulbs, computer monitors, and unshielded power supplies can actually leak radio waves that trick your Baofeng into opening its squelch gate.

- *The Fix:* Move the radio away from your computer desk or the running engine of your car. If the random static bursts continue, bump your Squelch (**Menu 0**) up from Level 1 to Level 2 or 3.

Quick Win Checklist

- [ ] I used the physical MONI button to verify my speaker actually produces sound.
- [ ] I checked Menu 0 (SQL) and ensured my squelch is reasonably low (Level 1, 2, or 3).
- [ ] I checked Menu 11 (R-CTCS) and Menu 10 (R-DCS) and made sure they are turned completely OFF.
- [ ] I set my Step Size (Menu 1) to 2.5K so the radio stops auto-changing my frequencies.
- [ ] I moved my radio away from computer monitors and LED lights to prevent random static bursts.

If you're stuck, go to...

- *Chapter 12: My Channels Won't Save* (if you applied these fixes in VFO mode, but the radio goes deaf again as soon as you switch to Memory Mode).

Narrative Interlude: When the Radio Goes Silent

The Silent Morning

CHARACTER: MARK JENSEN (MIDWEST Suburbs)

Skills reinforced:

- Triage flow
- Mode/Frequency/Tone checkpoints
- Squelch fixes
- Step-size sanity

Where to learn the steps: Chapters 8, 9, & 10

It was 6:15 AM on a Tuesday. Outside Mark Jensen's kitchen window, a fresh layer of Minnesota snow was furiously burying his driveway. Inside, the house was quiet, smelling of dark roast coffee and the comforting hum of the refrigerator.

Mark sat at the kitchen island, taking a sip of his coffee. This was his moment. He had his Baofeng UV-5R sitting proudly on the granite countertop. Over the weekend, he had successfully programmed his local repeaters, talked to Ron in the driveway, and felt like a certified communications expert. He knew there was a "morning commute net" on the Twin Cities repeater every weekday at 6:30 AM, where local hams swapped weather and traffic reports.

He reached out and confidently twisted the volume knob clockwise.

"BEEP. FREQUENCY MODE."

Mark waited for the familiar crackle of morning chatter.

Nothing.

Dead silence.

He frowned. He picked up the radio and held it to his ear like a seashell, hoping to catch even the faintest hiss of static. Nothing. The green LCD screen stared blankly back at him, displaying 146.700.

"Okay," Mark whispered. "Maybe nobody is awake yet. Let's check the weather."

He pressed the orange **VFO/MR** button to ensure he was in the manual typing mode. He started to punch in the National

Weather Service frequency he knew by heart: 1 6 2 5 5 0.

But as his finger hit the final zero, the radio gave a sharp, angry beep. The screen physically deleted his zero and replaced it with a five. It now read 162.555.

"What?" Mark muttered. He cleared it and tried again. 1 6 2 5 5 0.

Beep. The screen aggressively jumped to 162.555 again.

Mark's stomach dropped. The radio wasn't just deaf; it was actively fighting his inputs. He tried to type in a different local repeater, 146.520. The radio hijacked his typing and forced it to 146.525.

"It's broken," Mark said aloud, the panic rising in his chest. "I left it on the charger too long. Or the cold air by the window fried the motherboard. It's glitching."

His mind immediately went to the worst-case scenario. The cheap Chinese electronics had failed. His brief, glorious weekend as a radio operator was over. In a desperate bid to fix it, Mark started button-mashing. He pressed the UP arrow. The numbers went wild. He pressed the blue A/B button frantically. He hit MENU and started scrolling aimlessly through the 40 different options, looking for a magic "Fix Everything" button.

He landed on Menu 40: RESET.

He pressed MENU to unlock it. He scrolled to ALL.

He pressed MENU again. The radio flashed SOURCE?

Mark's thumb hovered over the MENU key. One more press, and he would nuke the radio from orbit. He would wipe out all his saved channels, his perfectly calculated offsets, and his transmit tones. He would be back to square one, sitting in the kitchen with a beeping brick.

"Having fun with the tactical walkie-talkie at six in the morning?"

Mark jumped. His wife, Jen, was standing in the kitchen doorway, wrapped in a heavy fleece robe.

"It's dead, Jen," Mark said miserably, pulling his thumb away from the nuclear reset button. "It won't let me type the right numbers, and the speaker is completely blown. I'm going to have to factory reset the whole thing."

Jen walked over to the coffee maker, poured herself a mug, and glanced at the yellow *Ghostwriter Guide* sitting on the counter next to Mark.

"Did you do the triage thing?" she asked, gesturing vaguely with her mug. "The book says not to panic. It says to check the 'Big Three' first."

Mark sighed. "Jen, it's fundamentally broken. The math chip is scrambled."

"Humor me, Mission Control," she said, leaning against the counter.

Mark pressed EXIT to escape the dangerous reset menu. He opened the guide to Chapter 8: *How to Diagnose in 60 Seconds.* He read the first rule. *Shift your mindset from "my radio is permanently broken" to "my radio just has a setting out of place."*

"Okay," Mark said, taking a calming breath. "Checkpoint one. Mode. Am I in VFO mode?" He looked at the screen. There were no tiny channel numbers on the right. "Yes. Mode is correct."

"Checkpoint two," Jen read over his shoulder. "Squelch and Monitor. See if the speaker actually works."

Mark pressed the small **MONI** button on the side of the radio.

SHHHHHHHHHHHHHHHHHH!

The rushing wall of white noise blasted out of the speaker, loud and clear. Mark almost dropped the radio.

"Okay!" Mark gasped, his eyes widening. "The speaker isn't blown. It works perfectly."

He let go of the button, and the radio went dead silent again. He flipped to Chapter 10. *If the radio is dead silent but the MONI button works, your squelch is too high.* "Wait a minute," Mark said, his brain finally engaging. He pressed MENU and typed 0.

The screen displayed SQL. Mark pressed MENU to unlock it. The squelch was set to 9. The absolute maximum setting. A literal brick wall.

"Why is this on 9?" Mark asked. Then he remembered. Yesterday afternoon, he had let Owen hold the radio to show him how bright the built-in LED flashlight was. Owen's teenage thumbs had been flying all over the keypad like he was playing a video game.

"Owen," Mark whispered, shaking his head. He changed the squelch back down to 2 and saved it.

"Alright, the gate is open," Jen said. "Now why wouldn't it let you type the numbers?"

Mark flipped the page to the next fix. *The Step-Size Rounding Error.* "The staircase," Mark read aloud. "If the radio won't land on a specific number and keeps rounding up, the step size is too large."

He pressed MENU and typed 1. The screen displayed STEP. He unlocked it. Sure enough, the step size was set to a massive 25.0K. Owen had scrambled that, too.

Mark smiled, feeling the anxiety completely melt away. He changed the step size down to a precise 2.5K and pressed MENU to save it. He exited back to the main screen. With a newfound swagger, he typed 1 6 2 5 5 0.

The radio accepted every single digit flawlessly. The squelch gate, now properly lowered to Level 2, instantly opened.

"...winter storm warning remains in effect until noon today. Expect heavy snowfall and hazardous driving conditions..." the robotic weather voice droned, loud and clear.

Mark set the radio down on the granite counter and took a long, victorious sip of his coffee. He looked at Jen.

"Fixed it," Mark said, tapping the side of his head. "Just took a little advanced technical diagnostics."

"Right," Jen smiled, taking her coffee toward the living room. "Next time, maybe check the settings before you initiate the self-destruct sequence."

Mark switched the radio over to Memory Mode and dialed up Channel 1. Right on cue, the morning commute net crackled to life, filled with local hams complaining about the snow. Mark listened, completely relaxed.

He hadn't needed to wipe his radio. He hadn't needed to buy a new one. He just needed to stop, breathe, and trust the troubleshooting tree. He had successfully diagnosed and cured a seemingly dead radio in under a minute.

"Checkpoints beat panic," Mark said to himself, watching the green screen glow.

Takeaway

When the radio fails or acts erratic, slow down. Do not immediately resort to a factory reset. The vast majority of "broken" radios simply have a setting out of place (like Squelch set to 9, or Step Size set too high). Use the 60-second triage flow to methodically check your Mode, your Speaker (MONI button), your Squelch, and your Step Size.

Bridge: Okay, Mark has his local settings dialed in perfectly. But why does one repeater work flawlessly, while another one completely ignores him?

Chapter 11

I Can Hit One Repeater but Not Another

The Symptom: You followed the instructions in Chapter 5 perfectly. You programmed "Repeater Alpha" into Channel 1, and it works flawlessly. You get signal reports, and everyone hears you. Confident, you use the exact same process to program "Repeater Bravo" into Channel 2. But when you switch to Channel 2 and press PTT... nothing. The tower ignores you.

You know your radio works, and you know you understand the menus. So why is one tower accepting you and the other shutting you out?

What you'll accomplish:

- Understand the difference between CTCSS and DCS tones.
- Check for "odd splits" (non-standard offset amounts).
- Learn the signs of a busy or "linked" repeater that requires different timing.

What you need:

- Your Baofeng radio.
- The RepeaterBook listings for *both* the working tower and the broken tower to compare them.

Fix Tree Step 1: The DCS vs. CTCSS Trap

Not all VIP passwords are created equal.

In Chapter 4, we learned about **CTCSS** (analog tones, like 100.0). This is what 90% of repeaters use. However, some newer or private repeaters use a digital password system called **DCS** (Digital Coded Squelch).

If you look at the repeater listing for Tower Bravo and the tone says something like D023N or 023, you are dealing with a DCS tower.

If you try to type 023 into the regular CTCSS menu (Menu 13), the Baofeng won't let you, or it will transmit the wrong type of password. The analog bouncer is looking for an analog password, and the digital bouncer is looking for a digital one.

- *The Fix:* If the tower uses DCS, leave Menu 13 (T-CTCS) completely **OFF**. Instead, press **MENU** and go to **Menu 12 (T-DCS)**. Enter your digital code here.

Fix Tree Step 2: The Non-Standard Offset (Odd Splits)

Most repeaters follow polite rules. In the 2-meter band, they almost always use a 0.600 MHz jump. In the 70-centimeter band, they almost always use a 5.000 MHz jump.

But sometimes, a local radio club has to put a tower in a crowded city where the standard frequencies are already taken. They have to use a weird, custom offset amount (like a 1.200 jump or a 7.600 jump). This is called an "Odd Split."

1. Look closely at the RepeaterBook listing for Tower Bravo.
2. Does the Offset exactly match the standard?
3. *The Fix:* If it is an odd split, you must go into **Menu 26 (OFFSET)** and manually type in that specific weird number (e.g., 001200) before you save the channel.

[FIELD NOTE] The Save Override: Remember, if you are fixing an existing memory channel, the Baofeng will not let you overwrite it easily. You must delete the channel (**Menu 28**) before you can re-save the new, corrected offset data into that slot (**Menu 27**).

Fix Tree Step 3: Linked Repeaters and Timing

Sometimes the settings are 100% correct, but the tower still seems to ignore you or cuts off the first half of your sentence.

Many modern repeaters are "linked" to other towers across the state via the internet. When you press the PTT button, Tower Bravo has to wake up, send a signal across the internet to Tower Charlie, wake that tower up, and open the gates. This process takes time.

If you press PTT and instantly say "This is K1ABC," you spoke while the internet link was still booting up. The tower missed your call sign entirely.

- *The Fix:* The "Press, Pause, Speak" rule becomes critical here. Press the PTT button. Count *"One-Mississippi"* in your head. *Then* speak. Give the digital network time to open the doors for you.

Quick Win Checklist

- [ ] I have compared the listings for Tower A and Tower B to look for differences.
- [ ] I checked if Tower B requires a digital DCS code (Menu 12) instead of a standard CTCSS tone (Menu 13).
- [ ] I checked if Tower B uses a weird, non-standard offset amount that requires me to change Menu 26.
- [ ] I am using the "Press, Pause, Speak" method to allow linked repeaters time to wake up.
- [ ] I remembered to completely delete the broken memory channel before trying to save my corrected settings into it.

If you're stuck, go to...

- *Chapter 14: CHIRP / Cable Problems* (if you have spent an hour fighting the menus and want to use a computer to just copy-paste the correct repeater data).

Chapter 12

My Channels Won't Save (Or They Disappear)

The Symptom: You painstakingly spend ten minutes programming a repeater into your radio. You carefully type the frequency, set the offset math, and dial in the correct tone password. You go to Menu 27, save it to Channel 001, and hear the robotic voice proudly confirm the save. But when you press the orange VFO/MR button to switch to Memory Mode and test your hard work, Channel 001 is completely empty. Or worse, it just shows the old, useless factory setting that was there yesterday.

You feel like the radio is intentionally gaslighting you. You might even wonder if the internal memory chip is physically defective.

Take a breath. The Baofeng memory system is not broken; it is just incredibly stubborn and highly literal. It follows a very rigid set of computer-logic rules dating back to its original engineering. If you violate even one of these invisible rules, it silently rejects your work without giving you a helpful error message or a warning beep. Let's outsmart it together.

What you'll accomplish:

- Understand the crucial "Sweep the Floor" rule to prevent silent overwrite failures.
- Diagnose the "Wrong Mode" error that tricks you into thinking your saves are working.
- Master the quirky "Double Save" sequence so your repeater channels actually stick permanently in the vault.

What you need:

- Your Baofeng radio.
- Your written notes for the specific channel you are trying to save.

Fix Tree Step 1: Confirm You Are in VFO Mode

You physically cannot build or save a new channel while you are in Memory (Channel) Mode. Think of Memory mode like a locked glass display case, you can look at the channels, and you can use them, but you can't alter them. The radio intentionally locks the keypad from making structural changes to protect your existing emergency presets from accidental bumps. If you try to save something while in Memory mode, the radio will pretend to accept the button presses, but absolutely nothing will happen behind the scenes.

- *The Fix:* Look at the right side of your screen. Do you see tiny channel numbers (like 001 or 127)? If yes, press the orange **VFO/MR** button once until those numbers completely disappear. You must build the channel on the blank, unlocked canvas of VFO mode before you can move it into the display case.

Fix Tree Step 2: The "Sweep the Floor" Rule

This is the number one reason channels fail to save, and it catches almost every beginner off guard. The Baofeng will not let you paint over a dirty canvas. Furthermore, it will *never* ask you, "Are you sure you want to overwrite this channel?"

If Channel 005 already has a frequency saved in it (even a random factory test frequency you didn't put there), and you try to save a brand new repeater into Channel 005, the radio's computer acts like a bouncer. It silently rejects the new data to protect the old data, leaving your new frequency lost in the void.

- *The Fix:* You must explicitly sweep the floor and delete the old channel before you save to it. Press **MENU**, type **2 8 (DEL-CH)**, unlock the menu, select your target channel number using the arrow keys, and press **MENU** to confirm the deletion. You will hear the robotic voice confirm the channel is cleared. Once the slot is truly empty, the radio will gladly accept your new data.

Fix Tree Step 3: The Double-Save Sequence

As we learned in Chapter 5, a repeater channel actually consists of two completely separate frequencies working together: the listening lane and the talking lane. Most modern, expensive radios automatically bundle these together for you. The Baofeng does not. It requires you to save them one at a time, manually, in a very specific, quirky order.

If you only save the listening lane, the channel will appear on your screen and you will hear traffic, but it will instantly fail when you press the PTT button because the radio doesn't know where to jump.

- *The Fix:* Let's review the exact rhythm and sequence in **Menu 27 (MEM-CH)**.

  a. Once your frequency, offset, and tone are set perfectly in VFO mode, go to Menu 27.

  b. Unlock it, pick your newly emptied channel slot, and press **MENU**. You will hear the voice proudly say "Receiving Memory."

[STOP] Do not press Exit yet! The job is only half done. You've only packed the listening half of the suitcase.

1. Press **MENU** again to unlock the menu a second time. The little black arrow will drop down.
2. Ensure the exact same channel slot is still selected, and press **MENU** a third time. You will hear "Transmitting Memory."
3. Now you can safely press **EXIT**. The complete package, listening lane, talking lane, and tones, has been successfully bundled and stored.

FIELD NOTE

Simplex Channels Only Need One Save: If you are saving a pure radio-to-radio "simplex" channel (like your family channel from Chapter 7) that has no offsets and no tones, your listening and talking lanes are the exact same frequency. In this specific case, you only need to perform the save sequence once. The "Double Save" is exclusively required for complex repeaters.

Quick Win Checklist

- [ ] I confirmed I am entirely in VFO mode before attempting to program any settings.
- [ ] I explicitly used Menu 28 (DEL-CH) to sweep the floor and delete my target memory slot before trying to save new data into it.
- [ ] I remember to perform the "Double Save" in Menu 27 when programming repeaters, ensuring I listen for both the "Receiving" and "Transmitting" voice prompts.
- [ ] I immediately switched back to Memory Mode (VFO/MR button) to visually test that my channel was actually stored.

If you're stuck, go to...

- *Chapter 14: CHIRP / Cable Problems* (if manual programming is causing you too much stress and you are ready to use a computer cable to simply drag-and-drop your channels).

Narrative Interlude: Why One Repeater Works and Another Won't

Character: Nate Brooks (Southeast)

Skills reinforced:

- CTCSS vs. DCS
- Tone-squelch pitfalls
- Deliberate channel overwrite
- Save discipline

Where to learn the steps: Chapters 11 & 12

The asphalt parking lot of the North Charleston Community Center was already radiating waves of intense morning heat. Nate Brooks sat in the driver's seat of his Tacoma, the air conditioning running on high, staring at a single piece of paper on his center console.

It was his CSCE, his Certificate of Successful Completion of Examination. He had just passed his Technician class amateur radio test. He was officially licensed. The government had assigned him a call sign, KN4XYZ, which meant his self-imposed ban on pressing the Push-To-Talk button was officially lifted. He was no longer just a listener; he was an operator.

But Nate wasn't the type to celebrate without finishing the job. He pulled out the bright yellow, laminated "Home Profile" index card from his tactical pouch. Channel 1, the Primary City Repeater, was locked in and working perfectly. However, for a hurricane preparedness plan to be truly viable, he needed redundancy. He needed Channel 2: The North County Backup Repeater.

He opened his notebook to the RepeaterBook printout he had generated earlier that week. *North County RPT. Frequency: 444.100. Offset: +5.000. Tone: D023N.*

"Alright," Nate muttered, slipping into his procedural mindset. "Let's build the backup route."

He picked up the Baofeng UV-5R, pressed the orange VFO/MR button to enter the manual typing mode, and keyed in 4 4 4 1 0 0. He went to Menu 25 and set the shift direction to positive. He went to Menu 26 and carefully typed 005000.

Next came the password. Nate went to Menu 13, T-CTCS, the transmit tone menu he had used for Channel 1. He scrolled through the numbers, looking for 023. It wasn't there. The lowest number was 67.0.

Frowning, Nate started pushing buttons. He went into Menu 11, R-CTCS, thinking maybe the backup tower required a receive tone instead. He selected a random tone just to see if it would force the radio to accept his input. He was guessing, and for a man who lived by checklists, guessing was a dangerous spiral.

Frustrated, he decided to just force the save and see what happened. He navigated to Menu 27, selected Channel 002, and pressed MENU. The robotic voice chimed, *"Receiving Memory."* He pressed it again. *"Transmitting Memory."*

Nate confidently pressed the VFO/MR button to drop back into his locked-down Memory mode. He clicked the UP arrow to Channel 2.

The screen read 155.500. It was a useless factory default frequency. His meticulously typed 444.100 was nowhere to be found.

"What?" Nate stared at the screen. He tried the entire process again, typing faster this time. He saved it to Channel 2. He checked Memory mode. Still 155.500.

The radio was silently rejecting his programming. Furthermore, when he clicked back down to his perfectly programmed Channel 1 to make sure the radio wasn't entirely broken, he was met with dead silence. He couldn't even hear the static tail of the primary repeater anymore.

His clean, disciplined communication plan was falling apart. He had introduced too many variables, changed too many settings without documenting them, and now he was locked out of his own system.

"Having trouble with the new ticket, KN4XYZ?"

Nate rolled his window down. Ms. Patel, the lead volunteer examiner who had just graded his test thirty minutes earlier, was walking toward her SUV. She was in her late fifties, possessed an aura of absolute calm authority, and had a massive VHF/UHF antenna mounted to the roof of her vehicle.

"I've compromised the baseline," Nate said flatly, holding up the Baofeng. "I'm trying to program the North County repeater into slot two. It refuses to save over the factory junk, and in trying to fix it, I've somehow deafened slot one."

Ms. Patel leaned against the door of his truck. "Let's look at the data. What's the RepeaterBook listing for North County?"

Nate read it off his notepad. "444.100. Plus offset. Tone is D023N."

"Stop right there," Ms. Patel said gently. "Look at that tone again. It doesn't say 100.0 or 141.3. It has a 'D' in front of it. That stands for Digital. You're trying to put an analog key into a digital lock."

Nate looked at the notebook. He had completely glossed over the letter 'D'.

"Menu 13 is for analog CTCSS tones," she explained. "When you couldn't find 023 in there, you started guessing. You went into Menu 11 and set a Receive Tone, didn't you?"

Nate checked the screen. There was a tiny CT icon illuminated on the left side of the display. "Yes. I changed a variable I didn't fully understand."

"And by doing that, you put a bouncer on your own speaker," Ms. Patel said, tapping the edge of his window frame. "Your radio is now ignoring every signal in Charleston because they don't have the secret password you just demanded. Turn Menu 11 off. Never use a receive tone unless you absolutely have to."

Nate immediately went to Menu 11 and set it to OFF. The tiny CT icon vanished.

"Now, for the memory failure," she continued. "The Baofeng is a stubborn machine, Nate. It will never let you paint over a dirty canvas. If Channel 2 already has 155.500 saved in it, the radio will silently reject your new save to protect the old data. You have to sweep the floor first."

"Deliberate overwrite," Nate realized, a wave of clarity washing over him. "I have to delete it before I can replace it."

"Exactly. Listen first. Verify receive. Then change one thing at a time," Ms. Patel said, giving him a warm smile. "Welcome to the airwaves, Nate. You'll get it." She turned and walked toward her SUV.

Nate took a deep breath. He looked at his radio and his notebook. Procedure. He could do procedure.

He went to **Menu 28 (DEL-CH)**. He selected Channel 002 and pressed MENU. *"Confirm,"* the radio said. The slot was now officially empty.

He went to VFO mode. He typed 4 4 4 1 0 0. Offset was already set to + and 005000.

Now for the password. He ignored Menu 13 entirely. He went to **Menu 12 (T-DCS)**. He scrolled through the digital codes until he found D023N. He pressed MENU to confirm it.

Finally, save discipline. He went to **Menu 27 (MEM-CH)**. He selected his freshly swept Channel 002. Press MENU. *"Receiving Memory."* Press MENU again. Arrow drops. Press MENU a third time. *"Transmitting Memory."*

Nate pressed the VFO/MR button to switch to his saved channels. He clicked up to Channel 2. The screen proudly displayed 444.100. It had saved perfectly.

He held the radio up to his mouth. It was time for a true, legal confidence check. He pressed the PTT button, watched the numbers automatically jump up by 5.000 to the transmit lane, and spoke clearly.

"KN4XYZ, testing."

He released the button. The numbers jumped back. A split-second later, a crisp burst of static erupted from the speaker, followed by a polite *beep* from the distant tower.

"Loud and clear, KN4XYZ. Good morning," an older gentleman's voice replied.

Nate smiled. He reached for his black permanent marker and placed a confident checkmark next to Channel 2 on his yellow index card. His redundancy was secured. He had two working repeaters, programmed correctly, saving reliably, and his comms plan was completely restored.

Takeaway

A good plan is small and clean. Random tweaks are how plans die. If a repeater listing uses a "D" before the tone (like D023N), it is a Digital Code. Leave the standard CTCSS menu alone and use Menu 12 (T-DCS) instead. Above all, the Baofeng will silently reject your programming if you try to save over an existing channel, you must explicitly use Menu 28 to delete the channel slot before attempting to save new data into it.

Bridge: If I'm going to do this, I need my programming and backups to survive reality.

Chapter 13

Scanning Doesn't Work (Or It Skips Everything)

The Symptom: You want to hear what is happening in your city during an event or emergency. You press and hold the Scan button (* / SCAN), hoping the radio will act like a high-end police scanner and automatically find local traffic for you to listen to. Instead, the frequency numbers blur past at lightning speed and never stop. Or, worse, it immediately stops on a channel that is just blasting pure, deafening static, and it gets stuck there forever, forcing you to turn the radio off in frustration.

Scanning on a Baofeng is famously quirky. Your device is a two-way radio first and a scanner second. It is not as smart, as fast, or as highly filterable as a dedicated $400 digital police scanner. But once you understand how its simple, mechanical "brain" actually works, you can make it a highly effective tool for gathering local intelligence and monitoring your neighborhood.

What you'll accomplish:

- Understand why your squelch setting is completely breaking your scan.
- Define and set the correct "Scan Resume" behavior in plain English so you stop missing the ends of conversations.
- Learn how to deal with "nuisance channels" that constantly hijack your scanner and interrupt you.

What you need:

- Your Baofeng radio.

Fix Tree Step 1: The Squelch Block

The Baofeng's scanning feature doesn't have a sophisticated computer brain; it relies entirely on the invisible Squelch gate that we set up way back in Chapter 2.

When the radio is in scan mode, it is rapidly flipping through frequencies. It only knows to stop when it physically collides with a signal strong enough to break through the squelch gate.

- **[STOP] Never scan on Level 0.** The squelch gate is completely removed. The radio will immediately "detect" the normal background static on the very first channel it checks. It will stop scanning instantly, thinking it found a real signal, and just blast static at you forever.
- If your Squelch is set to **Level 9**, the gate is a massive, impenetrable wall. The radio will scan right past legitimate, active human conversations because the signals aren't strong enough to break over that incredibly high wall. You will watch the scanner run in circles for hours and hear absolutely nothing.

- *The Fix:* Before you ever press the scan button, ensure **Menu 0 (SQL)** is set to **Level 2 or 3**. This provides the perfect Goldilocks balance, it allows the scanner to confidently skip over empty static, but immediately hits the brakes when a real human voice appears.

Fix Tree Step 2: The "Scan Resume" Mystery

The Symptom: Your scanner is working. It successfully finds a local repeater where two people are talking about a road closure. It stops so you can listen. But five seconds later, right in the middle of a critical sentence, the radio brutally cuts them off, beeps, and starts scanning again! It feels like someone grabbing the TV remote and changing the channel right at the climax of a movie.

This happens because the Baofeng has three entirely different "Scan Resume" modes, and the factory default is arguably the most annoying one for a beginner.

1. Press **MENU** and go to **Menu 18 (SC-REV)**. This stands for Scanner Resume.

2. Unlock it. You will see three obscure abbreviations: **TO**, **CO**, and **SE**. Here is what they actually mean in plain English:

- **TO (Time Operation):** The radio stops on an active signal, starts a strict 5-second countdown timer, and then blindly resumes scanning, *even if the person is still actively talking*. This is the annoying factory default. It is meant for just sampling the airwaves, not for actually listening to a conversation.

- **CO (Carrier Operation):** The radio stops on an active signal and patiently stays there for as long as the person is holding down their microphone. The moment they un-key their radio and the channel goes quiet, your Baofeng waits 3 seconds to see if anyone replies, and then politely resumes scanning.

- **SE (Search Operation):** The radio stops on the very first active signal it finds and stays there forever. The scanning function is permanently canceled the moment it finds a single voice.

- *The Fix:* Change Menu 18 to **CO (Carrier Operation)**. This is the absolute gold standard for emergency scanning. It lets you listen to the entire conversation naturally, and automatically goes back to hunting for new signals only when the conversation is actually over.

Fix Tree Step 3: Nuisance Channels

Sometimes, you have a local NOAA weather broadcast or a noisy digital dispatch frequency saved in your memory bank. Because weather stations broadcast continuously without ever taking a breath, and digital data streams sound like an angry 1990s dial-up modem playing on repeat, they will hold your scanner hostage. Every single time you run a scan, the radio stops on this noisy channel and refuses to leave.

- *The Fix:* You ultimately need to lock that specific channel out of the scan list. Unfortunately, you cannot easily do this from the radio's physical keypad. You can only skip these nuisance channels permanently by using a computer programming cable and the CHIRP software, which we will cover in the very next chapter.

GOOD ENOUGH

If you don't have a computer cable yet, or if you are out in the woods, there is a manual, split-second workaround. When the scanner inevitably stops and gets stuck on that noisy weather channel, just reach down and tap the **Up Arrow** key once while it is stuck. This manually forces the scanner to abandon that channel and immediately resume its hunt for the next signal. You might have to do this every time it loops around, but it keeps the scanner moving.

Quick Win Checklist

- [ ] I confirmed my Squelch (Menu 0) is set to Level 2 or 3 before initiating any scan.
- [ ] I understand how to start a scan by pressing and holding the * / SCAN key for two seconds.
- [ ] I have changed my Scan Resume setting (Menu 18) to **CO (Carrier Operation)** so the radio doesn't cut people off mid-sentence.
- [ ] I know to quickly tap the Up Arrow key to manually force the scanner past a stuck or exceptionally noisy channel.

If you're stuck, go to...

- *Chapter 10: I Can't Hear Anything* (if your scanner clearly stops on an active channel, but you hear absolutely no audio coming out of the physical speaker).

Chapter 14

CHIRP / Cable Problems (Optional but Common)

The Symptom: Programming your Baofeng by hand is a valuable rite of passage, but eventually, you will want to upgrade your workflow. You finally decided to buy the $20 USB programming cable. You eagerly want to use the free CHIRP software on your computer to type in custom channel names, import entire lists from RepeaterBook in seconds, and bypass the clumsy radio keypad entirely.

You plug the cable into your laptop, attach it to the radio, click the software, and instead of a seamless transfer, you get a terrifying red error box: **"Radio did not respond."** Instead of saving time and sanity, you are now locked in a frustrating battle with your computer's device manager.

What you'll accomplish:

- Understand the absolute golden CHIRP workflow (Download -> Edit -> Upload) to prevent data corruption.
- Perform a comprehensive "Sanity Check" on your physical cable connection.
- Diagnose and fix the dreaded "COM Port" error that plagues PC users.

What you need:

- Your Baofeng radio (fully charged).
- A Baofeng programming cable (the "BTECH PC03" FTDI cable is highly recommended over cheap knockoffs).
- A PC or Mac with the free CHIRP software installed.

Fix Tree Step 1: The "Radio Did Not Respond" Physical Check

The Baofeng has a notoriously tight, dual-pin accessory port. It requires the programming cable to be seated incredibly deeply into the side of the radio. If it is off by even a fraction of a millimeter, or if the rubber weather flap gets pinched underneath the plastic header, the metal data contacts won't touch. Your computer will stubbornly claim the radio doesn't exist.

1. Unplug the cable from the side of the radio entirely. Check to ensure the rubber port cover is pulled completely out of the way.
2. Push the two pins back in, and apply a surprising amount of firm, even pressure with your thumb. You want the hard plastic of the cable head completely flush against the plastic casing of the radio.
3. Ensure the radio is powered **ON**, and the volume knob is turned to maximum. (This sounds weird, but the

programming data actually travels through the physical speaker/microphone circuitry inside the radio as audio tones. The volume knob literally controls the data flow!)

4. Ensure the radio is tuned to a "dead" or quiet frequency with no static, active talking, or weather broadcasts. If the radio receives a radio signal and breaks squelch while the computer is trying to talk to it, the data transfer will instantly crash and fail.

Fix Tree Step 2: The COM Port Sanity Check

If the cable is firmly seated but CHIRP still gives you a stubborn error, you are likely playing "COM Port Roulette." Your computer physically sees the USB cable, but CHIRP doesn't know which specific digital port the cable is assigned to.

This is highly common with cheap, knockoff programming cables that use counterfeit "Prolific" microchips, which Windows actively tries to block. (This is why we highly recommend spending the extra five dollars for an authentic "FTDI" cable, they are true plug-and-play).

1. In CHIRP, click **Radio** > **Download from Radio**.

2. A small dialog box will pop up. Look closely at the **Port** dropdown menu at the top.

3. *Windows Users:* You should see a list of ports named something like COM3, COM4, or COM5. If the box is completely blank, or only shows COM1 (which is usually a reserved default system port), your computer hasn't installed the digital driver for the cable.

4. *Mac Users:* Look for a uniquely named port like /dev/cu.usbserial or /dev/tty.usbserial.

5. *The Fix:* Try selecting a different COM port from the dropdown menu and clicking OK. If it fails, try the next one on the list. If no other ports exist, completely unplug the USB cable, plug it into a different physical USB port on the opposite side of your computer, wait 30 seconds for the computer to recognize the new hardware, and try again.

Fix Tree Step 3: The Golden Workflow (Download First!)

The absolute biggest mistake beginners make with CHIRP is opening the software, creating a brand new, blank spreadsheet, typing in all their favorite frequencies, and trying to force-upload that blank document straight into the radio.

The radio will violently reject this. Every Baofeng has invisible, factory-specific firmware headers that must be present in the data file. If you force a blank spreadsheet into the radio, you overwrite the radio's "brain" with blank space, which can render the radio useless.

You must memorize and follow the Golden Workflow: **Download, Edit, Upload.**

1. **Download:** Always start your session by clicking *Download from Radio*. This pulls a safe "clone" of the radio's unique invisible system settings, along with its current channels, into CHIRP.

2. **Edit:** You will now see a spreadsheet of your radio's current memory on your computer screen. Make all of your changes safely here. Type in your frequencies, add convenient names (like "CITYRPT" or "FAMILY1", note that the Baofeng screen can only display up to 7 alphanumeric characters), and delete all the random factory junk.

3. **Upload:** Click *Upload to Radio*. Because you started with the radio's own native system file, it will gladly accept the beautifully edited version back.

STOP

Never unplug the cable during a transfer. When the green progress bar is moving across your computer screen, the radio's top LED light will flash rapidly to indicate active data flow. Do not touch the cable, do not bump the desk, and do not turn the radio off until that progress bar fully reaches 100%. Interrupting a data transfer mid-stream can "brick" the radio, scrambling its internal memory beyond repair.

Fix Tree Step 4: The Ultimate Safety Net (Save an Image)

Before you start getting creative, changing dozens of obscure menu settings in CHIRP, or importing 500 frequencies you don't actually need, you need to protect yourself.

As soon as you successfully *Download from Radio* for the very first time, go immediately to **File** > **Save As** in the CHIRP menu. Save this file directly to your computer's desktop as Baofeng_Factory_Backup.img.

This file is a digital time machine. It is an exact, bit-for-bit clone of your radio the day you bought it. If you ever severely mess up the internal settings, delete something critically important, or accidentally ruin your screen colors so badly you can't read the display, you can simply open this .img file in CHIRP and upload it to the radio to instantly rewind time and restore the radio to perfect factory-new condition.

Quick Win Checklist

- [ ] I physically pushed the programming cable into the radio until the hard plastic was completely flush against the radio body.
- [ ] I turned the radio ON and turned the volume knob to maximum before attempting a transfer.
- [ ] I selected the correct COM Port or USB Serial port in the CHIRP dropdown menu (avoiding COM1).
- [ ] I memorized the Golden Workflow: Download first, Edit the spreadsheet, then Upload back to the radio.
- [ ] I created a permanent .img backup file of my factory settings acting as my digital time machine.

If you're stuck, go to...

- *Chapter 15: Reset and Recover* (if you attempted a CHIRP upload, something went horribly wrong, and the radio screen is now blank, entirely unresponsive, or speaking in a different language).

Chapter 15

Reset and Recover (When You've Made a Mess)

The Symptom: You've spent the last two hours falling down a rabbit hole of confusing YouTube tutorials that used a slightly different model of radio. You pushed a sequence of buttons you didn't quite understand, and now your Baofeng is acting erratic and possessed. The screen is showing weird symbols you've never seen, it stubbornly refuses to transmit, the robotic voice prompts are suddenly speaking to you in Chinese, and absolutely none of the previous troubleshooting steps are working.

You feel a sinking wave of panic, thinking you have permanently destroyed your brand new radio and turned it into an expensive paperweight.

Take a deep breath and take your hands off the keypad. You haven't broken anything. The beauty of these inexpensive, utilitarian radios is that they are practically indestructible on a software level. There is no line of code you can type that will physically break the circuit board. Sometimes, the fastest and least stressful way to fix a deeply tangled, confusing mess is to simply stop fighting it, nuke it from orbit, wipe the slate completely clean, and start fresh.

What you'll accomplish:

- Understand the critical difference between a "VFO Reset" and an "ALL Reset."
- Execute a factory reset safely to clear out digital gremlins.
- Fix the default language settings that revert during a reset.
- Use a structured "Rebuild Checklist" to get your Home Profile back online in under 10 minutes.

What you need:

- Your Baofeng radio.
- Your physical paper backup card (from Chapter 7) detailing your Top 5 Channels.

Fix Tree Step 1: VFO Reset vs. ALL Reset

The Baofeng has two entirely different levels of amnesia you can trigger. It's incredibly important to choose the right one for your specific problem so you don't accidentally delete work you want to keep.

1. Press **MENU** and go to **Menu 40 (RESET)**.
2. Unlock the menu. You will see two options: **VFO** and **ALL**.

- **VFO Reset (The Soft Reset):** This resets all your basic menu settings (like your Squelch level, your Step Size, your display backlight colors, and battery saver timers) back to their original factory defaults. *However, it preserves your saved Memory Channels.* If your radio is acting a little weird or the screen looks wrong, but you definitely don't want to lose the 5 repeater channels you spent hours painstakingly programming, try this soft reset first.
- **ALL Reset (The Nuclear Option):** This deletes absolutely everything without mercy. It resets the menus, violently wipes out every single saved Memory Channel, and returns the entire radio to the exact pristine state it was in when it left the assembly line in China.

Fix Tree Step 2: Executing the Nuclear Option

If the radio is severely locked up, if you accidentally turned on a strange digital signaling feature you can't turn off, or if you bought the radio used and it is clogged full of the previous owner's random junk channels, you want the ALL Reset.

1. Ensure you have your physical index card from Chapter 7 with your frequencies written down directly in front of you. (Once you press this final button, those channels are gone forever).
2. Go to **Menu 40 (RESET)**.
3. Select **ALL** and press **MENU**.
4. The screen will say SOURCE? (This is the radio's poorly translated way of asking, "Are you absolutely sure you want to do this?")
5. Press **MENU** one more time to confirm your decision. **[STOP]** This is the point of no return. Once pressed, your memory channels are permanently gone.
6. The radio will display WAIT..., physically power itself off, and turn back on. **Do not remove the battery or turn the volume knob off during this reboot.** The screen will likely be in VFO mode, displaying factory test frequencies (like 155.500 or 136.000).

FIELD NOTE

Fixing the Language: When you perform a full factory reset, the Baofeng often defaults its robotic voice prompts back to Chinese. Don't panic. To switch it back to English, press **MENU**, type **1 4 (VOICE)**, unlock the menu, use the arrows to select **ENG**, and press **MENU** to confirm.

YOU'RE WINNING

The ghost in the machine is officially gone. All of your accidental menu mistakes, buried sub-tones, and confusing offsets have been wiped away. You now have a perfectly clean, predictable canvas to work with.

Fix Tree Step 3: The 10-Minute Rebuild

Now that the radio is thoroughly wiped, we need to get it combat-ready again. Because you practiced these exact skills in Part 1, what felt like an agonizing hour on your first day will now take you less than 10 minutes. You are going to build muscle memory right now.

Grab your paper backup card and follow this exact, ruthless order of operations:

Minute 1: The Hardware Baseline We need to lay the foundation so the radio behaves politely.

- Go to Menu 0 (SQL) and set your Squelch to 2 (so you aren't blasted by static).
- Go to Menu 1 (STEP) and set your step size to 2.5K (so you can type precise numbers).
- Go to Menu 2 (TXP) and ensure your transmit power is on HIGH.

Minute 2-5: Rebuild the City Lighthouses We need to restore your long-range repeater connections.

- Ensure you are in VFO Mode (no tiny channel numbers on the right). If you see channel numbers, press the orange **VFO/MR** button once to clear them.
- Program Channel 1 (Your Primary Repeater): Type the frequency, set your Offset Amount (Menu 26), set the Duplex Direction (Menu 25), and set your Transmit Tone (Menu 13). Use the careful "Double Save" sequence in Menu 27 to lock it in.
- Program Channel 2 (Backup Repeater) using the exact same methodical process.

Minute 6-8: Add the Neighborhood Simplex Lanes We need to restore your grid-down, direct-to-family channels.

- Program Channel 3 (Simplex 1): Type the frequency. Ensure Offset is OFF. Ensure Tone is OFF. Save it to Menu 27 using just a single save.
- Program Channel 4 (Simplex 2) identically.

Minute 9-10: The Final Vault Check

- Press the orange VFO/MR button to switch confidently back into Memory Mode.
- Use the arrow keys to flip through Channels 1, 2, 3, and 4. Visually verify they are all present and correct.
- Lock the keypad by holding the # key for two seconds. You will see a tiny padlock icon appear at the top of the screen, ensuring you don't accidentally ruin your fresh, beautiful rebuild.

The True Value of Backups

You just experienced firsthand exactly why we insisted on writing down the "Home Profile" on a physical index card in Chapter 7.

Under extreme stress, human memory is notoriously terrible. Adrenaline destroys your ability to recall fine details. You will absolutely not remember if your local repeater's tone was 100.0 or 110.9 when severe weather is rolling in or trees are coming down outside. If you have to spend an hour on your smartphone trying to look up RepeaterBook data while your neighborhood is out of power, your radio is a useless brick.

- **Level 1 Backup:** The waterproof, tape-laminated index card physically taped to the back of the radio.
- **Level 2 Backup:** Take a clear photo of that index card with your smartphone and "favorite" it in your photo album so it is always accessible, even without cell service.
- **Level 3 Backup:** The CHIRP .img file securely saved to your computer's desktop (covered in Chapter 14) for instant digital recovery.

Quick Win Checklist

- [ ] I understand the difference between a soft VFO Reset (saves my channels) and a hard ALL Reset (wipes absolutely everything).
- [ ] I have verified I possess a physical paper backup of my frequencies before triggering an ALL Reset.
- [ ] I successfully executed an ALL Reset to clear out corrupted, confusing settings.
- [ ] I used Menu 14 to change the robotic voice prompt back to English.
- [ ] I successfully used the 10-Minute Rebuild flow to restore my Squelch, Step Size, and Top 5 critical channels.
- [ ] I feel highly confident that I can completely "un-break" this radio anytime it starts acting erratic.

If you're stuck, go to...

- *Part 3: The Family Comms Plan* (Now that your radio is perfectly tuned, clean, and backed up, you have graduated from technical troubleshooting. You are ready to start running real-world communication drills with your family!)

Narrative Interlude: Cable Night at the Kitchen Table

Character: Mark Jensen (Midwest Suburbs)

Skills reinforced:

- Scan basics
- CHIRP workflow
- Cable/COM sanity
- Image backups
- Reset/rebuild

Where to learn the steps: Chapters 13, 14, & 15

"I'll just use CHIRP, how hard can it be?"

Those were the famous last words of Mark Jensen, spoken at 8:45 PM on a Thursday at his kitchen table.

Spread out before him was his laptop, a newly delivered twenty-dollar USB programming cable, and his trusty Baofeng UV-5R. Mark had successfully programmed his local repeaters by hand over the weekend, but his thumb was getting tired of the keypad. Plus, his scanner was misbehaving. Every time he tried to scan the local airwaves, it would immediately get stuck on the National Weather Service frequency and refuse to move, blasting automated forecasts about barometric pressure until he manually turned it off.

He knew CHIRP, a free, open-source radio programming software, could fix his scanner issues and let him name his channels. He had watched a five-minute YouTube video. He felt invincible.

He plugged the USB cable into his laptop, jammed the two-pronged end into the side of the Baofeng, and opened the software. A blank spreadsheet appeared. Mark spent the next twenty minutes meticulously typing in twenty different frequencies. He added local repeaters, simplex channels, and even a few fire dispatch frequencies he found online. He painstakingly typed in the offsets and the tones.

Finally, his digital masterpiece was complete. He clicked the button labeled *Upload to Radio*.

A little box popped up, loading for three agonizing seconds before flashing a harsh, red error message: **"Radio did not respond."**

"Excuse me?" Mark said to the screen. He clicked it again.

"Radio did not respond."

"I am right here," Mark muttered, tapping the radio. "Respond."

Assuming the USB port was acting up, Mark unplugged the cable. He moved it to the USB port on the other side of his laptop. He clicked Upload. *Error.* He unplugged it again, blew into the USB connector like an old Nintendo game cartridge, and plugged it into a dongle attached to the back of the computer. It felt like a modern technological rain dance.

Error. Radio did not respond.

"Is the tactical walkie-talkie rejecting your friend request?" Jen asked, walking into the kitchen to grab a glass of water.

"It's a COM port issue, Jen," Mark said defensively, rubbing his temples. "Or a driver conflict. The computer is speaking English and the radio is expecting Chinese. They are failing to negotiate terms."

"Did you read the guide?" she asked, pointing to the yellow field manual sitting under his coffee mug.

"I don't need the guide for a USB cable. You plug it in, it works. That's how computers have worked since 1998."

Jen just raised her eyebrows, took a sip of water, and walked back to the living room.

Mark sighed. He pulled the manual out from under his mug and flipped to Chapter 14: *CHIRP / Cable Problems*. He read the first section.

The Baofeng has a notoriously tight accessory port. If the plastic is not pushed entirely flush against the radio body, the data pins will not connect.

Mark looked at the side of his radio. The two pins were in, but there was a tiny, one-millimeter gap between the hard plastic of the cable and the radio casing. The rubber flap was slightly pinched. He pulled the cable out, moved the flap completely out of the way, and shoved the cable back in with a firm, aggressive push from his thumb. It seated with a dull *thud*. Flush.

Next, he read the volume rule. *Turn the radio on and turn the volume knob to maximum. The data travels through the audio circuits.* Mark twisted the knob all the way up.

He looked back at the book. *The Golden Workflow: Download first! Never force a blank spreadsheet into a new radio.*

Mark's stomach dropped. He had been trying to jam a blank, generic spreadsheet into the radio's brain. If the cable connection hadn't failed, he might have accidentally wiped the radio's internal firmware.

"Okay. Follow the flow," Mark whispered to himself.

He closed his masterpiece spreadsheet. He clicked *Radio > Download from Radio.*

A prompt asked him to select his COM Port. He looked at the dropdown list. It was currently set to COM1. The book explicitly warned that COM1 was usually a phantom system default. Mark clicked the dropdown and selected COM4. He selected his radio model (Baofeng UV-5R) and clicked OK.

A green progress bar appeared on the laptop screen. At the exact same time, the little LED light on the top of the Baofeng started flashing furiously. Data was flowing.

"We have a handshake," Mark grinned.

When the download finished, the screen populated with the radio's actual internal memory. There were the 5 channels from his Home Profile, exactly as he had programmed them by hand days ago.

Before he did anything else, Mark followed the ultimate safety rule in Chapter 14. He went to *File > Save As* and saved the file to his desktop as Baofeng_Factory_Backup.img.

"Time machine activated," he said. If he ever truly broke the radio in the future, he could upload that exact file and magically restore it to this moment. The fear of permanent failure completely evaporated.

Now, he could fix his scanner. He looked at the row for Channel 5, the automated National Weather Service frequency. He scrolled to the far right of the spreadsheet to a column labeled *Skip*. He clicked the box to change it to "S".

By doing that, he was telling the radio: *When I press the scan button, skip right over this noisy channel.*

He quickly typed in a few names for his channels, CITYRPT, FAMILY1, and then initiated the final step of the golden workflow. He clicked *Upload to Radio*.

The green progress bar sailed across the screen. The radio flashed, beeped loudly, and rebooted itself.

Mark unplugged the cable. The screen now proudly displayed alphanumeric names instead of just raw numbers. He pressed and held the * / SCAN key. The radio immediately began scanning through his memory channels, skipping perfectly over the annoying weather broadcast and hunting for real human voices.

It worked. He hadn't needed to try a fourth USB port. He just needed to follow the procedure: Seat the cable, pick the right COM port, download first, and save a backup.

He leaned back in his kitchen chair, extremely satisfied. He had a fully backed-up Home Profile saved to his hard drive. He had a working scanner. He had complete control over the machine.

"Cool," Mark said to the quiet kitchen, tapping his finger against the radio's plastic casing. "Now, can other humans in my house actually use this?"

Takeaway

Backups turn frustration into a speed bump. Never try to force a blank spreadsheet into a radio. Always push the physical cable in firmly, turn the volume up, and follow the Golden Workflow: **Download from Radio -> Edit -> Upload to Radio**. Save an .img file of your factory settings immediately so you can always safely reset and recover from any digital mistakes.

Bridge: Cool. Now, can other humans in my house use this?

Part 3:

The Family Comms Plan (Simple, Repeatable, Stress-Proof)

Chapter 16

The Family Comms Plan in One Page

In high-stress situations, human beings forget almost everything except the absolute simplest plan. Adrenaline is an incredible survival mechanism, but it completely destroys fine motor skills and complex memory recall. If cellular networks suddenly go down during a severe winter storm, a regional blackout, or a sudden neighborhood evacuation, your spouse or teenager will absolutely not remember how to navigate the Baofeng's VFO menu or manually program a CTCSS tone.

If they have to think about radio theory during an emergency, the plan has already failed. They need a communication strategy so painfully simple, so thoroughly rehearsed, that it fits entirely on a single index card.

What you'll accomplish:

- Create a one-page family communication plan that anyone in your household can follow under extreme stress.
- Assign clear, rigid roles so everyone knows exactly who is in charge of the radio network and who is checking in.
- Define strict check-in rules to conserve precious battery life and minimize airwave panic.

What you need:

- A notebook and a dark pen or marker.
- Your programmed Home Profile from Chapter 7.
- Your family members (or the core group of friends/neighbors you are planning with).

The Three Rules of the Family Plan

Before we write a single frequency down, your family needs to sit at the kitchen table and agree on three non-negotiable rules for using the radios.

1. **Keep it Simple:** The radio stays permanently locked in Memory Mode. Nobody is allowed to press the menu button, change the squelch, or type random numbers on the keypad. The instruction should be exactly three steps: Turn the top knob on, use the arrows to pick the channel number, and press the side button to talk. That is it.

2. **Keep it Legal:** Emphasize heavily that these are not walkie-talkie toys for casual chatting. They are serious tools for coordinating when normal grid methods fail. These airwaves are public, and during a crisis, they are shared with official emergency personnel. Everyone must follow the licensing rules we discussed in the front matter and behave professionally.

3. **Keep it Practiced:** A plan is just a worthless piece of paper until you drill it into reality. Under stress, you don't rise to the occasion; you default to your level of training. We will cover engaging, 10-minute weekly drills in Chapter 19 to build this muscle memory.

Assigning Roles

If an emergency strikes and everyone tries to grab a radio and talk over each other at the exact same time, you will have chaos. Assigning distinct roles creates a calm, military-style hierarchy.

- **Primary Operator (You):** You are the technical expert and the "quartermaster." You manage charging the batteries, programming the channels, and monitoring the wide-area repeaters for news and weather updates. You act as the central dispatcher.
- **Secondary Operator:** Usually a spouse, partner, or older teenager. This role is absolutely critical. If the Primary Operator is stuck at work across town or injured, the Secondary instantly takes complete control of the base station radio at home. They need to know exactly how to run the net.
- **Check-In Leads:** Anyone else holding a radio out in the field (kids at school, a partner commuting). Their only job is to turn the radio on, report their status to the Primary or Secondary operator at specific, pre-arranged times, and then turn the radio back off.

When to Listen vs. When to Transmit

A Baofeng battery is a finite resource. It will happily last for about 12 to 18 hours if you are just quietly listening to traffic. However, pushing your voice out across a city takes a massive amount of physical power. If you hold down the PTT button and talk non-stop, a fully charged battery will die in less than two hours.

STOP

The Walkie-Talkie Trap: Do not treat the radio like an open phone call. If the grid goes down, the psychological urge to constantly chatter for comfort is incredibly high. You must resist this. Establish a firm household rule: The radio is for listening 95% of the time. You only transmit to deliver a scheduled check-in, acknowledge a received message, or report an immediate, life-threatening emergency.

The "Say Less" Principle

When you do finally press the button to transmit, follow the "Say Less" principle. Under stress, people naturally tend to ramble. Rambling clutters the shared public channel, wastes precious battery power, and makes it impossible for the person on the other end to write down important details.

Instead of a breathless, rambling story about a flat tire and broken traffic lights, teach your family to use the highly structured **WHO / WHERE / WHAT / NEXT** format. It forces the brain to organize information before the mouth opens.

- **Who:** "Dad, this is Sarah." *(Establish identity immediately).*
- **Where:** "I am safe at the high school gym." *(Establish location).*
- **What:** "Roads are blocked by trees, I am staying here for now." *(Establish the situation).*
- **Next:** "I will listen for you at the top of the next hour. Over." *(Establish the future action).*

Deliverable: The One-Page Template

Grab your notebook. Write this template down in bold, clear letters. You will eventually copy this onto a small, sturdy piece of paper, completely laminate it with clear packing tape to make it waterproof, and put it in your family's emergency kit (or tape it directly to the back of the radio's battery).

FAMILY COMMS PLAN

- **Primary Operator:** *Name*
- **Secondary Operator:** *Name*
- **Home Channel (Simplex):** Channel 3
- **Backup Channel (Simplex):** Channel 4
- **City Info Channel (Listen Only):** Channel 1
- **Check-In Times:** Top of every hour strictly (e.g., 2:00, 3:00, 4:00).
- **Emergency Protocol:** If you are in immediate danger, say "Emergency, Emergency, Emergency" and give your location immediately.

YOU'RE WINNING

You just eliminated 90% of the sheer panic associated with suddenly losing cell service. Your family doesn't have to guess what to do. They now know exactly what channel to be on, who is supposed to talk, the format to use, and exactly what time to expect a reassuring message from you.

Quick Win Checklist

- [ ] I understand why the family plan must be kept incredibly simple to survive an adrenaline dump.
- [ ] I have clearly assigned Primary and Secondary operator roles in my household.
- [ ] My family understands the "Walkie-Talkie Trap" and knows the radio is for listening 95% of the time to save battery.
- [ ] I have taught my family the "WHO / WHERE / WHAT / NEXT" format for clear, punchy messages.
- [ ] I have filled out the One-Page Family Comms Plan template and prepared to laminate it.

If you're stuck, go to...

- *Chapter 8: Troubleshooting* (If a family member accidentally unlocks the radio keypad, changes a setting, and suddenly can't hear you, run them through the 60-second diagnostic flow).

Chapter 17

Simplex That Works (What You Can Expect and Why)

HOLLYWOOD MOVIES AND AGGRESSIVE walkie-talkie marketing boxes have lied to us. In the movies, a hero with a cheap handheld radio talks to a rescue helicopter three states away, achieving crystal-clear audio from inside a concrete basement. On the box of bubble-pack radios at the sporting goods store, you'll see bold claims like "36-MILE RANGE!"

In reality, pure radio-to-radio communication (referred to as "Simplex" or "Direct") is strictly bound by the unyielding laws of physics. If you hand your spouse a Baofeng and expect them to hear you clearly from ten miles away across a dense, hilly city without using a repeater tower, you are going to be severely disappointed. Understanding your actual limitations is the secret to building a plan that won't fail you. Let's set highly realistic expectations so your family doesn't panic the first time they hear a wall of static.

What you'll accomplish:

- Set realistic, field-tested range expectations for handheld radios.
- Learn how to instantly improve your results using location, environment, and body positioning.
- Choose simple, reliable family channels that actually work when the grid goes down.

What you need:

- Two programmed Baofeng radios.
- A partner to help you test.

Range Reality: The Line-of-Sight Rule

Your Baofeng transmits in the VHF (Very High Frequency) and UHF (Ultra High Frequency) bands. These radio waves behave almost exactly like a beam of light from a powerful flashlight.

If you shine a flashlight at a dense forest, the light scatters against the tree trunks and quickly stops. If you shine it at a concrete building, the light bounces off the wall. If you stand on top of a mountain and shine it down into a valley, the light travels unhindered for miles. Your radio waves act the exact same way.

- **Dense Urban City:** Expect 0.5 to 1 mile of reliable range. Steel-framed buildings, thick concrete walls, and massive electrical interference (the invisible "RF smog" created by thousands of Wi-Fi routers and power lines) will eat your 5-watt signal alive.

- **Suburban Neighborhood:** Expect 1 to 3 miles of range. Wood-framed houses are easier to punch through, but rolling hills and dense trees scatter the signal. Interestingly, you will often get slightly better range in the winter when the leaves have fallen off the trees, compared to the thick, wet foliage of summer.
- **Flat Open Country / Over Water:** Expect 3 to 5 miles of range. With nothing in the way to absorb the signal, your radio will reach out until the physical curvature of the earth literally gets in the way of the beam.
- **Mountaintop to Valley:** Expect 10 to 30+ miles of range. Height is absolutely everything in VHF/UHF radio. If your antenna can visually "see" the other antenna, the signal will make the trip.

Where Simplex Shines

If the range is so seemingly limited compared to a cell phone, why do we use Simplex channels at all? Because Simplex is entirely **grid-independent**.

If a massive winter storm knocks out power to your entire county, the cellular towers will go down within hours. A few hours later, the backup diesel generators running the local ham radio repeaters might run out of fuel. When absolutely all external infrastructure fails, Simplex is the only thing left. It requires nothing but your battery and their battery. It is absolutely perfect for:

- Coordinating between two cars in a convoy traveling through rural areas with zero cell service.
- Checking in with a spouse who is walking home from a neighborhood grocery store during a blackout.
- Running perimeter security around a rural property, farm, or campsite.

How to Instantly Improve Your Signal

If your spouse calls you on your designated Simplex channel and their voice is suddenly buried in scratchy, popping static, do not tell them to repeat themselves immediately. Tell them to change their physical environment first.

- **The "Step Outside" Rule:** Standard wood walls absorb about 30% of your radio signal. Brick, stucco with metal wire mesh inside (which acts like a giant cage), and modern energy-efficient windows with metallic coatings can absorb up to 80% of your signal. Simply walking out the front door, or stepping onto a wooden back porch, can turn a terrible, broken signal into a crystal-clear one instantly.
- **Get High:** Remember the flashlight analogy. Standing on a second-story balcony or at the top of a staircase is always vastly superior to transmitting from a ground-floor kitchen or a basement.
- **The "Two-Foot Shift":** UHF waves are incredibly short physically (about 70 centimeters long). Sometimes, radio waves bounce off a parked car or a metal fence and collide with each other, creating an invisible "dead spot" right where you are standing. Moving your body just two feet to the left or right can pull your antenna completely out of that dead zone and restore the audio.

Mini-Experiment: The Neighborhood Walk

It's time to prove these concepts in the real world. By doing this now, you eliminate the fear of the unknown later.

Step 1: Give one radio to your partner. Have them sit at the kitchen table (the base station).

Step 2: Ensure both radios are locked onto Channel 3 (your Family Simplex channel from Chapter 7) and the volume is turned up to a comfortable level.

Step 3: Take your radio and start walking down the street, away from your house.

Step 4: At 1/4 mile, press the PTT button. Pause for one second. Say: "Base, this is *Name*. I am at the corner of 4th and Main. How is my signal? Over." Have them reply. Both of your voices should be perfectly clear.

Step 5: Walk to 1/2 mile. Intentionally stand behind a large brick building, or walk down into a geographic dip in the road. Make the exact same call. Notice how the "hiss" of static starts creeping into the background of their voice.

Step 6: Now, step out from behind the building, or walk 50 feet up to the highest point on the street, and call again. Notice how the signal instantly clears up just by altering your line-of-sight.

FIELD NOTE

Build a Mental Map: Writing down the results of this walk on a piece of scratch paper gives you a permanent "mental map" of your neighborhood's radio terrain. If an emergency happens, you won't have to guess. You will already know exactly which street corners are total dead zones and which nearby hills are the perfect spots to stand when you need to transmit a critical message back to the house.

Quick Win Checklist

- [ ] I understand that VHF/UHF radios are strictly "line-of-sight" devices, not magic wands.
- [ ] I have highly realistic expectations for my neighborhood range (usually 1 to 3 miles).
- [ ] I understand that Simplex is a vital lifeline because it requires absolutely zero grid infrastructure to work.
- [ ] I know to step outside, get higher, or shift my position by two feet if an incoming signal is scratchy.
- [ ] I have conducted a "Neighborhood Walk" with my partner to map out the dead zones and high points around my home.

If you're stuck, go to...

- *Chapter 9: I Can Hear Them, But They Can't Hear Me* (if you do the neighborhood walk and you hear your partner perfectly, but they hear absolutely nothing but silence when you reply).

Chapter 18

Repeater-Assisted Family Check-Ins (When It Makes Sense)

SIMPLEX IS FANTASTIC FOR your immediate neighborhood, but what if you work ten miles away across the city in a concrete office park? What if your teenager's high school is on the other side of a massive, heavily wooded hill? If the cellular network suddenly crashes while your family is scattered during a severe weather event, pure radio-to-radio communication simply isn't going to punch through that much terrain and steel.

You need a bigger antenna, and you need a massive altitude advantage. This is where we bring the public repeater concepts you learned back in Part 1 directly into your family's practical emergency plan.

What you'll accomplish:

- Determine exactly how and when your family actually needs a repeater for emergency check-ins.
- Designate one official "Family Repeater" to act as your long-range rally point.
- Build a strict, disciplined schedule that works around the busy nature of public radio towers.

What you need:

- Your Baofeng radio.
- Your One-Page Comms Plan from Chapter 16.

When Repeaters Help (And When They Don't)

As we learned earlier, repeaters are the ultimate cheat code for handheld radios. They act like towering lighthouses, taking your weak, 2-mile signal from the ground and blasting it out over a massive 30-mile radius. However, when relying on them for survival communication, you must understand their one fatal vulnerability: **They belong to the grid.** These massive antennas sit on top of commercial buildings, water towers, or mountain peaks, and they run on standard commercial power. If a hurricane, earthquake, or severe winter freeze knocks out regional power, the repeater tower will automatically switch to a backup system. It might have a battery bank. It might have a diesel generator. It might have solar panels.

A good diesel generator might keep the tower alive for three days. A solar setup might struggle on day two of a dark blizzard. But eventually, if the grid stays down long enough, the repeater will inevitably die.

- **The Golden Rule of Thumb:** Use a repeater exclusively to coordinate getting your family home *during the first few hours* of a crisis. This is the "golden window" when anxiety is highest and people are actively trying to navigate blocked roads. Once everyone is safely back at the house, or within a 2-mile radius of each other, immediately switch off the repeater and use your private Simplex channels to conserve the tower's resources.

Choosing the "Family Repeater"

In Chapter 7, we saved two powerful repeaters into your Home Profile (Channel 1 and Channel 2). You now need to pick one of those to be your officially designated family rally point.

1. **Pick the tested, reliable one:** Do not just pick the repeater that is physically closest to your house. Pick the one that gave both you *and* your spouse the clearest, loudest signal during your tests in Part 1. A repeater on a mountain 15 miles away might actually perform better for both of your commute routes than a smaller repeater just 3 miles down the road.

2. **Make it official:** Write this specific channel clearly on your One-Page Comms Plan under "City Info Channel." Everyone in the family needs to know that this is the designated meeting spot in the sky.

3. **The Backup:** Technology fails. Antennas snap in high winds. If your primary repeater gets knocked offline by a storm and all you hear is dead static when you transmit, your family must automatically know to click up to Channel 2 (The Backup Repeater) and try to connect there.

The Check-In Schedule (Timing is Everything)

Here is the hard, unavoidable truth about using a repeater: **You do not own it.** It is a shared, public resource built and maintained by local amateur radio clubs using their own time and money.

During a major, wide-area emergency (like a spreading wildfire, a tornado touchdown, or a massive flood), official emergency volunteer nets (often called ARES or RACES) will immediately activate. They will take complete control of the primary repeaters to coordinate critical information with local fire departments, police, and hospitals.

[STOP] Stay Off the Button: If official emergency traffic is happening, you must stay off the button. Do not interrupt an active rescue operation just to tell your spouse you are stuck in traffic.

To work around this public shared space, your family check-in schedule needs to be incredibly precise and highly disciplined.

- **Top of the Hour:** In emergencies, staring at a silent radio waiting for a call will drive you crazy. Families should agree to only turn their radios on and attempt to contact each other exactly at the top of the hour (e.g., 1:00, 2:00, 3:00).

- **The 3-Minute Window:** This eliminates the anxiety of waiting. Both you and your spouse turn your radios on to Channel 1 at exactly 1:58 PM. You just listen. If the channel is clear of emergency traffic, you make your call at exactly 2:00 PM. By 2:03 PM, the check-in window is officially over, and you turn the radios completely off to save your batteries for the next hour.

GOOD ENOUGH

What if the repeater is totally busy with official emergency traffic at 2:00 PM? Don't panic and don't interrupt. You simply listen quietly to the traffic, which often provides incredible, life-saving clues about the disaster, like which roads are flooded or closed, and then try to make your family contact at the bottom of the hour (2:30 PM) instead.

The Quick-Strike Script

When you use a public repeater during a crisis, hundreds of other families might be waiting to do the exact same thing. A five-minute conversation about your flat tire or how scared you are is incredibly selfish and potentially dangerous to others waiting to speak. You must use the absolute shortest script possible.

(Wait for the channel to be totally clear of talking...) "This is *Your Call Sign* attempting to reach *Spouse's Call Sign*. Are you copying?" *(Spouse replies...)* "This is *Spouse's Call Sign*. I am safe at the office. Leaving in one hour. Over." "Copy that. See you at home. *Your Call Sign* clear."

Ten seconds. That is all it takes. You got the vital location information, you stayed legally identified with your call signs, and you immediately got out of the way so other people could use the tower.

Quick Win Checklist

- [ ] I understand that repeaters are vulnerable to long-term power outages and should be used primarily to get people home quickly during the initial hours of a crisis.
- [ ] I have officially designated Channel 1 (the most thoroughly tested tower) as my Family Repeater rally point.
- [ ] My family understands they must automatically switch to Channel 2 if Channel 1 is offline or unreachable.
- [ ] I understand that repeaters are public, and I must never interrupt official emergency or rescue nets.
- [ ] I have set a strict "Top of the Hour / 3-Minute Window" check-in schedule to conserve battery life and limit airtime.

If you're stuck, go to...

- *Chapter 11: I Can Hit One Repeater But Not Another* (If you successfully designate a Family Repeater, but your spouse can't seem to hit it with their Baofeng from inside their concrete office building, run the troubleshooting steps).

Narrative Interlude: Callsign Day

Character: Nate Brooks (Southeast)

Skills reinforced:

- One-page family plan
- Simplex expectations
- Repeater-assisted check-ins
- Short message format (WHO/WHERE/WHAT/NEXT)

Where to learn the steps: Chapters 16, 17, & 18

The email arrived at 7:14 AM on a Thursday. Nate Brooks was sitting at his kitchen island, sipping black coffee. He opened the FCC database link on his phone, typed in his last name, and there it was.

KN4XYZ.

It was a quiet moment, but Nate felt a distinct surge of serious pride. The letters and numbers weren't just a license; they were a certification of competence. He had done the reading, passed the test, and secured his legal right to transmit. He was fully operational.

Kara walked into the kitchen, tossing her keys onto the counter. "You've got that look," she said, eyeing him as she poured her own coffee. "The 'mission accomplished' look."

"Callsign day," Nate said, turning the phone screen toward her. "It's official. The grid-down comms plan is fully legal and ready to execute."

"Great," Kara said, taking a sip and looking him dead in the eye. "Now make this usable. Because right now, those radios are just intimidating plastic bricks sitting on your workbench."

Nate nodded. He knew she was right. A system is only as good as its weakest operator. That evening, after dinner, he called a mandatory family meeting in the living room. Kara sat on the sofa. Their sixteen-year-old daughter, Leah, slumped into the armchair, her eyes glued to a TikTok video playing silently on her phone.

Nate stood in front of the coffee table, where he had laid out three programmed Baofeng UV-5Rs.

"Alright," Nate began, slipping effortlessly into his professional briefing voice. "As we approach peak hurricane season, we need to establish our primary and secondary communication protocols. In the event of catastrophic cellular failure, these radios operate on VHF and UHF frequencies. Because UHF is strictly line-of-sight, we have to account for terrain masking and structural absorption. Now, if we need to reach across the county, we will utilize a duplex offset through the primary repeater, which requires a digital tone, "

"Dad," Leah interrupted, not looking up from her screen. "I have literally no idea what you are saying."

Nate stopped. He looked at Kara. She was tapping her foot, offering a thin, patient smile.

"You're lecturing, Nate," Kara said gently. "You spent three weeks learning the physics. We have jobs and high school. We are not passing an FCC exam tonight. We just need to know how to call you if the lights go out."

Nate looked at the radios, then at his family. He had fallen into the classic technical trap. He was trying to teach them *how* the watch was built, instead of just telling them the time.

"You're right," Nate said. He took a deep breath, resetting his posture. "Forget the physics. Let's look at the card."

He reached over to the primary base station radio and unclipped the bright yellow, laminated index card he had made in the garage. He handed a duplicate card to Leah, and one to Kara.

"This is the One-Page Comms Plan," Nate said, his voice dropping the rigid, tactical cadence and becoming calm and conversational. "Everything you need to know is on this yellow card."

Leah finally put her phone down and looked at the card.

"Rule one: Roles," Nate said. "I am the Primary Operator. If I am not home, Kara, you are the Secondary. You run the net from this house. Leah, your only job is to turn the radio on, check in, and turn it off."

"Okay," Leah said, sitting up slightly. "I can do that. Which button is it?"

"Top knob," Nate said. "Turn it until it clicks. That's it. Now, look at the channels on the card. Channel 3 is our Simplex channel. That means direct, radio-to-radio. It doesn't rely on cell towers or power grids. But there is a catch: it only works within a mile or two. If you are at the neighborhood entrance, it works perfectly. If you are downtown, it won't reach."

"So what do we do if we're downtown?" Kara asked, tracing the card with her finger.

"You click the arrow up to Channel 1," Nate explained. "That's the City Repeater. It's a massive tower that bounces our signals 30 miles. But, because it's public, it will be incredibly busy during an emergency. We cannot clog it up. So, we follow the Check-In Rule."

He pointed to the bottom of the yellow card.

"Top of the hour," Nate said firmly. "We do not sit around staring at silent radios. If the cell phones die, you turn your radio on at exactly 2:00 PM, 3:00 PM, or 4:00 PM. We have a three-minute window. We make contact, we exchange status, and we turn the radios off at 2:03 PM to save battery."

"What do I say?" Leah asked, looking at the intimidating keypad on the front of the radio. "Do I have to use police codes?"

"No codes," Nate said. "Plain English. Look at the back of the card. It's four words: Who, Where, What, Next."

He picked up a radio. "It sounds exactly like this: *Dad, this is Leah. I am at the high school. We are sheltering in place. I will check back in one hour. Over.*"

He set the radio down. "That is the entire system. Who you are, where you are, what your situation is, and what your next move is. Ten seconds, and you're off the air."

Kara nodded slowly. "Roles. Channel 3 for local. Channel 1 for long distance. Top of the hour. Who, where, what, next." She looked at Nate. "That's a plan I can actually remember."

"Let's prove it," Nate said. He looked at his watch. It was 7:45 PM. "Leah, take your radio. Walk down to the neighborhood clubhouse. At exactly 8:00 PM, I want you to initiate a Simplex check-in on Channel 3 using the four-step format."

Leah grabbed the radio, slipping the yellow card into her pocket. "Top of the hour. Got it."

For the next fifteen minutes, Nate and Kara sat in the kitchen. Nate forced himself not to touch his base station radio. At 7:59 PM, he turned the top knob on and verified the screen was on Channel 3. He set it on the island counter.

At exactly 8:00 PM, the squelch gate snapped open, accompanied by a faint hiss of static, proving they were operating on direct, line-of-sight simplex.

"Dad, this is Leah," her voice came through, completely clear. *"I am at the clubhouse. The pool is closed. Heading back to the house now. Over."*

Nate smiled. He picked up the radio, pressed the PTT button, and paused for one second.

"Copy that, Leah. Loud and clear. See you in five minutes. KN4XYZ clear."

He released the button and looked at Kara.

"She did it perfectly," Kara said, genuinely impressed. "No panic. No confusion."

"Because she didn't have to guess," Nate said, tapping the laminated yellow card on the counter. He had learned the most valuable lesson of emergency preparedness: the best gear in the world is useless if the procedure is too complex for a tired, stressed-out human being to remember.

He had his callsign. He had his hardware. But more importantly, he finally had a system his family could trust.

Takeaway

A plan that can't be explained in one page won't be followed. Do not burden your family with radio theory, physics, or complex menu operations. Give them a highly structured, laminated cheat sheet. Establish clear roles, realistic Simplex range expectations, strict check-in times to conserve battery, and enforce the short WHO/WHERE/WHAT/NEXT message format to eliminate rambling.

Bridge: Now we build the habit: 10 minutes a week.

Chapter 19

Drills: Build Muscle Memory in 10 Minutes a Week

A BEAUTIFUL, PERFECTLY THOUGHT-OUT plan written on a laminated index card is completely useless if nobody in your household actually knows how to execute it under pressure.

If the very first time your spouse or teenager tries to turn on the Baofeng and change a channel is during a terrifying neighborhood evacuation in the pitch black, it will be a complete disaster. Adrenaline is a powerful chemical that causes tunnel vision and destroys fine motor skills. Under that kind of intense stress, their hands will shake, their brain will freeze, they will press the wrong buttons, the radio will beep angrily at them, and they will give up in frustration, assuming the radio is broken.

Muscle memory is the only thing that reliably defeats panic. You don't rise to the occasion during a crisis; you default to your level of training. We need to turn operating this radio into a mindless, boring, automatic habit.

What you'll accomplish:

- Create highly repeatable, 10-minute drills for your family that don't feel like a chore.
- Turn basic radio operation into a stress-free, automatic habit that survives an adrenaline dump.
- Use gamification (like the "Lost Phone" drill) to keep kids and teenagers engaged in the process.

What you need:

- Your family members.
- Your programmed radios (fully charged).

Drill 1: The Blindfold Test (Basic Hardware)

In a real emergency, like a severe winter storm or a tornado, the lights are usually the first thing to go out. Before anyone worries about talking on the airwaves, they need to know how to manipulate the physical hardware without staring at the screen. You should be able to operate the basic functions of this radio in total darkness.

The Drill:

1. Hand the radio to your family member completely turned off and with the volume knob rolled all the way down.
2. Ask them to close their eyes (or sit together in a completely dark room).

3. Say: *"Turn the radio on, set the volume to a low, comfortable level, and prove the speaker actually works."*

4. **Pass Condition:** Without opening their eyes, they successfully click the top knob on. They physically slide their fingers down the left side of the radio to find the MONI button (distinguishing it from the larger PTT button by feel), press it to hear the rushing static, adjust the volume knob to a comfortable level, and release the button.

Drill 2: The Channel Switch

Under stress, people tend to grip the radio tightly, accidentally crushing the keypad with the palm of their hand. This drill ensures they can reliably jump between your City Repeater and your Family Simplex channel without accidentally pressing the wrong keys and getting lost in the menus.

The Drill:

1. Put the radio in Memory Mode on Channel 1 (with the keypad officially locked).

2. Say: *"The city repeater just went down and all we hear is static. Switch to the Backup Family Channel."*

3. **Pass Condition:** They calmly hold the # key for two seconds until the tiny padlock icon disappears, use the Up Arrow to click up exactly three times to Channel 4, and press the # key again to re-lock the pad.

STOP

Correct gently. If they accidentally press the orange VFO/MR button and suddenly get lost in frequency mode, do not yell or get frustrated. Panic kills the desire to learn. Calmly remind them: "Take a breath. If you see a screen full of raw numbers without a tiny channel number on the right, you just accidentally bumped the door open. Press the orange button once to get back home."

Drill 3: The "Clean Call"

Now we practice the "Say Less" principle from Chapter 16. It is incredibly common for beginners to start talking *before* they press the button, which cuts off the first half of their sentence.

The Drill:

1. Put one family member in the living room with a radio on Channel 3. Go out to the backyard with your radio so you can't see each other.

2. Text them on their phone: *"Initiate a check-in."*

3. **Pass Condition:** They must execute the perfect "Push, Pause, Speak" rhythm. They press the PTT button, pause for a deliberate half-second, state who they are, state where they are, state their status, and cleanly say "Over," then immediately release the button. *(Good Example: "Dad, this is Leo. I am in the living room. Everything is secure here. Over.") (Bad Example: "...ead, it's Leo, I'm just sitting here on the couch, what do you want me to do next?")*

Drill 4: The "Lost Phone" Scenario (Advanced)

Teenagers might roll their eyes at sitting in the living room practicing radio calls. Once a month, run a practical, gamified scenario that simulates a minor crisis in the real world. This proves to them that the tool actually works when the cell network doesn't.

The Drill:

1. Tell your family: *"For the next 30 minutes, all cell phones are dead."* Have everyone physically put their smartphones in a drawer.
2. Send a family member to the neighborhood grocery store, a friend's house, or a nearby park (somewhere you have already proven is within your Simplex range) with a Baofeng radio.
3. Have them coordinate their pickup, or report when they are walking home, using *only* the radio on the strict Top of the Hour check-in schedule.
4. **Pass Condition:** They successfully contact you at the designated time, deliver the required information cleanly without rambling, and return home. The feeling of success when they hear your voice without using a cell tower is a massive confidence booster.

Tracking the Habit

Consistency beats intensity every single time. Don't try to run all four drills on the same day for an hour straight. People will get incredibly bored, annoyed, and view the radios as a chore.

Run just one drill for ten minutes on a Saturday morning while making coffee. Keep a simple physical checklist on the fridge. Once everyone in the household can pass Drill 1 and Drill 2 flawlessly without hesitating or looking at the buttons, your family is officially better prepared than 99% of your neighborhood for a local emergency.

[YOU'RE WINNING] You have successfully transitioned your family from passive bystanders to capable radio operators. The panic is gone.

Quick Win Checklist

- [ ] I understand that muscle memory is the only reliable tool during a stressful emergency or adrenaline dump.
- [ ] I have successfully run the "Blindfold Test" with my family members to ensure they can find the volume and monitor buttons by feel alone.
- [ ] I have verified my family can confidently unlock the keypad, change channels precisely, and re-lock it.
- [ ] I have practiced "Clean Calls" to enforce the Push-Pause-Speak rhythm and the WHO/WHERE/WHAT protocol.
- [ ] I have planned one simple, 10-minute radio drill for this upcoming weekend to start building the habit.

If you're stuck, go to...

- *Chapter 12: My Channels Won't Save* (If a family member accidentally deletes a channel while fumbling with the keypad during a drill, go back and rebuild it using the Double Save method).

Chapter 20

Go-Bag Setup (Power, Spares, and What Actually Matters)

You wake up at 2:00 AM to the terrifying wail of tornado sirens. The power grid in your entire neighborhood goes completely black. Your heart rate spikes. You calmly reach into the hallway closet, grab your trusty Baofeng radio, twist the volume knob on, fully expecting to hear the comforting voice of the local weather dispatcher... and the screen stays black.

No beep. No static. Nothing. The lithium battery slowly died three months ago while sitting neglected in the bottom of a drawer.

A radio without a rock-solid power and storage strategy isn't a lifeline; it's just a tactical-looking plastic brick. In an emergency, you will not have the time, the electricity, or the presence of mind to hunt for a charging cable. Preparedness means ensuring your equipment is in a permanent, organized "ready state," so it functions perfectly the exact second you reach for it in the dark. Let's cut through the internet hype and build a foolproof, budget-friendly Go-Bag setup that actually works when the lights go out.

What you'll accomplish:

- Develop an automated power strategy so your radio is never dead when you need it most.
- Understand which antenna upgrades actually matter (and which "tactical" traps are a waste of money).
- Build a consolidated, grab-and-go kit for your emergency bag that protects your gear.
- Learn the secret to safely storing a backup radio in your vehicle without ruining it.

What you need:

- Your Baofeng radio.
- The original desktop charging cradle.
- A dedicated small padded pouch or bag.

The Power Strategy (Rule of Two)

Lithium-ion batteries are fantastic, but they have a fatal flaw: they naturally lose their charge over time, even when the radio is completely turned off. Furthermore, the Baofeng has a tiny amount of "parasitic drain", meaning the internal circuitry sips a microscopic amount of power even when the knob is clicked off. If you fully charge your radio in January and put it in a closet, it might be completely dead by July.

- **The Rotation (Avoid the Silent Killer):** Never, ever store the radio sitting inside the plugged-in charging cradle. The Baofeng charging docks are famously "dumb." They constantly trickle a tiny bit of voltage into the battery, which degrades the chemical cells over time and can cause the battery pack to physically swell or fail. Instead, charge the battery to 100%, take it entirely off the charger, and put it in your bag. Set a recurring calendar reminder on your phone for the 1st of every month to plug it in and "top off" the charge for an hour.

- **The Rule of Two:** "Two is one, and one is none." You must own at least one spare, extended battery (like the incredibly popular 3800mAh extended battery pack). The extended battery makes the radio slightly taller and heavier, but it effectively doubles your listening time from 14 hours to almost 30 hours. Keep the extended battery physically attached to the radio, and keep the smaller stock battery fully charged sitting right next to it in the bag. Use a silver marker to write A and B or the date of the last charge on them.

- **The AA Battery Shell (The Ultimate Grid-Down Backup):** You can buy a special, empty plastic battery shell designed for the UV-5R that holds 6 standard AA batteries. If the power grid is down for a massive two-week blackout and you absolutely cannot plug your charger into a wall, popping in standard AA batteries from your kitchen drawer is an absolute lifesaver.

STOP

Not all AAs are created equal. Do not put standard alkaline batteries (like cheap Duracells or Energizers) into your backup shell and leave them in a hot bag for three years. They will leak corrosive battery acid and permanently destroy the plastic shell. You must use Lithium AA batteries (like Energizer Ultimate Lithium). They cost twice as much, but they have a 20-year shelf life, they perform beautifully in freezing weather, and they absolutely never leak.

Antennas: Keep It Simple and Avoid the "Tactical" Trap

New radio operators love to jump on Amazon and immediately buy massive, 15-inch "tactical" whip antennas that fold in half with a velcro strap. They look incredibly cool, but they introduce a massive vulnerability.

While these massive antennas *do* slightly improve your transmit range, they make the radio incredibly top-heavy and clumsy to hold. More importantly, they create leverage. If you drop the radio on the kitchen floor, or if the long antenna snags on a car door as you are getting out, that 15-inch lever will instantly snap the fragile brass SMA connector completely off the internal circuit board. Your radio is now effectively useless and requires complex micro-soldering to repair.

- **The Go-Bag Standard:** The 6-inch stock "rubber duck" antenna that came in the box is perfectly fine for 90% of neighborhood emergencies. It is short, flexible, and extremely difficult to break. Keep it on the radio.

- **The Smart Upgrade:** If you desperately want a range upgrade, skip the heavy tactical folding whips. Look for a "Signal Stick" (made of a special memory-metal that can be tied in a knot without breaking and snaps right back into perfect shape) or a genuine Nagoya NA-771. However, if you attach an upgraded antenna, you must keep the original stock antenna tucked away in your bag as a permanent backup. Antennas break easily in chaotic environments, and a radio without an antenna is just a plastic brick.

The Grab-and-Go Kit

Do not leave your radio components scattered around the house. The charger is plugged into the wall in the home office, the radio is sitting on the kitchen counter, and the spare battery is buried in a junk drawer. When the lights violently go out and panic sets in, you will never find all three pieces in the dark.

Buy a small, inexpensive padded electronics pouch, a tactical admin pouch, or a small Pelican-style hard case. Inside this single, dedicated container, pack:

1. **The Radio:** With the extended battery and stock antenna physically attached, ready to turn on.

2. **The Spare Battery:** Fully charged.
3. **The AA Battery Shell:** Pre-loaded with 6 Lithium AA batteries.
4. **Your Comms Plan:** The printed, waterproofed, laminated One-Page template from Chapter 16.
5. **A Discreet Earpiece:** In an emergency shelter or a crowded room, you might not want a loud dispatcher's voice broadcasting news to everyone around you. A cheap earpiece provides critical privacy and prevents panic from spreading. It also prevents drawing unnecessary attention to your supplies if you are trying to remain low-profile.
6. **A Headlamp & Pencil:** A small, cheap LED headlamp (so you can read the radio screen and use both hands to operate it in the dark) and a small carpenter's pencil (which will never dry out or freeze like an ink pen) to log important information.

The "One Extra Radio" Concept

Baofengs are incredibly cheap, often less than the cost of a large pizza. Once you have built your perfect "Home Profile," run your family drills, and packed your Go-Bag, the absolute smartest thing you can do is buy one more radio.

Program this second radio exactly like yours. Turn it off, lock the keypad, and put it in the glovebox of your car or the bottom of your spouse's daily work bag.

If an emergency happens while you are physically separated at work or running errands, you don't have to stress about whether or not they remembered to bring their radio today. You already know they have one, you know it's programmed with the exact same frequencies, and you know exactly which channel to find them on.

FIELD NOTE

Surviving the Hot Car: If you decide to store a backup radio in your vehicle, do not leave the lithium-ion battery attached to it. The interior of a car parked in the summer sun can easily reach 140°F (60°C). Extreme heat will completely destroy a lithium battery, causing it to swell or even catch fire. For a dedicated vehicle radio, remove the lithium battery entirely. Power it exclusively using the AA Battery Shell loaded with Energizer Ultimate Lithium AA batteries, which are chemically designed to safely withstand extreme temperature swings without degrading or exploding.

YOU'RE WINNING

Take a moment to look at your organized, fully charged, programmed kit. You have successfully graduated. You started this book with a confusing, beeping, stressful piece of plastic, and you have methodically transformed it into a trusted, tested, family-integrated emergency lifeline. You are officially ready for whatever the world throws at you.

Quick Win Checklist

- [ ] I understand the danger of parasitic drain and have set a monthly calendar reminder to top-off my battery charge.
- [ ] I have permanently removed my radio from the desktop charging cradle so I don't degrade the battery chemistry over time.
- [ ] I have avoided the "tactical antenna" trap to protect my radio's fragile antenna connector.
- [ ] I have consolidated my radio, spare battery, earpiece, and Comms Plan into a single, easy-to-grab padded pouch.
- [ ] I understand how to safely store a backup radio in a hot vehicle using Lithium AA batteries.
- [ ] I have completely finished Part 3! My family is fully prepared, practiced, and equipped to communicate in an emergency.

If you're stuck, go to...

- *Part 4: Quick Cards and Templates* (Now it's time to print out the final, concentrated cheat sheets so you have all of this critical information securely on paper for your Go-Bag).

Narrative Interlude: Ten Minutes a Week

Character: Mark Jensen (Midwest Suburbs)

Skills reinforced:

- Weekly drills
- Go-bag readiness
- Calm check-ins
- Labeling/ready state

Where to learn the steps: Chapters 19 & 20

The local meteorologist on the morning news was pointing to a massive, angry red blob sweeping down from Canada. It was a classic Minnesota "Alberta Clipper," promising white-out conditions, plummeting temperatures, and a very high probability of downed power lines by nightfall.

Mark Jensen stood at the kitchen island, absolutely vibrating with preparedness.

Spread across the granite countertop was a tactical, olive-drab backpack with a dozen molle webbing straps. Next to it sat three Baofeng UV-5R radios, a tangle of extended antennas, battery charging cradles, a throat microphone he'd ordered late one night on Amazon, and a notebook filled with tiny, scribbled frequency tables.

Jen walked into the kitchen, holding her empty coffee mug. She stopped dead in her tracks, staring at the military-grade garage sale covering her counter. Owen shuffled in behind her, wearing pajama pants and earbuds.

Jen looked at the red blob on the TV, and then looked at Mark. She gave him *the look*. It was a look of deep, loving caution that explicitly meant: *Do not make this complicated.*

"The storm is tracking south," Mark announced, adopting his best Mission Control voice. "We have a highly probable grid-down scenario. I have spent the morning establishing a multi-tiered communications web. I've programmed fourteen local repeaters, a cross-band simplex relay, and I've configured dual-watch so we can monitor county fire dispatch while simultaneously tracking the weather node."

Owen slowly pulled one earbud out. "Dad, are we going to war with the snow?"

"No, Owen, we are ensuring continuity of operations," Mark said, picking up the throat microphone. "Now, for today's drill, I want us to practice switching between the primary and tertiary repeater offsets while wearing these tactical headsets. It will take about an hour, "

"Mark," Jen interrupted gently, sliding her coffee mug onto a small sliver of empty counter space. "You're doing the NASA thing again."

"I am establishing a robust protocol," Mark protested mildly.

"You have a throat mic," she pointed out. "I am not wearing a throat mic to tell you the power went out. You spent two weeks making that beautiful, simple yellow card. Why are we suddenly trying to run a CIA safehouse?"

Mark looked down at the sprawling mess on the counter. He looked at the giant, floppy 15-inch tactical antennas he'd bought. He looked at the charging cradles that were slowly degrading his lithium batteries. He realized she was entirely right. In his excitement to be "ready," he had completely abandoned the core principle that got him here: *Keep it simple.*

He had fallen back into the trap of thinking more gear meant more safety.

"Okay," Mark sighed, a smile breaking through. "You're right. I'm over-engineering. Let's dial it back."

Mark took the throat mic and the massive antennas and shoved them back into a cardboard box. He took the radios completely off their charging cradles, remembering the 'parasitic drain' warning from the guide, and placed them in a neat row.

He pulled a silver metallic Sharpie out of the kitchen drawer. On the back of the first radio's battery, he wrote **BASE**. On the second, he wrote **JEN**. On the third, he wrote **OWEN**.

"Labeling. Good," Jen nodded approvingly. "What's the actual plan?"

"Ten minutes. Two drills. That's it," Mark said, pointing to the laminated yellow Quick Card taped to the Base radio. "Owen, you're up first. Drill 1. The Blindfold Test."

Owen rolled his eyes, but he put his phone down. "I don't have a blindfold."

"Just close your eyes," Mark said, handing Owen his labeled radio. "The power is out. It's pitch black in the house. The cell networks are down. Turn the radio on, find the volume, and prove to me the speaker works."

Owen closed his eyes. Mark watched his son's hands. Owen fumbled for a second, then his fingers found the top knob. *Click.* The radio beeped. His thumb slid down the left side of the casing, bypassing the large PTT button, and found the smaller MONI button below it. He pressed it.

SHHHHHHHHH! A loud rush of static filled the kitchen. Owen rolled the volume knob down until it was a quiet hiss, then let go of the button.

"Target acquired," Owen said, opening his eyes and smirking. "Too easy."

"Perfect," Mark said. "Muscle memory. You don't need to see the radio to use it."

"My turn," Jen said, grabbing her radio and slipping on her winter coat. "Drill 3. The Clean Call. I'll go out to the cold car in the driveway."

She walked out the front door. Mark and Owen stayed in the kitchen. Mark turned the Base radio on and verified they were securely locked on Channel 3, their neighborhood simplex channel. He set the radio on the counter and stepped back.

A minute later, the squelch gate snapped open.

"Mark, this is Jen," her voice came through, completely clear and steady, untainted by the howling wind outside. *"I am in the frozen minivan. Status is very cold, but secure. Ready to come back inside for coffee. Over."*

Mark grinned. She had used the exact WHO / WHERE / WHAT / NEXT format without rambling. He pressed the PTT button, paused for a full One-Mississippi count, and replied.

"Copy that, Jen. Coffee is hot. Proceed to the kitchen. Mark clear."

When Jen walked back inside, brushing a few stray snowflakes off her shoulder, Mark was already packing the olive-drab Go-Bag. But this time, he wasn't packing a chaotic mess.

He packed the Base radio with the extended battery attached. He packed one spare stock battery, fully charged. He packed the empty AA battery shell, pre-loaded with six Energizer Ultimate Lithiums for a grid-down backup. He tucked the laminated Quick Card and a pencil into the front zipper.

It was light. It was organized. It was in a permanent 'ready state.'

"There," Mark said, zipping the bag closed and setting it by the backdoor emergency kit. "Ten minutes. We proved we can operate in the dark, and we proved we can make a clean transmission without panicking. That's our weekly routine."

"I like this routine," Jen said, pouring her coffee. "It doesn't involve any three-hour lectures on frequency math."

"Or throat mics," Owen added, picking his earbuds back up.

Mark leaned against the counter, listening to the wind start to rattle the kitchen windows as the Alberta Clipper finally arrived. He didn't feel the urge to press buttons or test the squelch again. He didn't feel anxious. He had successfully transitioned from being a guy playing with a tactical walkie-talkie to a guy with a trusted, reliable system.

"You know," Mark said, tapping the Go-Bag with his foot. "I think we're actually ready for the snow."

Takeaway

Skills beat features. Repetition beats panic. Do not overwhelm your family with overly complex gear, tactical accessories, or massive channel lists. Strip your Go-Bag down to the reliable essentials, ensure your radios are physically labeled, and store them *off* the chargers to protect the batteries. Stick to a simple, 10-minute weekly drill routine to keep the muscle memory sharp.

Bridge: Mark and Nate have both built their systems. Now, there's only one final test...

Part 4:

Quick Cards and Templates (Print & Laminate)

Print It. Keep It. Use It When It Counts.

Scan the QR code to get the printable quick cards, templates and appendix pages from this book. You'll get practical tools for setup, repeater programming, first calls, troubleshooting, and family comms planning... ready to print and keep with your radio.

https://www.morsecodepublishing.com/bootcamp

Chapter 21

Quick Card: First Hour Checklist

GOAL: From unopened box to your first transmission. *If you check every box, you are successfully on the air.*

1. Hardware Setup

- [ ] **Attach Antenna:** Screw on the stock rubber antenna until snug. (STOP: Never press the talk button without an antenna or you risk permanent hardware damage!)
- [ ] **Seat Battery:** Slide on the battery and press firmly until it physically clicks.
- [ ] **Power On:** Turn the top knob clockwise. Set volume to roughly 50%.
- [ ] **Test Speaker:** Press the side **MONI** button to hear static. Adjust volume to comfort.

2. Basic Configuration

- [ ] **Set Squelch:** Press MENU -> Type 0 -> MENU -> Select 2 (this is the Goldilocks setting, quiet enough to block static, but sensitive enough to hear people) -> MENU -> EXIT.
- [ ] **Set Step Size:** Press MENU -> Type 1 -> MENU -> Select 2.5K -> MENU -> EXIT.
- [ ] **Clear Receive Tones:** Check MENU 10 and MENU 11. Both must be **OFF**.

3. VFO Mode (The Canvas)

- [ ] **Enter VFO Mode:** Press the orange VFO/MR button until channel numbers on the right disappear.
- [ ] **Select Top Line:** Press the blue A/B button until the tiny arrow points to the top line (A). The radio only listens to and obeys the line with the arrow!

4. Finding a Signal (Listen First)

- [ ] **Tune to Weather:** Type 1 6 2 . 5 5 0 (or local equivalent). Listen for the robotic voice.
- [ ] **Tune to Repeater:** Type a local Repeater Receive Frequency. Wait to hear a conversation.

5. Programming a Repeater

- [ ] **Clear Slot:** MENU 28 -> Select 001 -> MENU to delete old data.
- [ ] **Enter Frequency:** Type the Repeater Receive Frequency.
- [ ] **Set Offset Direction:** MENU 25 -> Select + or - -> MENU.
- [ ] **Set Offset Amount:** MENU 26 -> Type 0 0 0 6 0 0 (for 0.600) -> MENU.
- [ ] **Set Transmit Tone:** MENU 13 -> Select Tone (e.g., 100.0) -> MENU. *(Leave OFF if none required).*

6. The Double-Save

- [ ] **Save Receive:** MENU 27 -> Select 001 -> MENU *(Hear: "Receiving Memory").*
- [ ] **Save Transmit:** Press MENU again -> Select 001 -> MENU (Listen for the radio to say: Transmitting Memory, this is the proof your settings are saved!) -> EXIT.
- [ ] **Test It:** Press VFO/MR to enter Memory Mode. Go to Channel 001.

7. First Transmission

- [ ] **Check License:** Verify you hold a valid Amateur Radio License.
- [ ] **Listen:** Wait 30 seconds to ensure the channel is totally clear.
- [ ] **Talk:** Press PTT, pause for a full One-Mississippi count, say: "This is *Your Call Sign*, monitoring." Release PTT.

IF YOU ARE STUCK: Go directly to Chapters 8–12 (The Troubleshooting Tree).

Chapter 22

Quick Card: Repeater Setup Decision Chart

USE THIS FLOWCHART WHEN looking at RepeaterBook or a local radio club listing to know exactly which menus to change.

Q1. Does the listing have an Offset or Shift (+ / -)?

- **YES:** 1. Press **MENU**, type **25**, press **MENU**, select + or -, press **MENU** to confirm. 2. Press **MENU**, type **26**, press **MENU**, use the keypad to type the digits (e.g., **000600**), press **MENU** to confirm.
- **NO:** * Leave Menu 25 OFF. Leave Menu 26 at 000000. This is a simplex (direct) frequency.

Q2. Does the listing require an "Uplink Tone" (CTCSS / PL Tone)?

- **YES, it lists a standard number (e.g., 100.0 or 141.3):**
    - Go to **Menu 13 (T-CTCS)** and set it to that exact number.
- **YES, but it lists a "DCS" or "Digital Code" (e.g., 023 or D023N):**
    - Go to **Menu 12 (T-DCS)** and set it to that exact code.
- **NO, it says CSQ or lists no tone at all:**
    - Leave Menu 13 and Menu 12 **OFF**.

Q3. Does the listing show a "Downlink Tone" or "Receive Tone"?

- **DEFAULT ADVICE: IGNORE IT.** * Leave **Menu 11 (R-CTCS)** and **Menu 10 (R-DCS)** completely **OFF**.
    - *Why?* Setting a receive tone puts a lock on your own radio's speaker. Unless you are experiencing severe local interference, keeping your receive tones OFF ensures you hear everything on the channel and never accidentally lock yourself out.

THE "SAVE IT" REMINDER

- Did you sweep the floor? (**Menu 28: DEL-CH**). You must delete the channel slot first if it has an existing number, or the radio may not save your new offset settings correctly.
- Did you double-save? (**Menu 27: MEM-CH**). 1. Save RX: Press **MENU**, select channel, press **MENU** (Hear: *Receiving Memory*). 2. Save TX: Immediately press **MENU** again (while still on Menu 27), press **MENU** once more (Hear: *Transmitting Memory*).*

Chapter 23

Quick Card: First On-Air Script + Etiquette

TAPE THIS NEAR YOUR radio base station until muscle memory takes over.

The 3 Starter Scripts

1. The "I'm Just Listening" Script *(Use this to announce your presence without demanding a conversation.)*

"This is *Your Call Sign*, monitoring."

2. The "Radio Check" Script *(Use this to verify your antenna and audio are working clearly.)*

"This is *Your Call Sign*, testing a new radio setup. Is anyone available for a quick signal report?" (When they reply: "Thanks for the report, you sound great as well. *Your Call Sign* clear.")

3. The "Quick Question" Script *(Use this to ask for help or directions.)*

"This is *Your Call Sign*. Does anyone on frequency have a recommendation for *Your Question*?"

The 7 Golden Rules of Etiquette

1. **Listen First:** Never turn on a radio and immediately talk. Wait 30 to 60 seconds. Check your screen for a green RX light or signal bars, which indicate the channel is busy even if you hear silence.
2. **Press, Pause, Speak:** Press the PTT button, count "One-Mississippi" in your head, and *then* speak. This ensures the first word of your sentence isn't cut off while the repeater wakes up.
3. **Use Plain English:** Drop the CB slang. Speak naturally and clearly. Don't worry about the Q-codes (like QSL or QSY) you hear others using, plain English is the gold standard for beginners.
4. **Identify Properly:** Use your legal, government-issued call sign. By law, you must state it every 10 minutes during a long conversation and when you sign off for the last time.
5. **No "CQ" on FM Repeaters:** Do not yell "CQ, CQ, CQ." Just say your call sign and that you are monitoring.
6. **Don't Hog the Channel:** Keep transmissions under 45 seconds. Always wait for the 'repeater tail' (the click or beep after someone else stops talking) before you start your next turn.
7. **No Profanity or Music:** The airwaves are public, unencrypted, and monitored. Keep it family-friendly.

Chapter 24

Quick Card: Family Channel Plan (Fill-in Template)

INSTRUCTIONS: FILL THIS OUT with a dark marker, completely laminate with clear packing tape, and store with your radio battery.

FAMILY COMMS PLAN

- **Primary Operator:** ____
- **Secondary Operator:** ____
- **Home Channel (Simplex 1):** Channel 003
- **Backup Channel (Simplex 2):** Channel 004
- **City Info Channel (Listen Only):** Channel 001

CHECK-IN TIMES (Strict): [ ] Top of the Hour (e.g., 2:00, 3:00) [ ] Bottom of the Hour (e.g., 2:30, 3:30) *Rule: Turn radio on 2 minutes before. Turn radio off 3 minutes after to save battery.*

MESSAGE FORMAT: "WHO / WHERE / WHAT / NEXT"

- **WHO:** "Dad, this is Sarah."
- **WHERE:** "I am at the high school."
- **WHAT:** "Roads are blocked. I am staying here."
- **NEXT:** "I will check back in one hour. Over."

EMERGENCY PROTOCOL: If in immediate, life-threatening danger, break protocol. Say: **"Emergency, Emergency, Emergency"** and give your specific location immediately.

Chapter 25

Quick Card: Weekly Drill Tracker

Instructions: Keep this on the fridge. Run only ONE drill per week. Keep it under 10 minutes. Muscle memory defeats panic.

The Drills:

- **Drill 1 (Blindfold):** Turn radio on, set volume, use side MONI button to test speaker with eyes closed.
- **Drill 2 (Channel Switch):** Unlock keypad, switch from Channel 1 to Channel 4, re-lock keypad.
- **Drill 3 (Clean Call):** Go to different rooms. Execute a flawless "Press, Pause, Speak" using the WHO/WHERE/WHAT format.
- **Drill 4 (Lost Phone):** One person goes to a local park/store without a cell phone and checks in at the Top of the Hour via radio.

Week	Date	Drill #	Pass / Fail	Notes (What went wrong?)
Week 1	_____	Drill 1	[ ] Pass [ ] Fail	______________________________
Week 2	_____	Drill 2	[ ] Pass [ ] Fail	______________________________
Week 3	_____	Drill 1	[ ] Pass [ ] Fail	______________________________
Week 4	_____	Drill 2	[ ] Pass [ ] Fail	______________________________
Week 5	_____	Drill 3	[ ] Pass [ ] Fail	______________________________
Week 6	_____	Drill 3	[ ] Pass [ ] Fail	______________________________
Week 7	_____	Drill 4	[ ] Pass [ ] Fail	______________________________
Week 8	_____	Drill 4	[ ] Pass [ ] Fail	______________________________

Note for Parents: Start with Drill 1 and Drill 2 *only* until they are completely automatic and boring. Do not rush to Drill 4 until basic hardware mastery is achieved.

Chapter 26

Quick Card: Fix-It Fast Cheats (Top 12 Symptoms)

When it fails, don't panic. Check this list.

Symptom	Fastest Fix	Reference
Keypad ignores me (beeps angrily)	You are in Memory mode. Press **VFO/MR** to switch to manual typing.	*Ch. 8*
I transmit, but friends don't hear me	Check the **A/B** arrow. It's probably pointing to the wrong line.	*Ch. 8*
Radio restarts when I press PTT	Your battery is loose. Remove it and slam it back on until it clicks.	*Ch. 8*
Tower ignores my transmission	Your **Menu 13 (T-CTCS)** password tone is wrong or missing.	*Ch. 9*
Tower ignores me (Frequency doesn't jump)	Your Offset is OFF. Set **Menu 25** (+/-) and **Menu 26** (Amount).	*Ch. 9*
My buddy is talking, but radio is silent	You accidentally set a Receive Tone. Turn **Menu 10 & 11 OFF**.	*Ch. 10*
Random blasts of loud static	Your squelch is too low, or you are near a PC. Bump **Menu 0** to 2 or 3.	*Ch. 10*
I can't type the exact frequency (rounds up)	Your step size is too large. Change **Menu 1 (STEP)** to 2.5K.	*Ch. 10*
One tower works, the other doesn't	The broken tower uses a digital code. Turn off Menu 13, use **Menu 12**.	*Ch. 11*
My saved channel disappears	You didn't "sweep the floor." Use **Menu 28** to delete it before saving.	*Ch. 12*
My saved channel works, but won't transmit	You forgot the Double-Save. Redo **Menu 27** twice (RX and TX).	*Ch. 12*
Scanner won't stop on active traffic	Squelch is on 0 (stuck open) or 9 (stuck closed). Set it to 2.	*Ch. 13*

Appendices

Appendix A: Common Terms (Plain English)

- **Baofeng:** A highly popular, inexpensive brand of handheld two-way radios manufactured in China.
- **Call Sign:** A unique alphanumeric identifier (e.g., K1ABC) issued by your government once you pass your amateur radio test. It is your official, legal name on the airwaves.
- **CTCSS/ PL Tone:** An invisible, sub-audible hum added to your voice transmission that acts as a "VIP password" to open up a repeater tower.
- **DCS:** The exact same concept as a CTCSS tone, but it uses a digital code instead of an analog hum.
- **Duplex:** The ability to transmit on one frequency and receive on a different frequency at the same time (how repeaters work).
- **Offset:** The exact mathematical difference between a repeater's listening lane and its talking lane (e.g., jumping 0.600 MHz).
- **PTT (Push-To-Talk):** The large button on the side of the radio. You hold it down to broadcast, and let it go to listen.
- **Repeater:** A large radio tower placed on a high elevation that listens for weak handheld signals and re-broadcasts them at massive power over a wide area.
- **Simplex (Direct):** Pure radio-to-radio communication without using a repeater tower.
- **Squelch:** The radio's invisible "gate" that mutes background static, only opening when a sufficiently strong human signal is detected.
- **VFO (Variable Frequency Oscillator):** The mode on your radio that allows you to manually type in raw numbers and explore the airwaves.

Appendix B: Recommended Default Settings (Beginner-Friendly)

If you get lost in the menus, verify these settings. This configuration creates the path of least resistance for a beginner.

- **Menu 0 (SQL):** 2 *(Filters static but catches weak signals)*
- **Menu 1 (STEP):** 2.5K *(Allows precise frequency typing)*
- **Menu 2 (TXP):** HIGH *(Maximum transmit power)*
- **Menu 3 (SAVE):** 3 *(Good battery saving ratio)*
- **Menu 7 (TDR):** OFF *(Dual Watch off prevents confusing channel jumping while learning)*
- **Menu 8 (BEEP):** ON *(Provides auditory feedback when keys are pressed)*
- **Menu 10 (R-DCS):** OFF *(Prevents locking yourself out of hearing traffic)*
- **Menu 11 (R-CTCS):** OFF *(Prevents locking yourself out of hearing traffic)*
- **Menu 14 (VOICE):** ENG *(Keeps voice prompts in English)*
- **Menu 35 (STE):** OFF *(Squelch Tail Elimination - turn off to fix weird repeater clicks)*
- **Menu 39 (ROGER):** OFF *(Nobody likes the obnoxious "roger beep" on ham bands)*

Appendix C: Channel Naming Conventions (Examples)

When using CHIRP software to program your radio via computer, you are limited to exactly 7 alphanumeric characters for channel names. Keep them incredibly descriptive so you know what they are in a crisis.

- **CTY-RPT** (Primary City Repeater)
- **NTH-RPT** (North Backup Repeater)
- **FAM-DIR** (Family Direct / Simplex)
- **CONVOY1** (Vehicle-to-Vehicle Travel Channel)
- **WX-CH1** (Weather Broadcast 1)
- **FIRE-DS** (Local Fire Dispatch - Receive Only)

Appendix D: Getting Local Repeaters (Where to Look)

Do not guess frequencies. The easiest way to find accurate, up-to-date repeaters in your specific neighborhood is:

1. **RepeaterBook:** Download the free "RepeaterBook" app on iOS or Android, or visit www.repeaterbook.com. It uses your GPS to show you exactly how many miles away the nearest tower is.

2. **Local Clubs:** Google "[Your County] Amateur Radio Club." They almost always list their primary repeater on the homepage.

What to Copy Down on Your Notepad:

1. The **Downlink / Receive** Frequency (e.g., 146.700).

2. The **Offset** Direction (+ or -).

3. The **Uplink Tone** (e.g., 100.0). Ignore the Downlink tone if they list one.

Appendix E: Legal Basics (What You Can / Can't Do)

- **Listening is Free:** In the US (and most countries), it is 100% legal to listen to almost any unencrypted frequency without a license.
- **Transmitting is Governed:** To press the PTT button on Amateur Radio (Ham) bands, you **must** hold a valid license from your government (like the FCC).
- **The Exception:** Fines are steep for illegal transmitting. The *only* exception generally recognized is an immediate, genuine threat to life and limb where no other method of communication is available. A flat tire does not count. A life-threatening medical emergency during a total grid collapse does.
- **Get Licensed:** The entry-level "Technician" license in the US is a 35-question multiple-choice test. It requires zero morse code and costs about $15 to take. Spend the time to study, take the exam, and unlock the full legal power of your Baofeng.

Appendix F: Emergency Reference (Stay Calm)

If the grid goes down right now:

1. **Do not transmit immediately.** Turn the radio on, tune to your City Repeater (Channel 1), and simply listen.
2. **Gather Intel.** Local emergency nets will establish a "control station." Write down what they are saying (road closures, shelter locations).
3. **Switch to Simplex.** If you need to find your family, switch to your Family Direct channel (Channel 3).
4. **Use WHO/WHERE/WHAT.** Keep transmissions under 10 seconds.
5. **Conserve Power.** Turn the radio off when not actively listening for scheduled check-ins. Keep your radio warm (inside your jacket) if you are in a freezing environment to preserve battery chemistry.

Narrative Interlude: Two Weather Watches

Characters: Mark Jensen (Midwest) & Nate Brooks (Southeast)

Skills reinforced:

- Quick cards under stress
- Triage mindset
- Family plan execution

Where to learn the steps: Parts 2, 3, & 4

At 6:14 PM in the Minnesota suburbs, the neighborhood transformers began to fail. It sounded like a string of distant, muffled fireworks popping off under a thick blanket. Mark Jensen stood in his living room, staring out the front window as the heavy, blinding curtain of an Alberta Clipper furiously buried his driveway in white.

With a final, loud metallic *pop* from the pole at the end of the street, the regional power grid surrendered. The comforting, continuous hum of the furnace spun down to a halt, and the house was instantly plunged into absolute, freezing darkness. The silence that followed was heavy and immediate.

"Wi-Fi is down!" Owen shouted from the basement, his voice laced with the sudden, primal panic of a disconnected teenager whose digital world had just vanished. "Dad, the router is completely dead!"

"Everyone stay put! Nobody move until we get lights!" Mark called out.

He didn't scramble in the dark. He didn't frantically tear through the kitchen junk drawers looking for loose AA batteries or half-dead flashlights. He walked with measured, practiced steps to the hallway closet. He reached in, unzipped the front pouch of his neatly staged olive-drab Go-Bag, and pulled out his Baofeng radio and a small, high-lumen LED flashlight. He clicked the light on, instantly sweeping away the darkness.

Twelve hundred miles to the south, the situation was much louder and infinitely wetter.

Nate Brooks stood in his garage in South Carolina, the air thick with barometric pressure and the smell of wet pine needles. He was listening to the outer bands of a Category 3 hurricane actively punishing the reinforced aluminum of his garage door. The wind didn't howl; it roared like a freight train idling right in his driveway. The power had been out for three agonizing hours. The cell networks, overwhelmed by a million people trying to stream weather updates, had overloaded and died thirty minutes ago. His phone screen simply displayed the useless, terrifying phrase: *SOS Only*.

Kara had gone down the street to check on an elderly neighbor right before the cell towers failed, intending to be back in ten minutes. That was half an hour ago.

Nate stood at his workbench, his face illuminated by the harsh, white glare of a single battery-powered lantern. Sitting perfectly centered next to the lantern was his Baofeng UV-5R. The bright yellow, laminated Quick Card was tied securely to its pouch, an anchor of procedure in a chaotic environment.

Back in Minnesota, Mark clicked the top knob of his radio.

"BEEP. CHANNEL MODE."

"Dad," Owen said, appearing at the top of the basement stairs, squinting against the flashlight beam with his own radio clutched in his hand. "I think I broke mine in the dark. I bumped the keypad when the lights went out, and now it's just hissing at me. I can't get it to stop."

Mark felt a brief, phantom spike of his old beginner's anxiety, the deep-seated fear of the beeping brick, the assumption that cheap electronics had failed exactly when he needed them most. But then his training kicked in. He remembered the triage flow.

"Bring it here," Mark said, aiming his flashlight at the green LCD screen of Owen's radio. He ran the 60-second mental checklist, refusing to panic. *Mode?* He saw the tiny numbers on the right. Memory mode. *Arrow?* It was pointing to the top line. *Squelch?* Mark looked closely at the screen. The tiny "SQL" icon wasn't showing, meaning the radio wasn't receiving a signal. Then, he looked at the side of the chassis. He noticed the small rubber button below the PTT was physically wedged against the edge of Owen's nervously gripping thumb.

"You're holding down the Monitor button, bud," Mark chuckled. He felt a wave of profound relief, mostly because he knew he would have assumed the radio was fried just two weeks ago. He gently pulled Owen's thumb off the side of the radio. The rushing waterfall of static instantly stopped, plunging the hallway back into quiet. "The invisible gate was open. You just bypassed the squelch. You're fine. The radio works perfectly."

Mark checked his own radio. He was locked onto Channel 5, the National Weather Service frequency. He listened for thirty seconds to confirm the blizzard warning hadn't escalated, then confidently clicked the Up Arrow down to Channel 3, their designated Family Simplex channel.

"Okay, Owen," Mark said, switching seamlessly back into Mission Control mode. "Take your radio. Go back down and check the basement egress windows for snow drafts or leaks. I'll stay up here and check the front doors. Let's do a clean call in two minutes."

In South Carolina, Nate Brooks checked his tactical watch. It was 14:58. Two minutes to the top of the hour.

He didn't touch the radio. He didn't pace the garage floor. He simply stood still and waited. The urge to break protocol and call Kara immediately was a physical weight in his chest, but he resisted it. Radio discipline was the line between control and chaos. The plan was built to eliminate the anxiety of the unknown, and he trusted the procedure above his own nerves. He knew that Kara had her radio in her jacket pocket, perfectly tuned to Channel 3, and he knew she wouldn't waste her battery trying to call him outside of the designated window.

At exactly 15:00, Nate reached out, his hand steady, and clicked the top knob on. He verified the green screen read Channel 3. He set it back down on the bench.

At 15:01, the green LED indicator light on the top of the radio flashed. A split-second later, the squelch gate snapped open. It was a loud, aggressive burst of static that cut cleanly through the muffled roar of the hurricane outside.

"Nate, this is Kara," her voice came through the small speaker. It was slightly scratchy, battling the dense, wet foliage of the storm, but perfectly legible. *"I am at Mrs. Gable's house. Everyone is secure. Waiting out this band of rain before walking back. Over."*

Nate picked up the radio. He pressed the PTT button, counted a deliberate *One-Mississippi* in his head to let the hardware wake up and open the channel, and replied with a steady, flat cadence.

"Copy that, Kara. Stay put until the rain slacks. I am monitoring Channel 3. KN4XYZ clear."

He set the radio down. The transmission had taken eight seconds. It was entirely devoid of panic, perfectly legal, and completely off the grid. The feeling of absolute, quiet confidence that washed over Nate in that dim garage was profound. He had taken a thirty-dollar piece of plastic and turned it into a reliable lifeline. The system wasn't just theoretical anymore. It worked.

In Minnesota, Mark's radio chirped cheerfully from the kitchen counter, cutting through the eerie silence of the dead house.

"Dad, this is Owen," the voice said, sounding surprisingly professional and composed for a sixteen-year-old in a blackout. *"I am in the basement. Windows are dry, no snow getting in. It's really cold down here, though. Over."*

Mark grinned in the dark, incredibly proud of his son for sticking to the script. He picked up his radio, pressed the PTT button, and paused.

"Copy that, Owen. Come back upstairs and grab a heavy blanket from the closet. Good work. Mark clear."

Jen walked into the kitchen, wrapped in a heavy wool sweater, holding a flickering, battery-operated candle. She looked at Mark standing tall in the dark, then at the radio resting next to the printed, laminated Comms Plan.

"Are we officially surviving the snowpocalypse?" she asked, raising an eyebrow with a fond smile.

"We have continuity of operations, Jen," Mark said, unable to hide his grin. "The NASA control room is fully functional. We have eyes on the basement."

"It's a thirty-dollar walkie-talkie, Mark. You are standing in a coat closet."

"It's a verified, multi-tiered communication node," Mark corrected gently, locking the keypad with a precise two-second press of the # key. "And more importantly, it works."

Hundreds of miles apart, battling completely different elements on opposite sides of the country, both men sat in the dark and listened to the ambient noise of their respective storms. The fear of isolation, that creeping, helpless dread that usually accompanies a sudden loss of the cellular grid and the internet, simply wasn't there.

They weren't electrical engineers. They didn't have to memorize complex physics, build massive backyard towers, or understand the intricate internal wiring of their digital devices. They had just followed the steps, built their simple Quick Cards, and practiced their lines before the crisis hit.

They had taken a confusing, beeping piece of plastic and turned it into a trusted lifeline.

They were ready.

Takeaway

The ultimate goal of amateur radio and emergency preparedness is peace of mind. You don't need a hundred programmed channels, a trunk full of expensive tactical gear, or an advanced degree to protect your family. You need a charged battery, a simple, laminated one-page plan, and the discipline to practice it before the lights go out.

You can do this. Turn that beeping brick into a lifeline. Stay prepared, and stay on the air.

The Baofeng Radio Revolution

The Beginner Guerrilla's Guide to Break Through the Complexity, Secure Communications, and Prepare for Disaster With Prepper Tactics

MorseCode Publishing

Introduction

Baofeng Survival

WHEN ALDEN JONES PACKED his backpack for a solo hike along Vermont's picturesque Long Trail, his expectations were set on breathtaking vistas and serene solitude amidst nature's embrace. Little did he know, this journey would thrust him into a harrowing ordeal, where a simple device – a Baofeng radio – would become his lifeline (Tate, 2021).

Stricken by a sudden diabetic emergency, Alden's situation turned critical. Miles away from the nearest town, deep in the wilderness, his cell phone was as good as a brick. The sun dipped below the horizon, bringing the chilling prospect of a night spent helpless in the wild. But Alden wasn't alone in his plight as a fellow hiker came to his aid. Alden's foresight in bringing along his Baofeng handheld radio, a modest yet powerful tool, was about to pay off.

Activating his radio, Alden transmitted a desperate mayday call across a local amateur radio frequency. Miraculously, his call was answered. Two dedicated ham radio enthusiasts, equipped with their Baofeng radios, picked up his distress signal. They mobilized swiftly, coordinating a helicopter rescue that would whisk Alden to safety. This incident wasn't just a testament to Alden's survival instinct; it underscored the critical role a small, battery-powered Baofeng radio could play in emergencies, even in the most secluded corners of our planet.

Inspired by Alden's real-life survival tale, I embarked on my own journey into the world of radio and preparedness. As a novice, the realm of radio communication was daunting, filled with technical jargon and complex operations. Yet, I was convinced of its invaluable role in disaster scenarios where conventional communication methods fall short. This book is my foray into understanding Baofeng radios, focusing on the popular UV-5R model.

My aim is straightforward: to introduce beginners to the fascinating world of Baofeng radios. Through my experiences, I've gained insights into selecting, setting up, and troubleshooting these devices. This book is my conduit for sharing this knowledge, simplifying the technical jargon of ham radios, and empowering anyone to operate a Baofeng radio confidently. There is no better feeling of safety in a disaster than knowing you have functioning communications.

The Baofeng Revolution

In these pages, you will find a comprehensive guide to Baofeng radios, tailored specifically for beginners. While I discuss the Baofeng product line broadly, I use the UV-5R model as a practical example to illustrate setup instructions and operational tips. My goal is to equip you with the knowledge to tune into your first frequency – your local NOAA Weather Radio station – and, from there, embark on a thrilling journey into the world of amateur radio.

This guide covers various topics, from choosing the right model, initial setup, understanding ham radio lingo, and programming your device to enhancing signal clarity, understanding legal aspects, exploring advanced features, and applying your knowledge in real-world scenarios. Furthermore, I delve into emergency communication strategies, routine maintenance, and effective troubleshooting techniques. By the end of this book, you'll possess the knowledge to use your Baofeng radio independently and develop a deep appreciation for its capabilities.

While numerous guides cater to seasoned radio enthusiasts, this book fills a critical gap, offering practical and accessible guidance for newcomers. My mission is to empower you, the Baofeng beginner, with the competence and confidence to handle your radio proficiently, regardless of your prior experience. I aspire to ignite in you the same passion for Baofeng radios that now fuels my enthusiasm.

So, I invite you to join me in uncovering the tactical potential of the little Baofeng radio that could. This journey is not just about mastering a piece of technology; it's about discovering how this technology can simplify, secure, and even save our lives in times of need. Whether standing before an array of radios in a store, feeling overwhelmed, or simply curious about communication technology, this book is your beacon.

Baofeng radios are the ideal entry point for beginners venturing into the realm of communication technology. Comparable to a guide for novice mountain climbers, these radios ensure safety, simplicity, and enjoyment as you navigate the peaks and valleys of radio communication. Their appeal lies in their affordability, performance, and versatility. The real question isn't why you should choose Baofeng radios, but why wouldn't you?

As we progress through this book, we will demystify the jargon, untangling the complex web of frequencies, channels, and signals. Fear not; the world of amateur radio is far less intimidating than it appears. Remember, every expert was once a beginner. I penned this book initially to aid my learning journey and soon realized that in doing so, I could help others embark on their own path.

The Baofeng radio is more than a mere gadget; it's a steadfast companion. It stands by you in emergencies, offering a crucial connection when other means falter. It accompanies you in your leisure, linking you to communities across the globe. This book sets the stage for your adventure into the Baofeng universe. We'll guide you through setting up your first radio, maintaining it, programming it, and resolving basic operational issues. You don't need to be a tech whiz; you only require curiosity and a zeal for learning.

As you turn each page, prepare to unearth a new passion, hobby, and skill set. Welcome to the Baofeng Radio Revolution – your journey begins now.

Chapter 1: The Baofeng Blueprint - An Introduction to Handheld Radio & Choosing A Baofeng Model

In the world of emergency preparedness, the ability to communicate effectively can be the difference between safety and peril. The journey into two-way radio communication is not just a hobby; it's a critical aspect of ensuring resilience in the face of emergencies. This chapter aims to guide you through the diverse landscape of Baofeng two-way radio models, which are renowned for their reliability and affordability. Choosing the proper radio is a strategic decision, pivotal in establishing dependable communication lines during crises when traditional networks might fail.

Understanding the nuances of Baofeng radios is crucial, whether you're coordinating with a community network during a natural disaster, staying in touch with your group during off-grid adventures, or simply ensuring a line of communication when other systems are down. This chapter will help you navigate the options available, aiding you in selecting a device that aligns with your specific preparedness goals and scenarios. It will also ensure you're well-equipped to communicate effectively in any situation.

The Very Basics of Radio

In choosing your ideal Baofeng radio model, it helps to have a fundamental understanding of radio communication. In simple terms, two-way radios allow the wireless transmission of radio waves between a transmitter and a receiver (Ricky, 2022). The transmitter converts electrical signals into radio waves, which are picked up by the receiver's antenna. The receiver then converts the radio waves into electrical signals (*Problem With Baofeng BF-888 Radios,* n.d.).

Understanding radio waves is also crucial. Radios mainly use these types of waves (W0SJF, n.d.):

- VHF or very high-frequency waves (136–174 MHz range): Excellent outdoors over open terrain.
- UHF or ultra-high-frequency waves (400–520 MHz range): Great for penetrating obstacles like buildings and forests.

Here's my trick for remembering these differences: For VHF, focus on the 'V' for very long distances. For UHF, the 'U' reminds me of Urban, so I remember it's better at penetrating through buildings.

Baofeng radios can transmit on both VHF and UHF frequencies, a feature known as dual-band operation. This versatility is part of what makes them so popular.

The Radio Spectrum

To understand how radios work, one must have a basic grasp of the radio spectrum. This refers to the full range of radio frequencies used for wireless communications (*Baofeng Cheat Sheet*, n.d.).

Why is this information important? Understanding frequency fundamentals lets you choose the right radio equipment for your intended application.

The spectrum is divided into different bands based on frequency and wavelength, such as AM, FM, VHF, UHF, etc. Each band has unique propagation characteristics (the way radio waves travel through the air).

VHF refers to very high-frequency signals in the 136–174 MHz range. The wavelength ranges from 1–2 meters. VHF provides excellent propagation outdoors for services like TV, FM radio, and two-way land mobile radios.

UHF refers to ultra-high frequency signals from 400–520 MHz with wavelengths around 0.5 meters, or you may see it called 50-70 centimeters. Major uses include cell phones, WiFi, Bluetooth, two-way radios, satellite communication, and microwave ovens.

It is also important to note that the full VHF and UHF ranges are larger than the practical range used in radio service. The full VHF range is 30–300 MHz, and the full UHF range is 300–3000 MHz (3 GHz).

Higher frequency bands allow faster data transmission rates but have shorter transmission ranges. Lower frequencies travel further but with less bandwidth.

Scan the QR code or click on the link to receive 'The Baofeng Radio Revolution Command Packet,' which includes an overview of the full radio spectrum.

www.MorseCodePublishing.com/Command

Next, let's get to know more about the primary radio services.

Primary Radio Services

Two-way radios utilize radio services, such as FRS, GMRS, MURS, and Ham radio. These differ based on licensing requirements, frequencies, transmission power, and a few other differences (Best Baofeng Radios, 2022).

FRS—Family Radio Service

The Family Radio Service (FRS) is a set of frequencies the FCC has designated for personal communication, much like an express lane reserved for family vehicles. These channels require no license and can be used for both routine and emergency communication. Key attributes include (Family Radio Service, 2011):

- FRS is a short-distance two-way radio service that anyone in the US can use without a license.
- Operates on 22 UHF channels within 462 and 467 MHz ranges.
- These channels are in the UHF band (great for close urban areas or dense forests).
- Limited power output requirements max out at 0.5 watts ERP (effective radiated power).
- Because of the UHF and limited power, the unobstructed range is generally under 2 miles.
- These are ideal for activities like hiking, skiing, and camping, where groups want to coordinate, and no licenses are needed.

FRS radios are the "walkie-talkies" you can purchase off the shelf at any sporting goods store and immediately use with limited prior knowledge. While highly accessible, their limited range may mean communicating only within your immediate vicinity.

GMRS—General Mobile Radio Service

The General Mobile Radio Service (GMRS) is another set of frequencies the FCC has designated for personal use. Unlike FRS channels, GMRS channels require a license, allowing for higher power levels and greater range. Highlights include (*General Mobile Radio Service [GMRS]*, n.d.):

- GMRS provides longer-range licensed two-way radio communication.
- It requires a simple FCC license (currently $35), which covers you and your immediate family for five years.
- Power output is limited to 50 watts for GMRS channels, with power limited to 0.5 watts for the channels in which GMRS and FRS overlap.
- Operates on 30 channels near 462 MHz and 467 MHz, with eight channels dedicated to repeaters exclusive to GMRS.
- The typical handheld range is under 5 miles, but repeaters can extend it further.

The GMRS license offers affordable, licensed, long-range communication for families and small businesses. The license gives you legal access to higher power levels and repeater infrastructure, allowing expanded coverage compared to unlicensed bands. For example, GMRS could enable clear communication across a large campground or construction site. It provides an ideal emergency channel for coordinating with family. The licensing also avoids interference issues common on crowded, unlicensed frequencies.

FRS vs GMRS – What's the Difference?

The FRS (Family Radio Service) and GMRS (General Mobile Radio Service) are radio services operating in the same 462-467MHz range of the UHF band. They share 22 simplex channels, meaning that a channel on an FRS radio corresponds to the same channel on a GMRS radio across different brands. This compatibility allows FRS and GMRS users to communicate directly with each other.

However, there are key differences. FRS radios don't require a license and are restricted to handheld models with non-detachable antennas and a maximum of 2 watts of power on certain channels. In contrast, GMRS radios need a license. GMRS radios can be handheld, mobile, or base station units with detachable antennas and higher power limits – up to 5 watts on channels 1-7 and up to 50 watts on channels 15-22. GMRS also offers 8 additional repeater channels, which FRS doesn't have.

While FRS and GMRS share channels and frequencies, their distinct features, like radio types, wattage, antenna flexibility, and repeater capability, cater to different user needs. FRS is more suited for personal, casual use without the licensing hassle. In contrast, GMRS offers enhanced features and power for a more robust communication experience, albeit with a licensing requirement.

MURS—Multi-Use Radio Service

MURS, or the Multi-Use Radio Service, is a private, two-way, short-distance voice or data communications service for personal or business activities in the United States. MURS facilitates two-way communication without needing any license. Key details are (*GMRS*, n.d.):

- Operates on 5 VHF frequencies near 151–154 MHz.
- Limited to a maximum of 2 watts transmitter power output.
- It is intended for sporadic, short-distance exchanges.

- It has a typical range of 1–3 miles.
- MURS is used for personal and business communications by families, small businesses, and recreational activities. It's also used for remote control of devices and telemetry applications.

MURS provides an easy entry point into VHF radio operation for personal or business use without any license requirement. MURS allows doubled power levels, extended range, and business usage compared to FRS radios. The frequencies are less crowded than FRS as well. However, the range is still limited compared to licensed services like GMRS or commercial radios. MURS is a good option for simple, short-range communication at low cost and complexity. However, a licensed service may work better for heavy daily use or wide coverage.

Marine VHF Channels

These frequencies are expressly set aside for maritime communication, similar to dedicated boat lanes in a bustling harbor. These channels are used for various purposes, including communication between vessels, shore stations, and emergency distress calls.

If you're using your Baofeng radio in a marine environment, knowing how to access these Marine VHF channels can be a lifesaver. Whether calling for assistance in an emergency, receiving weather updates, or simply coordinating with other vessels, these channels are your lifeline in the open sea.

Aircraft Radio Service

The Aircraft Radio Service, essential for aviation safety, operates primarily in the VHF band (108-137 MHz) for communication and navigation. Managed by federal authorities, it facilitates crucial two-way communication between pilots and air traffic control, vital for flight operations, especially during takeoff, landing, and controlled airspace. This service includes emergency frequencies (121.5 MHz and 243.0 MHz). Operators require licensing, often included in pilot certifications, underscoring the service's role in coordinating global air traffic and maintaining high safety standards.

Amateur Radio Service

Amateur Radio, or "ham radio," is a hobby in which enthusiasts use various communication modes, such as voice, digital, Morse code, and slow-scan television, to connect worldwide. It's not just for casual chatting; ham radio operators also provide vital communication during emergencies when conventional networks fail. One must pass a licensing exam covering radio theory and regulations to participate. It's a diverse and engaging hobby, regulated by national authorities like the U.S. FCC to ensure responsible and safe use of the radio spectrum.

Here are some basics (*Amateur Radio Service*, 2016):

- It requires passing an exam to obtain an FCC license such as Technician, General, or Extra Class (more on these later). Each license level requires passing an examination, with the complexity and depth of the test material increasing with each higher level. The higher the class of license, the more operating privileges the holder has.
- It privileges larger frequency bands and power limits, dependent on license class.
- An estimated 730,000 licensees are in the US.

While licensing is a hurdle, ham radio offers extensive customization and long-range communication capabilities for electronics hobbyists. Compared to GMRS, ham radio allows greater technical experimentation and broader frequency access, but passing exams is required to unlock these privileges. GMRS provides a simpler entry point for affordable licensed long-range communication without customization.

In summary, choosing a radio service involves balancing licensing needs, intended use, power, and range requirements.

Radio Technology Fundamentals

Now that we have covered radio frequency basics and services let's demystify some key radio technology terms to round out your knowledge foundation.

Radio Frequency (RF)

- Radio Frequency (RF) is vital in wireless communications. It encompasses electromagnetic waves for data transmission, with frequencies ranging from 20 kHz to 300 GHz. These waves enable technologies such as mobile phones, radios, and Wi-Fi.

- RF works by emitting waves that carry information through space, similar to ripples from a stone thrown in water. This principle allows voice, data, and video to be transmitted wirelessly over distances.

- RF is ubiquitous in modern life. It's essential for mobile communications, where voice signals are transformed into RF waves, connecting people without physical wires. For ham radio enthusiasts, like Baofeng radio users, understanding RF's properties is key for effective communication and troubleshooting.

In summary, RF is the cornerstone of wireless connectivity, crucial for the functioning of various modern technologies.

Frequency Modulation (FM)

FM radio works by changing or modulating the radio signal's frequency to encode information. It is like tuning a radio dial up and down to different stations. The frequency shifts up and down based on the loudness of the audio signal. FM encodes sound differently than AM radio.

Compared to AM, FM has some significant advantages:

- Reduced noise and interference since only one frequency carries the whole signal.

- Allows transmitting higher fidelity audio than AM modulation.

- Resists signal distortion from lightning, electrical motors, and power lines.

FM provides excellent audio clarity and resistance to external distortions for portable two-way voice communications, so services like FRS and GMRS use FM.

CTCSS and DCS These systems help filter out unwanted conversations on a radio channel. They use unique low-frequency tones or codes sent along with the voice signal.

- The Continuous Tone-Coded Squelch System (CTCSS) transmits one of 50 super low, sub-audible tones. Sub-audible means below what humans can hear—under 20 Hz. You can't hear these tones but can feel them as vibrations! Each tone is unique (Vivian, 2021).

- Digital Code Squelch (DCS) uses 1 of 105 special digital codes sent with the voice signal (Vivian, 2021).

- Both radios must use the same CTCSS tone or DCS code to talk to each other. This blocks out other random conversations on that channel that do not match your tone or code. It's like a password or filter to keep communications more private.

These extra tones and codes help prevent unwanted chatter from cutting in. You only hear radios using the right tone or code to match yours. It clears up the channel, so you get the calls that matter to you.

Antennas and Transmission Antennas take the incoming radio waves and convert them into electrical signals, which are fed into the receiver (Ho, 2023). They also take transmission signals from the transmitter and convert them into radio waves. Key aspects are:

- Efficient power transfer between transmitter/antenna and antenna/receiver is critical for maximum range.
- Antenna gain in the desired direction results in higher signal strength.
- Matching the antenna design parameters like frequency, impedance, and radiation pattern to the radio frequency is vital for optimal performance—for example, a VHF antenna would be poorly matched to a UHF radio.
- Environmental factors like terrain and structures impact the propagation of transmitted signals.

So, upgrading to a better antenna significantly improves performance. We will cover optimal antennas for Baofeng radios later in the guide.

Understanding how antennas radiate signals and interact with the environment further demystifies radio wave transmission and reception.

We will build on these radio basics extensively in later chapters. However, you are now equipped with a foundational understanding of radio. Next, look at what features are important when buying a Baofeng.

Deciphering the Features of Radios

With the wide range of Baofeng two-way radio models available, it can be overwhelming trying to understand all the technical specifications and features. However, understanding the key capabilities and differences between models is essential for making the right purchase decision tailored to your needs. We will break down the most critical features to evaluate across popular models so you can zero in on your ideal device.

Range

One of the most vital features to consider is the expected transmission range (Encyclopedia Britannica, Inc., 2022). After all, the range determines the maximum distance your messages can cover and receive replies from other users. Baofeng radios support frequency bands like VHF and UHF, which have innate differences in propagation characteristics that impact range (Chen, 2021):

- VHF signals can travel further distances in outdoor open environments thanks to line-of-sight propagation. However, the range is more limited indoors due to attenuation (the reduction in strength of a radio signal as it travels through things such as walls and objects).
- UHF signals, while having a reduced range in open areas compared to VHF, can better penetrate through foliage, buildings, and other obstructions. This makes UHF well-suited for dense urban settings.

Within VHF and UHF bands, several factors like transmission power output and the antenna used significantly influence the range (Encyclopedia Britannica, Inc., 2022). Most handheld Baofeng models output 5W of power, but high-powered units like the BF-F8HP can transmit 8W for extended reach. Similarly, upgrading to an aftermarket high-gain antenna can dramatically increase the range compared to the standard rubber duck antenna.

However, when comparing the different Baofeng models for the expected range, the tested max range under optimal conditions provides a helpful indicator.

Entry-level models like the UV-5R are rated for up to 5 miles under ideal circumstances using the stock antenna (*Baofeng UV-82HP vs UV-5R*, 2022).

Upgraded models like the BF-F8HP can achieve over 15 miles of range thanks to their higher transmit power and better internal components (*Best Baofeng Radios*, 2022).

Due to power restrictions, commercial variants like the BF-888S have a more limited range of just 1–2 miles (TalkieMan, 2019b).

Consider your typical usage—a basic entry-level model may suffice if you need short-range communications within a few blocks in urban areas. However, the extended reach of high-powered radios like the BF-F8HP will prove invaluable for remote locations with challenging terrain.

Battery Life

Baofeng handhelds rely on rechargeable Li-ion battery packs for power. Enough battery capacity to last through the day is essential when using your radio for outdoor activities or emergencies. Entry-level Baofeng models often have limited battery life, especially on high-power transmissions. Upgraded models feature improved batteries offering enhanced runtime between charges (Knoji Staff, n.d.; Superior Support, 2022):

- The UV-5R ships with a standard 1,800 mAh battery, providing only 8–10 hours of battery life. Heavy use will drain it within half a day.
- In comparison, the UV-82HP has a beefy 2,800 mAh battery, providing 15-18 hours of operation and allowing you to use the radio from sunrise to sunset without recharging.
- Some premium models, like the UV-9R Plus, boast extended 2,500 mAh batteries despite being compact. These models strike a balance between battery life and portable size.

If you need your radio for all-day outdoor excursions or camping trips or as an emergency preparedness tool, prioritize those Baofeng models from the factory that offer the highest capacity battery packs. Most batteries can be replaced with third-party options. However, it's convenient when the stock battery provides sufficient operating time before needing spares.

Durability and Water Resistance

Baofeng radios are inherently durable, given their compact polycarbonate casing and sturdy controls. But you can expect specific models to withstand the elements better when used outdoors:

- Entry-level models like the UV-5R lack any water resistance. Exposure to heavy rain or submerging them can permanently damage the internals (Rock Networks, n.d.-a).
- Upgraded variants like the UV-9R Plus offer an IP67 rating, making them fully dustproof and waterproof up to 1-meter depth for 30 minutes. Their rubber gaskets provide a sealed design (*Best Baofeng Radios*, 2022).
- The UV-82HP features a more robust chassis than the UV-5R and uses metal alloy in high-wear parts. However, it lacks a sealed design, so water should be avoided (Superior Support, 2022).

If your radio will likely get wet when used in inclement weather or take the occasional splash during water sports, opting for one of the IP67-rated waterproof models is wise. Similarly, choose a dustproof-rated option if it will be exposed to dirty or dusty environments on your adventures. However, the added weather resistance comes at a higher price. If you plan to use your radio only in fair weather, an entry-level model like the UV-5R will likely suffice since it will be affordable to replace if damaged.

Audio Quality and Volume

Being able to hear messages clearly is vital to effective communication. Some environments, like noisy construction sites, require loud and clear audio. Evaluating the speaker quality and volume levels across Baofeng models is essential if audio clarity is paramount (Rock Networks, n.d.-a):

- The speaker on basic models like the UV-5R maxes at around 700 milliwatts (mW), a relatively low power level. In noisy surroundings, received voices can be nearly unintelligible, even at full volume.
- Premium models, like the UV-82HP, boast 2-watt (W) front-facing speakers. Combined with the dual push-to-talk (PTT) feature, they excel in noisy workplace conditions with loud incoming audio.

- For discrete communications, personal receiver earpieces allow hearing transmissions clearly without blaring the message aloud. Most Baofeng radios support external microphones and earpieces.

Speaker capabilities may be less critical if you only plan to use your radio occasionally in quiet indoor settings. However, opt for a loud, clear audio model for outdoor group hikes or crowded public events to avoid miscommunications.

Ease of Use

All Baofeng radios feature a standard control interface with a PTT button (Push to Talk), volume knob, keypad, and LCD screen. However, some models make the experience more ergonomic and intuitive (Rock Networks, n.d.):

- Models like the UV-5R have a smaller monochrome display and buttons that can feel cramped.

- Newer designs like the UV-9R Plus feature a larger 1.5" color display, oversized PTT, and a more ergonomic grip for one-handed use.

- The UV-5R Plus simplifies the interface by adopting an Android smartphone-style menu UI (user experience) rather than the traditional icon-based menu navigation.

If you are less technically inclined or want an intuitive radio that even beginners can operate, evaluate the ease of use. Handling the device and comparing the button sizes, display visibility, menu navigation, and grip ergonomics in person can be worthwhile. While all Baofeng radios share a common architecture, subtle refinements across models impact usability.

Programming Capability

One of the top advantages of Baofeng radios is their programmability using a computer and programming cable. This allows configuring them for your desired frequencies and channels beyond the default settings. However, specific models are more accessible to program than others:

- The UV-5R is notoriously tedious to program manually, requiring combing through confusing menus. Software like CHIRP simplifies the process, but some users still experience hiccups (Centers, 2022).

- Newer options like the UV-9R Plus can be programmed directly on the device. Adding channels with the intuitive interface is more manageable, reducing reliance on a computer.

- Some premium models even include built-in touchscreens allowing programming from the radio without cables.

Those who plan to minimize manual programming and use default frequencies may find even basic Baofeng models sufficient. But if you intend to program memory banks spanning multiple channels, opt for a model that streamlines and simplifies the process.

Accessories

One benefit of the immense popularity of specific Baofeng models is the wide range of compatible accessories (Encyclopedia Britannica, Inc., 2022). You can enhance your radio with add-ons like:

- Aftermarket antennas: Upgrade your whip or duck antenna for increased range with options like the Nagoya NA-771.

- Headsets: Discreetly receive transmissions using earpieces. Specific models have built-in headset jacks.

- Microphones: Use extended microphones for long-range or outdoor use, keeping the radio itself protected.

- Batteries: Extended batteries help keep the radio operational for multi-day trips.

- Belt clips: Keep the device securely attached and instantly accessible.

Well-known models like the UV-5R, BF-F8HP, and UV-82HP have a thriving accessory ecosystem. Some key points when evaluating accessories:

Antennas

- Ensure any aftermarket antennas are designed for your radio's frequency bands with appropriate connectors. Also, match the antenna impedance to the radio's rated impedance (the resistance of the circuit - impedance is crucial in radio systems for matching the transmitting antenna and the receiver to ensure efficient power transfer and minimize signal reflection and loss).
- Higher gain (directional antennas) antennas boost range but require compatible power levels supported by the radio.
- Select antennas suited for your usage, like flexible rubber models for outdoors or small magnetic mounts for vehicular use.

Cases and Clips

- Hard cases offer the most protection, while silicone and TPU (Thermoplastic Polyurethane) cases are more grippy. Soft pouch cases make radio storage and transport convenient.
- Belt clips must match the radio's body in tightness, width, and mounting style to stay securely attached.
- Commonly available belt clips usually work for the UV-5R and UV-82. However, the BF-F8HP requires specially sized clips that fit its thicker body.

For a safe, customized setup, ensure all add-ons and replacements meet the technical specifications of your particular radio model.

Understand the Different Baofeng Models

Having covered the key factors, like frequency, power, and programming, let us explore some specific Baofeng models with those features to help clarify choosing the right radio. With so many Baofeng models available, it can get overwhelming trying to determine which one is right for you. Each model has its unique blend of features, performance factors, and price points. Evaluating your specific needs and usage scenarios will help narrow down the ideal option. We will provide an in-depth feature comparison across some of the most popular Baofeng radios to simplify your purchase decision.

Baofeng UV-5R

The UV-5R is the classic Baofeng model, the gold standard of affordability and functionality. It operates on VHF 136-174MHz and UHF 400-520MHz frequencies with a power output of 5 watts. The UV-5R is compact and lightweight (about the size of a deck of cards), making it a great companion for outdoor adventures. It includes features such as dual-band, built-in flashlight, and an FM radio, which come in handy when you least expect it.

The UV-5R is feature-packed, given its low cost, making it a versatile starter radio. But the maximum 4W power output and short battery life limit its utility for long-range or all-day use. Still, it remains a popular "first Baofeng" choice.

Baofeng UV-82HP

The UV-82HP model builds on the UV-5R, improving some drawbacks for a modest price increase. The UV-82HP is a model of choice for many radio enthusiasts. It operates on the same frequencies as the UV-5R and BF-F8HP but has a power output of 7 watts. The higher capacity 2800 mAh battery provides 18+ hours of battery life, over double the UV-5R (Superior Support, 2022). What sets the UV-82HP apart is its ergonomic design. The radio's button layout is designed for ease of use, and the built-in speaker delivers crisp and loud audio.

So, suppose you need a rugged radio with extended runtime suitable for outdoor conditions or workplace use. In that case, the UV-82HP is worth the slight cost increase over the UV-5R. Just lack of weatherproofing is a limitation.

Baofeng BF-F8HP

Where the UV-82HP improves battery life over the UV-5R, the BF-F8HP boosts power output. The BF-F8HP is an upgrade over the UV-5R, boasting a power output of up to 8 watts. The increased power means better range and signal clarity, especially in challenging environments. The BF-F8HP also operates on VHF 136-174MHz and UHF 400-520MHz frequencies. It shares many features with the UV-5R, but its standout quality is its improved battery life, thanks to the 2100mAh battery. It also has an IP55 rating (see the glossary for an expanded definition of IP ratings, but it measures protection from solid objects and liquids), which provides some dust and splash resistance (*Baofeng - Baofeng Troubleshooting*, 2022)

On the other hand, the BF-F8HP does not have an illuminated keypad or shortcut keys. Moreover, it is about 25% heavier than other handheld models (Best Baofeng Radios, 2022). If you prioritize extended-range capability, the BF-F8HP is a top choice despite its larger form factor.

Baofeng UV-9R Plus

The UV-9R Plus is Baofeng's attempt at a more premium model, boasting an IP67 waterproof and dustproof rating. It operates on VHF 136-174MHz and UHF 400-520MHz frequencies with a power output of 8 watts and a longer-running 2500 mAh battery. The UV-9R Plus is a fantastic choice for outdoor adventurers, as it can withstand harsh weather conditions. This model also features a dual push-to-talk button, allowing you to switch between two frequencies easily.

Due to the upgrades, the UV-9R Plus is more expensive than the UV-5R. However, if you need a rugged radio with excellent battery life, the UV-9R Plus checks all the boxes.

Baofeng BF-888S

The BF-888S is Baofeng's entry in the compact FRS/GMRS walkie-talkie category (Rock Networks, n.d.; TalkieMan, 2019). Attributes include a smaller, compact design ideal for a backpack or pocket. It supports 2W output power and a basic range of 2–5 miles. The 16 channels in FRS/GMRS frequencies of 462 and 467 MHz are great for something that needs to work right out of the box with no license required.

This micro-sized option is designed for casual short-range communication. It provides an affordable, lightweight walkie-talkie option.

Make Your Selection

Like choosing between different types of vehicles, choosing the suitable Baofeng model depends on your specific needs and circumstances. For example, if you're going for a hike on a clear day, the UV-5R might be enough. But if you're planning a week-long camping trip in unpredictable weather, the weather-proof UV-9R Plus could be your best bet.

Remember, the perfect radio doesn't exist, but the ideal radio for you does. Understanding the features and capabilities of each Baofeng model can help you make an informed decision.

Key Factors in Selecting a Baofeng Radio:

- Range: Models like the BF-F8HP offer extended reach due to higher power output.
- Battery Life: Consider models like the UV-82HP and UV-9R Plus for longer battery endurance.
- Durability: The UV-9R Plus stands out with its IP-rated water and dust resistance.
- Audio Quality: The UV-82HP is notable for its superior speaker output.
- Ease of Programming: While all models are programmable, some may offer a more user-friendly experience.

- Portability: The BF-888S is compact and lightweight, perfect for on-the-go use.

With this knowledge, you can confidently select a Baofeng model that delivers the right mix of functionality and value tailored to your communication needs.

Making the Right Purchase: When buying your chosen Baofeng radio, you have several options, each with advantages and considerations.

Buying Online:

- Online platforms like Amazon offer a wide selection of new and used Baofeng radios.

Baofeng Specialty Retailers:

- Direct purchases from sites like BaoFengTech.com guarantee genuine products and full warranty support.
- Specialty retailers like Buy Two Way Radios and STS Telecom offer expert advice and a focused selection of Baofeng products.

General Electronics Stores:

- Stores like Best Buy or Walmart may carry Baofeng radios, offering the advantage of in-person purchases.
- The drawback is that staff may lack in-depth knowledge of Baofeng radios, and selection might be limited.

Local Ham Radio Outlets:

- Local stores often have knowledgeable staff with practical experience in ham radio.
- Finding a well-stocked outlet can be challenging, especially in smaller towns.

Purchasing Considerations:

- Check for warranty coverage and ensure it's from an authorized dealer.
- Understand the return policy.
- Confirm if the seller offers post-purchase technical support.

Avoiding Counterfeit Baofeng Radios:

- Buy only from authorized dealers listed on official Baofeng websites.
- Confirm the FCC ID on the radio matches the FCC's database. Counterfeits usually lack proper FCC certification.
- Check for FCC certification on the radio and verify it on the FCC ID search website (https://www.fcc.gov/oet/ea/fccid).
- FCC Equipment Authorization
    - Just as a new car must meet specific safety and emissions standards before it can hit the road, radio transmitting equipment like your Baofeng radio must meet specific technical standards the FCC sets before it can be marketed and sold in the United States. This is known as FCC equipment authorization.
    - As a user, you can check if your Baofeng radio is FCC-authorized by looking for an FCC certification label on the device or by searching the FCC's Equipment Authorization Database.

- FCC ID and Labeling Requirements
  - Imagine walking into a library and finding all the books without titles or authors. It would be nearly impossible to find what you're looking for. Similarly, the FCC requires all authorized radio devices to have an FCC ID. This unique identifier helps the FCC keep track of all devices in the market and allows consumers and the FCC to retrieve the device's certification information.
  - The FCC ID is typically found on a label affixed to the device. This label must be permanently attached and readily visible. It includes the FCC logo, the FCC ID, and a statement confirming the device's compliance with the FCC's technical regulations.

Retaining Important Documentation:

- Keep all provided documentation, including the user manual, warranty card, packing list, and FCC grant.
- These documents are essential for setup, warranty claims, and verifying FCC compliance.

Understanding these aspects of purchasing a Baofeng radio will help you navigate the market effectively. Whether you choose a specialty retailer for expert advice or an online marketplace for convenience and selection, ensuring the authenticity and suitability of your radio is paramount. With careful consideration, you can acquire a Baofeng radio that perfectly fits your communication needs and skill level.

Final Considerations: Add Ons

Choosing the best Baofeng radio model is just the beginning of your journey. To truly maximize the potential of your new radio, it's important to consider investing in essential accessories. These additions can significantly enhance the functionality, durability, and overall experience of using your Baofeng radio.

Essential Baofeng Radio Accessories:

1. **Spare Batteries:**
   - Crucial for extended use, especially in areas without charging facilities.
   - Vital in emergencies where power sources may be scarce.
   - Cold weather can drain batteries faster, so having spares is beneficial.
2. **Battery Chargers:**
   - Multiple chargers ensure you can recharge batteries wherever you are – at home, in your vehicle, or in a bug-out bag.
   - Owning more than one charger provides backup in case one fails.
3. **Carry Cases:**
   - Protects your radio from physical damage.
   - Waterproof cases guard against rain and moisture.
   - Features like belt clips and storage pouches enhance convenience and organization.
4. **Microphones and Headsets:**
   - Ideal for hands-free operation and discreet communication.

  - Improves audio clarity, especially in noisy environments.
  - External PTT buttons facilitate easier transmission.

5. **Upgraded Antennas:**
  - Enhances reception and transmission range.
  - Select an antenna compatible with your radio's frequency range and suitable gain rating.
  - Durability and size are important considerations based on your usage.

6. **End Fed Half Wave Antennas:**
  - For long-range communication in emergencies.
  - Portable and quick to deploy, they can significantly extend your radio's reach.
  - Dual-band models are versatile for various frequency needs.

7. **Programming Cable:**
  - Facilitates easy programming of channels via software like CHIRP.
  - Essential for backing up settings and updating firmware.
  - Opt for genuine Baofeng cables for the best compatibility.

8. **AA/AAA Battery Pack:**
  - Adds flexibility when standard batteries are more readily available.
  - Useful in prolonged scenarios where recharging Li-ion batteries is not an option.

9. **Go Bag:**
  - A tactical bag for storing and transporting your Baofeng and accessories.
  - Features like MOLLE webbing and internal organization cater to various gear.
  - Ensures readiness and quick deployment in any situation.

Once you've acquired your Baofeng radio and essential accessories, the next step is unboxing and setting it up. The upcoming chapter will guide you through this process in a beginner-friendly manner, covering:

- Initial unboxing and inspection of the radio and its components.
- Basic setup procedures, including battery installation and charging.
- Programming your first channels using the keypad.
- Testing key functions to ensure your radio is ready for use.

Even if you're new to radio communications, this next chapter will provide clear, step-by-step instructions to get your Baofeng radio up and running. Get ready to embark on an exciting journey of discovery and communication with your new Baofeng radio!

Chapter 2: Breaking into Baofeng - Unboxing and Basic Operations

THE ARRIVAL OF NEW gear, particularly something as crucial as a communication tool, is a moment charged with anticipation and excitement. Unboxing a Baofeng radio isn't just about revealing a new gadget; it's about unveiling a key component of your preparedness arsenal. As you tear through the packaging, it's not just the thrill of a new possession but the realization that you're a step closer to being better prepared for emergencies.

This Baofeng radio represents more than a communication device; it's your lifeline to the outside world in times of crisis, your tool for connecting with members during drills or actual emergencies. Its significance is profound for someone dedicated to preparedness. Every dial, button, and antenna is a component in your readiness strategy, ensuring that you have the means to stay informed and connected whether you're facing a natural disaster, a power outage, or any unforeseen event.

So, as your heart races with excitement over this new acquisition, know that it's more than just an addition to your tech collection; it's a vital step in fortifying your preparedness plan.

The First Encounter: Unpacking Your Baofeng Radio

The excitement of unboxing a new gadget! As you unwrap your Baofeng radio packaging, here's what you'll typically find inside (Superior Support, 2022):

- **The radio device itself**: Most Baofeng radios have a compact, rectangular shape that fits nicely in your palm. Popular models like the UV-5R are about 5.5 inches tall and 2 inches wide (again, about the size of a deck of cards).
    - **Keypad:** The keypad is your control center, your command bridge. It's where you input frequencies, navigate menus, and access the various features of your radio. Each key on the keypad has a specific function beyond just number input. For instance, the [#] key will lock the keypad if pressed and held for 2 seconds (press and hold for 2 seconds to unlock it). The more comfortable you become with the keypad, the smoother your radio operation.
    - **Display Screen:** The display screen is your window into the workings of your Baofeng radio. It's where you can see the current frequency or channel, monitor transmission status, and access menus. It also displays battery status and other settings like power and squelch levels. The display screen is backlit, making it easy to read even in low-light conditions.
    - **Speaker/Microphone**: Are your radio's voice and ears. The speaker broadcasts the incoming signals, turning them into audible sounds. Conversely, the microphone captures your voice, converting it into a signal that can be transmitted. These components are usually combined into a single unit in handheld radios like Baofeng radios.
- **Battery:** Your radio's battery is its heart, pumping the power needed to function. Baofeng radios typically come with a rechargeable Lithium-ion battery, known for their long lifespan and high energy density. This means you can operate your radio for a considerable time before recharging. To install, match up the alignment guides on the

battery to the radio and slide the battery upwards until it clicks securely. To remove it, turn your radio unit to the back. You'll notice a small latch on the back at the top. Push the latch down and slide the battery down and out.

- **Battery charger**: While Li-ion batteries are great, they eventually drain. The battery charger restores the health of your battery heart! The battery charger typically comes with a USB cord that can be inserted into either a wall socket or a car charger.
    - Batteries can be charged alone in the battery charger. You do not need the battery attached to the radio to charge it. This becomes important when you use your radio and have a backup battery in the charger, ready for future use.
- **Antenna**: Think of the antenna as the periscope of a submarine. The component reaches into the vast ocean of frequencies, catching signals transmitted in its range. A standard Baofeng radio comes with a rubber-duck antenna, which is versatile, portable, and suitable for most situations. It's crucial to remember that the antenna is the lifeline of your radio, the primary tool for receiving and transmitting signals. Treat it with care, and it will serve you well.
    - Attach the supplied rubber antenna to the SMA (SubMiniature version A) female connector protruding from the radio's top. Screw it on gently but firmly until snug (Rick, 2016).
- **User Manual**: This short and sweet document provides instructions for setup and operation. Keep it handy as you get started. While it is limited in scope and explanations, it is still helpful and convenient.
- **Belt Clip**: A plastic clip lets you attach the radio to your belt or strap. If you regularly carry your radio and like the feel of it on your belt, I recommend upgrading to a sturdier carrier.
- **Wrist Strap**: A loop that attaches to the radio and helps prevent dropping while carrying the radio. There's not much else to say about this!
- **Programming Cable** (with some models): Connects your radio to a PC for programming. Imagine trying to input a hundred different contact numbers into your smartphone manually. Tedious, isn't it? That's where the programming cable comes in. This accessory connects your Baofeng radio to your computer, allowing you to use software to input frequencies, name channels, and customize other settings. Say goodbye to the monotony of manual input and hello to the efficiency of the programming cable.
- **Earpiece with Microphone** (with some models): Allows hands-free talking. The external microphone is an accessory that you may find helpful for hands-free communication. Think about those times when you're multitasking, maybe cooking dinner while coordinating a neighborhood watch. An external microphone allows you to relay messages without holding the radio. Some models even come with a built-in speaker, enhancing the audio clarity.

Once you've inventoried the contents, installed the battery, and connected the antenna, your radio is ready for that exciting first power-on! So go ahead and explore your Baofeng radio. Hold it in your hands and feel its weight and texture. Familiarize yourself with its parts and features because this is not just a radio. It's your companion on a journey of discovery and into the world of radio communication. And as with any journey, the first step is always the most exciting. So, let's take that step together. Your Baofeng radio is ready. Are you?

Powering On: Initial Setup of Your Baofeng Radio

After attaching the battery and antenna, the moment you've been waiting for has arrived —you can power on your radio for the first time (*How to Use Baofeng Radios*, 2022).

Turn the radio over and locate the volume/power knob on the top. Turning this knob clockwise powers the radio on. You'll hear a beep and see letters or graphics displayed on the screen, indicating it is powered on. If the voice prompt is turned on, you will hear if you are in frequency or channel mode. The LED indicator will also light up, confirming the radio is now on.

On some models, you can customize the initial setup's radio name, owner name, power-on message, and menu language directly from the keypad. However, other models require you to connect to a computer using the programming cable.

Getting Familiar: Understanding the Basic Operations

Now that your Baofeng radio is powered on and initialized, it's time to understand the basic controls and functions. Getting familiar with buttons and knobs will help you operate the device confidently (*How to Use Baofeng Radios*, 2022).

On the top, you'll find the volume/power knob, which turns the radio on and off and controls its audio volume. Turn clockwise to increase loudness.

The Push-to-Talk (PTT) button is on the left side and has an indent for easy identification by touch. Pressing and holding this button transmits your voice at the set frequency.

- Also, on the left side is a bright orange CALL button. Go ahead and push that button – but push it and let go. Please don't hold it unless you want to set off the alarm (more on the alarm later)! You turn on the FM radio by just pushing and releasing the button! Then, tune into your favorite FM radio music station using the number buttons. Push and release the button to turn off the FM radio.

- The Final button on the left is MONO. For now, push that button and see the flashlight come on. Press again for a strobe light, and press again for off. The other function of this button will be covered later in the book.

The numeric keys for inputting frequencies and text are on the front of your radio. The keys are also used for selecting menu options and programming the radio.

- The menu button activates the radio's menu system. Use it to access configuration options and settings. The "MENU" key also acts as the "enter" function. When you want to select a menu option, click the menu button. The arrow keys help navigate between menus and settings. The basic UV-5R has 41 menu sections (menu sections 0-40)!

- Also seen on the front is the transmit/receive LED indicator. This is activated when transmitting or receiving signals. It blinks green when receiving and turns solid red when transmitting. Within the menu, these are slightly customizable, with the ability to select between a few color options.

- The bright orange button is the VFO/MR. This button switches between 'frequency mode' and 'channel mode.' VFO (Variable Frequency Oscillator), or frequency mode, allows you to enter a frequency manually. MR (memory recall), or memory mode, will enable you to access your saved channels easily.

- Also standing out on the front in a nice blue color is A/B. This button switches between the two frequencies displayed on the screen. Go ahead and press A/B, and watch as the arrow moves between the top and bottom frequencies. We'll cover their usage in more detail later. For now, familiarize yourself with button placement and functionality.

Let's Talk: Setting Up Your First Communication

Once you have a basic grasp of the controls, it's time to set up your first actual two-way communication using your Baofeng radio. This helps confirm everything is working correctly and builds confidence in using your radio (*How to Use Baofeng Radios*, 2022).

First, ensure the battery is fully charged—the icon on the display indicates the battery level. Then, turn the radio on!

Set your radio to a frequency like FRS channel 15 (462.550 MHz). Use the numeric keypad to input the 6-digit frequency. Have a second person ready with their own FRS or GMRS radio set to the same channel or frequency.

Press and hold the PTT button on your Baofeng radio, speak into the mic to transmit your voice message, and then release PTT when done. The other radio should receive your audio through its speaker. Have the other person press their PTT and respond to confirm two-way communication.

Start by testing at close range, then slowly increase the distance between the two radios while communicating to determine the maximum range. Try moving between, in, and around different structures or buildings. Keep your Baofeng's antenna vertical for best results. This test confirms everything is working and teaches you about practical radio communication.

Now, let's say you don't have someone or another radio. Then, hold tight, and we will connect you with the NOAA Weather station for your area.

Find your local NOAA weather radio frequency:

- Go to www.weather.gov/nwr and click the "Find My Station" button.
- Enter your zip code into the 'NWR Station Search' section and press 'Enter.'
- Note the frequency in the yellow box. In my area, the frequency is 162.550 MHz.

On your radio, first, ensure you have selected the correct band. Click the BAND button until you are on VHF (136–174 MHz range). To help me, I think of 'V' in VHF as standing for very long distances (excellent outdoors over open terrain). Since your weather station is probably a long way off, you want to be on VHF.

- Type in the frequency of your weather station.
- You may have to wait a minute to hear the transmission.
- You may also hear static or breaks in the communication. If that is the situation, hold MONO on the left side of the radio press. We will explain this function later, but it will help you hear the transmission for now.

You did it! You are now on the radio. You have taken your first step to becoming a radio operator and setting yourself up for future preparedness. Congratulations!

3 or 4 Decimal Places

When programming frequencies into your Baofeng, you might encounter a situation where the frequency you need to input has more decimal places than your radio seems to support. Most Baofeng radios, including the UV-5R, typically display and operate with frequencies up to three decimal places (e.g., 162.555 MHz).

If you need to access a frequency with four decimal places (like 162.5575 MHz), here's what you can do:

- Round to the Nearest Frequency: In your case, you could round 162.5575 MHz to 162.558 MHz or down to 162.557 MHz. The Baofeng UV-5R should accept either of these frequencies. The slight difference is often within the acceptable tuning error margin for most amateur radio and scanning purposes.
- Consider the Bandwidth: The bandwidth of your transmission (or the receiver's filter bandwidth) may be wide enough to accommodate slight frequency discrepancies. For instance, if you're using a typical FM mode with a bandwidth of 5 kHz or more, being off by a few hertz will generally not be significant.
- Check Real-World Performance: Test the radio in real-world conditions after rounding the frequency. If you're trying to access a specific repeater or listen to a particular signal, see if the radio successfully receives or transmits as expected. Often, you'll find that the minor discrepancy doesn't impact functionality.
- Use Software for Programming: Programming software like CHIRP might allow you to enter the frequency more precisely. However, the radio's actual tuning accuracy and display will still be limited to hardware capabilities.
- Hardware Limitations: Remember that the Baofeng UV-5R and similar models are budget-friendly radios. Their

frequency resolution and accuracy might not be as high as expensive amateur radio equipment.

For most amateur radio applications, especially on VHF and UHF bands, rounding to the nearest available frequency setting on your Baofeng should be sufficient. However, if you frequently need to operate with greater precision, you might consider investing in a radio with finer tuning capabilities.

Radio Etiquette

Now that you are "on the radio," let's review radio etiquette. This section will guide you through some of the most important ones.

Stay Legal

First and foremost, know what frequencies you are on and which ones you are legally allowed to transmit. If you are unsure, for now, just listen.

Listening First

Before you hit that transmit button, it's essential to listen first. Check if the frequency or channel is currently in use. Interrupting ongoing communications is considered rude in the radio community. Take the time to understand the flow of the conversation, and when you're sure the channel is clear, proceed with your transmission.

Call Sign Usage

Imagine attending a formal event where everyone is addressed by their titles and names. It lends a sense of identity and respect to the conversations. In the realm of radio communication, call signs serve a similar purpose.

A call sign is a unique identifier essential in radio communication, assigned to stations and operators for organized and legal frequency use. In amateur radio, the call sign, issued by authorities like the FCC in the U.S., comprises a region-indicating prefix, a local area numeral, and a unique suffix.

Operators must use their call signs at the start and end of transmissions, ensuring clear identification and responsible frequency use. Managed internationally by the ITU, these call signs facilitate the global identification of transmission origins. Some operators personalize their call signs, adding elements like initials. As a vital component of radio communication, call signs ensure order and accountability across radio networks.

Remember, your call sign is your identity on the airwaves. Use it with pride and respect.

Brevity and Clarity

Keep your communications short and to the point when speaking over the radio. Remember that the airwaves are shared, and lengthy transmissions can tie up a frequency. Speak clearly and slowly enough to be understood, but try not to hold the airwaves longer than necessary.

Using Standard Phrases and Protocols

Radio communication has its language and protocols. Using standard phrases and codes can make your transmissions easier to understand.

Staying Polite and Respectful

Remember to always be polite and respectful in your communications. Avoid using offensive language or engaging in arguments over the air.

Radio Language: Q-Codes and Phonetic Alphabet

Use of Q-Codes

Imagine being in a foreign country, unable to speak the local language, relying solely on a handful of universally understood signs to navigate your way. In the world of radio communication, Q-codes serve a similar purpose. They're a set of standardized abbreviations used to convey specific messages, transcending language barriers and ensuring clear communication.

Your Baofeng radio allows you to communicate using these Q-codes. For example, if you're in an emergency and need assistance, you can use the Q-code 'QRR' to signal that you're in distress and require immediate help.

Phonetic Alphabet

Think back to your school days when you first learned the alphabet. Each letter had a name and a corresponding sound that helped you remember it. In radio communication, the phonetic alphabet functions in a similar way. It's a set of words used to represent each letter of the alphabet, ensuring clear and accurate communication over the airwaves.

The phonetic alphabet is particularly useful when communicating complex information, such as coordinates or license details, over your Baofeng radio. For example, instead of saying 'B,' which might be misheard as 'D' or 'V,' you would say 'Bravo.' This ensures that the information you're transmitting is received accurately, even in challenging conditions.

A printable version of Q-codes and the Phonetic Alphabet can be found here in the 'The Baofeng Radio Revolution Command Packet.'

Common Issues: Navigating the First Roadblocks

As with any new technology, there is a learning curve. You may encounter minor issues initially when using your Baofeng radio. Here are some common problems beginners face, along with potential solutions (*Baofeng - Baofeng Troubleshooting*, 2022; Rick 2016).

- Difficulty powering on: First, ensure the battery pack is charged and correctly inserted. Check that the power/volume knob is turned clockwise to switch it on. If it still doesn't power up, contact the dealer.
- No sound: Check if the volume is turned up adequately by turning the volume knob. Inspect the speaker to ensure it is clean and not blocked or damaged.
- Battery drains quickly: Turn down the brightness if it is too high. Disable features like VOX if not required to save battery life. Check the battery icon - if it shows just 1 bar, charge the battery.

These basic troubleshooting tips will help you resolve common initial hiccups. Don't be discouraged by minor issues. Regular usage will make you increasingly adept at using your Baofeng radio.

Learning Curve: Becoming Comfortable With Your Baofeng Radio

Developing familiarity and comfort with your new Baofeng radio takes regular usage. Here are some tips to quickly become proficient:

- Practice operating all the buttons and looking through the menu until the layout and functioning becomes second nature. Muscle memory helps build reflexive skills.
- Initially, focus on key features like changing frequencies, transmission, and reception. Don't get overwhelmed trying advanced options before mastering the basics.

- Another quick and easy thing to start with is to scan frequencies. Here is how:
    - Go to Frequency Mode – click the VFO/MR button.
    - Click BAND and select either VHF or UHF. Try them both, but start with VHF.
    - On the UV-5R, click the * button with blue writing on it 'SCAN.' Press and hold until you hear 'Scanning Begin,' or if your voice prompts are off, you see the frequencies start changing. Then release.
    - Your radio will scan through all available frequencies.
    - Listen and see what you pick up. Then, write down that frequency for future use.
    - Don't worry if you don't hear much. It can be hit or miss at times. But if you are scanning VHF, you should hear your NOAA weather station again.

In this chapter, we walked through unboxing your new Baofeng radio, powering it on for the first time, understanding the essential functions, setting up your first communication, and getting comfortable with operating it. Don't worry if you don't feel like you know everything yet—the key is just getting hands-on experience and becoming familiar with the basics. Now, you are ready to dive deeper.

The next chapter will help cut through the noise by explaining common radio terms and jargon you'll encounter with your Baofeng. Understanding these key terms and concepts will make mastering your radio much more manageable. You've completed the first step of unboxing and basic setup—get ready to boost your radio knowledge even further in the next chapter! Let's get started learning essential radio lingo.

Chapter 3: Decoding Baofeng - A Beginner's Guide to Radio Terminology

MASTERING THE LANGUAGE OF your tools is as crucial as owning them. Baofeng radios, a staple in any prepper's kit, come with unique technical jargon and terminology. Understanding these terms is not just about technical knowledge; it's about ensuring effective communication during critical situations. Whether coordinating with a group during a drill, communicating during a community event, or navigating through an emergency, the clarity and precision afforded by proper terminology can make all the difference.

This chapter is dedicated to demystifying the language of Baofeng radios. It's designed to help you confidently operate your radio and communicate with precision and authority. By grasping these key terms, you're enhancing your ability to use your Baofeng radio as an efficient tool in your preparedness arsenal, ensuring that your communication is clear, concise, and effective when needed.

The Basics: The Importance of Proper Radio Terms and Language

You frequently encounter technical terms and acronyms specific to two-way radios, whether reading the user manual, browsing online forums, or communicating with other radio operators. Familiarity with these terms is vital; reading about two-way radios can feel like learning a foreign language (*Baofengradio*, 2021)!

Knowing the proper definitions helps you effectively configure and operate your radio, troubleshoot any issues, and research problems through online knowledge bases. Understanding their terminology allows you to tap into the broader radio communication community (*Amateur Radio Terms*, 2022).

Don't be overwhelmed by the jargon! Start with the most essential terms covered in this chapter. You will steadily pick up more terminology with practical experience. Refer to the glossary for help.

Tuning In: The Basics of Frequency and Channel

Two key terms you'll continually encounter are frequency and channel. When operating your radio, understanding the distinction between entering a frequency versus selecting a channel is important.

In your Baofeng radio, frequency is the specific radio wave it sends out and picks up, akin to choosing a particular station on your car radio. It's quantified in Hertz (Hz), which counts the number of wave cycles occurring every second. Essentially, a greater frequency value corresponds to a higher frequency of the radio wave, much like tuning to different stations on your radio involves different frequency numbers.

Channels let you save groups of related frequencies under shortcut numbers for quick recall. For example, you may have channel 1 set to 154.570 MHz. So, when you select channel 1 on your radio, it automatically tunes to that saved frequency. Channels are an easy way to store and access the frequencies you use often. It's like setting your favorite music channels in your car for quick and easy access.

Baofeng radios can function in either frequency mode or channel mode (*General Mobile Radio Service*, 2023):

- **Frequency mode:** You directly enter the desired frequency using the numeric keys to access any frequency within the radio's range. In addition, if you enable the scanning feature, your radio will scan through each frequency.
- **Channel mode**: You select from pre-programmed channels configured with specific frequencies. This is quicker but restricts you to only those already set-up channels. When enabling scanning, your radio will scan through only your preset channels.

Making Connections: The Role of Repeaters

Repeaters play a vital role in two-way radio communications. A repeater is a device that receives radio signals on one frequency and retransmits them at a higher power on another frequency. This allows the signal to cover longer distances by boosting the radio signal strength. Amateur radio operators typically use repeaters to extend the range of their VHF and UHF band radios. For effective repeater operation, you need to understand terms like input frequency, output frequency, and offset (*Ham Radio Licenses*, n.d.):

- **Input frequency**: This is the specific frequency on which your Baofeng radio transmits to communicate with a repeater. When setting up your radio to use a repeater, you'll need to know this frequency to ensure your radio is sending signals that the repeater can receive. It's like dialing the correct number to make a phone call – you need to use the right frequency so the repeater can 'answer' your radio's call.
- **Output frequency**: Once your signal reaches the repeater, it doesn't just relay your exact signal. Instead, it rebroadcasts, or 'retransmits,' your signal on a different frequency, known as the output frequency. This is the frequency other radios in the area will tune into to hear your transmission. Using a different output frequency allows for efficient and clear communication, as it prevents your transmitted signal from directly interfering with the signal being received by others.
- **Offset**: The offset is a critical concept in repeater operation. It's the difference between the input and output frequencies. For example, if a repeater has an input frequency of 145.270 MHz and an output frequency of 145.870 MHz, the offset is 600 kHz. This separation is essential to avoid interference; it ensures that the repeater's transmissions don't interfere with the signals it's trying to receive. Think of it as a conversation where one person speaks at a time to avoid confusion – the offset allows the repeater to 'speak' and 'listen' at different 'tones,' avoiding a conversational clash.

When programming your Baofeng with a repeater, you must input the proper transmit and receive frequencies with the correct offset between them. Getting this configuration right is essential for making repeaters work smoothly.

In a later chapter, we'll provide step-by-step instructions on how to program frequencies and offsets for repeaters. For now, remember that repeaters operate using separate frequencies for receiving signals from your radio and retransmitting them.

Speaking the Same Language: CTCSS and DCS

CTCSS (Continuous Tone-Coded Squelch System) and DCS (Digital Code Squelch) are sophisticated systems used in ham radio to ensure clearer and more exclusive communications. These systems are particularly valuable in filtering out unwanted conversations on busy frequencies, acting like a selective hearing mechanism for your radio.

- **CTCSS**: Imagine being in a crowded room where everyone is talking, but you only want to hear conversations from your friends. CTCSS works similarly. It embeds one of 50 low-frequency tones into your transmitted audio signal. These tones are like secret handshakes; they are inaudible during transmission but are essential for communication.

When you set a specific CTCSS tone on your Baofeng radio, it 'listens' exclusively for signals with that tone. This means your radio will remain silent (or squelched) until it detects a transmission with the matching tone, effectively filtering out all other chatter on the frequency.

- **DCS**: DCS is a more advanced version of CTCSS, using one of 105 digital codes instead of tones. Think of it as a digital password added invisibly to your communication. When you configure a specific DCS code on your radio, it acts as a gatekeeper, only allowing transmissions with the correct digital code to be heard. This is particularly useful in environments with a lot of radio traffic, as it ensures you receive only relevant communications.

In summary, CTCSS and DCS are integral features of your Baofeng radio, enabling you to participate in clutter-free and focused radio communications. By setting up these systems, you can ensure that your radio conversations are exclusive to your intended group, eliminating irrelevant transmissions from others on the same channel. Once configured in your radio's menu, this automated process enhances your communication experience by providing a more tailored, interference-free listening environment.

Modes of Communication: Simplex and Duplex

Two fundamental modes of communication in ham radio are simplex and duplex. Understanding the distinctions and applications of these modes is crucial for any amateur radio enthusiast.

- **Simplex Communication:** This mode is akin to a one-lane bridge, allowing two-way communication but only one direction at a time. For example, if a frequency of 146.520 MHz is set on a simplex channel, your Baofeng radio uses the same frequency to transmit and receive signals. It's ideal for straightforward, short-range conversations, as seen in walkie-talkies and basic ham radio setups. The main limitation is the inability to talk and listen simultaneously.

- **Duplex Communication:** Duplex, on the other hand, is like a two-lane bridge, enabling simultaneous two-way communication. For example, your radio might receive signals at 146.520 MHz but transmit at 147.120 MHz. Duplex can be either full (simultaneous talking and listening) or half-duplex (both functions, but not at the same time).

In essence, simplex and duplex are the building blocks of radio communication in ham radio. Simplex, with its one-at-a-time communication style, is straightforward and effective for short-range conversations. Duplex offers simultaneous interaction, which is more complex but essential for continuous dialogue. These modes shape how ham radio operators connect, each serving distinct purposes and scenarios.

Listening Carefully: Squelch and VOX

Two crucial features in Baofeng radios that significantly enhance audio reception and transmission are squelch and VOX (Voice-Operated Transmission). Understanding and properly configuring these functions can dramatically improve your radio communication experience.

- **Squelch:** This function acts as an audio gatekeeper. It mutes the speaker output when no signal is present, eliminating background static noise. The squelch threshold is adjustable to suit the RF noise levels in your environment. Properly setting the squelch level filters out weak signals, allowing only strong and clear communications to be heard. If set too low, it lets in too much background noise; set too high, and it might block desired signals. Thus, finding the right balance in squelch settings ensures crisp and uninterrupted reception.

- **VOX:** VOX automates the transmission process by detecting your speech and automatically triggering the transmit mode. This hands-free operation is beneficial when pressing the PTT (Push-To-Talk) button is inconvenient, like handling equipment or climbing. VOX sensitivity should be adjusted to activate with your voice while reliably avoiding unintended transmissions. This feature allows for smooth, natural conversation flow in radio communications without manual switching.

In summary, mastering the squelch and VOX features on your Baofeng radio can dramatically improve your communication experience. Squelch eliminates unnecessary background noise, ensuring that only relevant transmissions are heard. VOX, on the other hand, offers convenient hands-free operation, allowing for seamless conversations. Both features ensure efficient, straightforward, and user-friendly radio operation when optimally configured, making them indispensable tools for any Baofeng radio operator.

Staying Within Bounds: Understanding RF Gain and Bandwidth

In the world of ham radio, two critical settings that can significantly impact the performance of your Baofeng radio are RF Gain and Bandwidth. Mastering these settings can significantly enhance your radio's range and audio quality.

- **RF Gain:** This controls how sensitive your radio is to incoming signals (Noonan., n.d.). Increasing RF Gain is akin to increasing the volume but for signal strength, not audio. It strengthens weaker signals and makes them more intelligible. It's beneficial for capturing distant or faint transmissions. However, too much RF Gain on strong signals can cause distortion. The key is to start with a moderate level and adjust upwards only as needed to improve the reception of weaker signals.

- **Bandwidth:** This refers to the width of the frequency range your radio can receive or transmit at any given time (*Choosing an Antenna*, n.d.). Wider bandwidth allows your radio to monitor a broader range of frequencies simultaneously, which helps scan busy bands. However, a narrower bandwidth provides better filtering from interference, focusing on a tighter frequency range. Choosing the right bandwidth depends on your situation: wider for monitoring more activity, narrower for clearer, more focused communication.

Understanding and correctly adjusting RF Gain and Bandwidth can significantly enhance your Baofeng radio experience. RF Gain helps you fine-tune the radio's sensitivity to pick up distant or weak signals, while Bandwidth allows you to control the range of frequencies you can access at once. Balancing these settings optimally will enable you to maximize your radio's performance, whether trying to reach far-off stations or need clarity amidst a crowded frequency band. These settings are vital tools in the ham radio operator's kit, enabling more effective communication.

Safety Features: Time-Out Timer (Transmission Timer) and Emergency Alert

Baofeng radios have essential safety features like the Time-Out Timer and the Emergency Alert, designed to enhance the user's safety and the radio's functionality. Understanding and utilizing these features can be crucial in various situations.

- **Time-Out Timer (TOT):** Also listed as 'Transmission Timer' in user manuals, it is a built-in safety mechanism that prevents the radio from transmitting too long. The timer typically has a preset limit, often around 60 seconds (*How Far Can I Talk?*, n.d.). If you transmit continuously past this duration, the Time-Out Timer automatically stops the transmission. This is particularly useful in preventing potential damage to the radio, such as overheating, which could occur from an accidentally pressed or stuck microphone button. The TOT is a safeguard, ensuring that transmissions are brief and the radio remains functional. To adjust the TOT:

  - Press the MENU key, followed by '9' on the keypad. This takes you to the 'TOT' (Time-out Timer) setting. Press MENU to enter.

  - Use the UP and DOWN keys to set the desired timeout duration. This is the maximum length of time you can transmit continuously. It will range from Off to 600 seconds.

  - Press the MENU key again to save your selection.

  - Hit EXIT to leave the menu. Your timeout timer is now set.

- **Emergency Alert:** This feature is a critical safety tool in emergencies. The radio triggers a loud siren and activates flashing LEDs by pressing a designated button, usually labeled "Alarm" or the 'Call' button (*How to Use Baofeng Radios*, 2022). This function is designed to attract attention in emergencies, especially when the user is in danger but unable to speak. The alert broadcasts a siren to nearby radios and provides a visual signal, making it an effective way to summon help. It's a vital feature for public safety, with personnel often trained to respond to these alerts. However, using this function responsibly is essential to avoid false alarms, which can undermine its effectiveness in real emergencies.

In conclusion, the Time-Out Timer and Emergency Alert are invaluable features in Baofeng radios. The Time-Out Timer ensures safe and responsible use by limiting continuous transmission and protecting the radio from damage. The Emergency Alert function provides a critical means of signaling for help in dangerous situations. Both features enhance the safety and utility of Baofeng radios, making them more reliable in various scenarios.

Expanding Your Reach: Ham Radio and Dual Band

Baofeng radios offer advanced features like ham radio operation and dual-band functionality, significantly enhancing their versatility for amateur radio enthusiasts. Understanding these features and their implications is crucial for anyone looking to maximize their Baofeng radio's capabilities.

- **Ham Radio Operation:** Refers to amateur radio, a hobby and service enabling long-range wireless communication and experimentation. You need a valid FCC-issued amateur radio license to access certain frequency bands and higher power operations (*Ham Radio Licenses*, n.d.). Baofeng radios cover various ham bands, so having at least a Technician-level license opens up a wider range of operating privileges. It's important to remember to only transmit on ham bands for which you're licensed.

- **Dual-Band Functionality:** A dual-band Baofeng radio can operate on two different frequency bands, typically VHF and UHF. This feature is like having two radios in one, allowing you to monitor a local UHF repeater while also scanning a VHF channel, for instance (*How Far Can I Talk?*, n.d.). Dual-band functionality, especially with the dual watch feature, enables simultaneous monitoring of two channels, expanding your communication possibilities.

While a ham radio license is required to fully utilize certain frequencies, Baofeng radios' dual-band capability makes them highly versatile, even for beginners. Understanding and using these features can significantly enhance your radio experience.

Venturing into the world of radio communications introduces many new terms and concepts. It's natural to feel a bit overwhelmed initially, but don't worry—gradually familiarizing yourself with these concepts is vital. The glossary section is a handy resource whenever you need to clarify terms or refresh your memory. It's designed to make your journey into radio communications smoother and more enjoyable.

With a solid grasp of the basic radio terminology in your toolkit, we're poised to dive into the exciting next chapter: programming your Baofeng radio. This is where the real adventure begins, as you unlock the full potential of your radio by tailoring it to your specific needs. From installing necessary software to inputting channels, we'll guide you through every step of the process, ensuring a beginner-friendly experience. You'll soon discover the impressive range of advanced features your Baofeng radio offers. So, let's roll up our sleeves and delve into the hands-on experience of radio programming!

Chapter 4: Baofeng Preparedness - Mastering Radio Programming

A Baofeng radio isn't just a gadget; it's a lifeline. Whether traversing rugged landscapes, coordinating community safety initiatives, or keeping in touch during emergencies, the ability to program your Baofeng radio is a crucial skill in your preparedness toolkit.

Programming your radio equips you with the agility to adapt to changing situations, ensuring that you have quick access to essential frequencies, from local emergency channels to weather alerts. This chapter goes beyond the basics of radio use; it's about empowering you with the knowledge to tailor your Baofeng to your unique needs.

Whether manually programming channels or using software like CHIRP, these guidelines are tailored to help you maintain a crucial edge in readiness. You'll learn to navigate the radio spectrum with ease, ensuring that in any scenario – be it adventure, community service, or emergency – your Baofeng radio is an asset you can rely on.

Demystifying the Programming Process

Programming a complex communications device like a Baofeng radio can seem intimidating for a beginner. However, it simply involves entering frequencies you want to use into the radio's memory. This allows you to quickly tune in to programmed channels instead of manually entering a frequency each time (Centers, 2022).

Programming is like creating a speed-dial phonebook on your radio. You save significant time and effort by saving your most-used frequencies as channel entries rather than manually inputting them repeatedly.

Software programming requires a computer and cable to interface with your Baofeng and configure the settings. However, you can also manually input frequencies directly on the radio. This chapter will cover both methods.

Manual Programming: The Old-School Approach

Manually programming your Baofeng involves directly inputting frequencies using the numeric keypad. Let's walk through it step-by-step with a real-world example:

1. Find your local NOAA weather radio frequency:
   - Go to www.weather.gov/nwr and click the "Find My Station" button.
   - Enter your zip code into the 'NWR Station Search' section and press 'Enter.'
   - Note the frequency in the yellow box. In my area, the frequency is 162.550 MHz.
2. Determine what memory channel you want to store this on. **Deleting a Saved Channel:**
   - If your radio came programmed with channels already saved, you may need to delete a saved channel to open it

up for your use. To delete a channel on a UV-5R:

- Press 'MENU'.
- Enter '28' or use the arrows to find DELCH (delete the memory channel)
- Press 'MENU' to confirm and enter.
- Find the channel you want to delete and free up. Let's use channel #5 in this example. Arrow up to channel 5.
- Press 'MENU' to confirm and delete the saved channel.
- We now have a free channel to use.
- Press 'EXIT'.

Voice Prompts:

1. Press the VFO/MR button to enter frequency mode.
    - To hear if you are in channel or frequency mode (and hear other operations), turn on 'Voice Prompts.'
    - Press 'MENU'.
    - Enter '14' or use the arrows to find VOICE (Voice Prompt).
    - Press 'MENU.'
    - Use the up arrow to switch to ENG (English) or CHI (Chinese). I assume you will want English (ENG) and then press 'MENU.' I will say it's fun to change it to Chinese and then scroll through all the menu options to hear Chinese!
    - Press 'EXIT.'
2. Again, press VFO/MR; you should hear if you are on Channel or Frequency Mode. Ensure you are in frequency mode.
3. Use the numeric keys to enter the NOAA frequency in MHz format.
    - For example, in my area, I will enter the frequency as 162.550 MHz and enter 162550 on the keypad.

Save a Channel

1. To save this frequency into a memory channel, press 'MENU.'

- Enter '27' or scroll up to MEMCH (stored in the memory channel)
- Press 'MENU.'
- Use arrows to go up to channel 5.
- Press 'MENU'
- You have now saved this frequency to channel 5.
- Press 'EXIT.'

- To test this out, press VFO/MR until you are on Channel Mode
- Use the arrows to select channel 5.
- Is the correct frequency displayed?

Always double-check your frequencies to avoid typos. Manual input takes patience, but it works in emergencies if your computer software is unavailable or you are deep in the woods. Manual programming will serve you well if you are starting with radio. However, as the years pass and your list of frequencies expands, upgrading to software is a significant time saver.

Manual Programming: Repeaters

Programming a repeater into your Baofeng can be a bit complex for beginners, but you can do it successfully with step-by-step instructions. Here's how to manually program a repeater frequency into your Baofeng UV-5R:

Step-by-Step Manual Programming

1. Switch to Frequency Mode.
    - Turn on the radio.
    - Press the 'VFO/MR' button to switch to Frequency Mode (VFO).
2. Enter the Repeater's Frequency.
    - Use the keypad to enter the repeater's frequency. For example, if the repeater's frequency is 146.940 MHz, type 146940.
3. Set the Transmit Offset.
    - The standard offset for a 2-meter band VHF (144-148 MHz) repeater is usually 600 kHz, while it is often 5 MHz for a 70 cm band UHF (420-450 MHz) repeater.
    - Determine if the offset is positive or negative based on the repeater specifications.
    - Your repeater resources will typically give you both of these.
4. Set the Offset Direction and Amount on your radio.
    - Press the 'MENU' button.
    - Enter **25** (for frequency direction) SFT-D, and press 'MENU.'
    - Use the up/down arrow keys to select the offset direction (N for simplex, + for positive, - for negative). For most repeaters, you'll select + or -.
    - Press 'MENU' to confirm.
    - Press 'EXIT' to return to the main screen.
    - Press 'MENU' again.
    - Enter '26' (for offset frequency) OFFSET, and press 'MENU.'
    - Enter the offset amount (e.g., 000600 for 600 kHz) using the keypad.

- Press 'MENU' to confirm.
- Press 'EXIT' to return.

5. Set the CTCSS/DCS Code (If Required)
   - If the repeater requires a CTCSS/DCS tone for access, you must set it. Your repeater resource will provide this information.
   - Press 'MENU'.
   - Enter '13' (for CTCSS) T-CTCS or '12' (for DCS) T-DCS, and press 'MENU.'
   - Use the keypad or the up/down arrows to select the correct tone.
   - Press 'MENU' to confirm.
   - Press 'EXIT' to return.
6. Save the Channel following the instructions from the manual programming section.

Final Steps

- Turn the radio off and then back on.
- Press the VFO/MR button to switch to Channel Mode.
- Use the up/down arrows to find your new channel.

To verify if you have correctly set up your Baofeng to communicate with a repeater, you can follow these steps:

- **Check Your Settings**:
  - Ensure the frequency and the offset are correctly entered.
  - Confirm that the offset direction (plus or minus) is set according to the repeater's requirements.
  - Verify the CTCSS/DCS tone is correctly set if the repeater uses one.
- **Listen to the Repeater**:
  - Before transmitting, listen to the repeater's frequency to see if you can hear other conversations. This can confirm whether you're on the right frequency and whether the repeater is within range.
- **Perform a Radio Check**:
  - When the frequency is clear, key up (press the transmit button) and briefly state your call sign along with a request for a radio check. For example, "This is [Your Call Sign] requesting a radio check."
  - If your setup is correct and someone is listening, you might receive a response.
  - Only do this step if you are legally licensed for this frequency.
- **Check for a Repeater's Automated Response**:
  - Some repeaters have an automated voice or a beep tone that indicates you have successfully accessed the repeater.

  - After transmitting, if you hear this automated response, it's a good sign that you're set up correctly.

- **Ask for Confirmation from Local Hams**:
  - If you're part of a local ham radio club or have ham friends nearby, ask them to listen and confirm if they can hear you through the repeater.

- **Check the Transmission Quality**:
  - If you do make contact with someone, ask for feedback on the quality of your transmission. They can tell you if your signal is clear, too weak, or distorted.

- **Adjust Antenna or Location if Necessary**:
  - If you cannot hit the repeater or receive poor reports on your signal, consider adjusting your antenna or moving to a location with less obstruction.

- **Use a Different Radio to Monitor**:
  - If you can access another receiver or a different radio, you can listen to the repeater's frequency while transmitting from your Baofeng. This can help you hear what your transmission sounds like.

Remember, success in accessing a repeater can depend on various factors, such as the power output of your radio, your distance from the repeater, the quality of your antenna, and geographical obstructions.

Software Programming: Preparing for the Process

Software programming allows you to configure your Baofeng radio's settings and memory channels using a computer. Let's break down the programming process using CHIRP, a popular open-source software program (Centers, 2022). Here are some tips to get prepared:

1. Download programming software:
   - Go to https://chirp.danplanet.com/projects/chirp/wiki/Home and click "Download CHIRP for your platform."
     - You can also go to Google and search for "Chirp danplanet," and it will be the first thing that comes up.
     - Warning: there are ads on this site that have a 'Download' button. DO NOT click on those! Click only in the text or paragraphs where it says 'Download.'
   - Select the version for your operating system. CHIRP is free and works with most Baofengs. The version for your computer will usually be highlighted in green.
   - Follow your computer prompts to download the software.
2. Install CHIRP on your computer following standard software installation steps. Ensure you download the software version compatible with your specific Baofeng model. Then, open the program.
3. Connect the USB programming cable from your radio to your computer.
   - If you are like me, your programming cable came with a CD to install the driver. Do this first so that you install the driver from the CD. Before I did this, CHIRP didn't recognize my cable or radio.
   - Note: CHIRP will ask you what computer port your cable is connected to. If you are like me, you may not know,

so I always choose the 'HELP ME' option in the drop-down and follow the instructions.

4. You should have obtained a list of frequencies you want to program.
    - Search databases like RadioReference.com to find local repeater and public safety frequencies.
    - Refer to your local radio club's recommended frequencies.
5. Consider your intended use—emergency, local communications, private channels, etc.- as this will determine your channels, settings, and how you want to group them.
    - Pro-Tip: Organize your thoughts on paper or a spreadsheet. Group things out here before doing the work in the software.

These steps will help ensure a smooth programming experience using CHIRP.

Programming With Software (CHIRP) in Memory Editor

1. Launch CHIRP.
2. Connect your Baofeng radio to your computer using the supplied programming cable.
3. In CHIRP, go to the 'Radio' menu at the top of the screen and click 'Download from Radio' to download your current radio memory contents as a .csv file. Use 'Help Me' to find your port.
    - Save this to your computer as a backup copy.
4. Your memory editor will display and look like the image below.

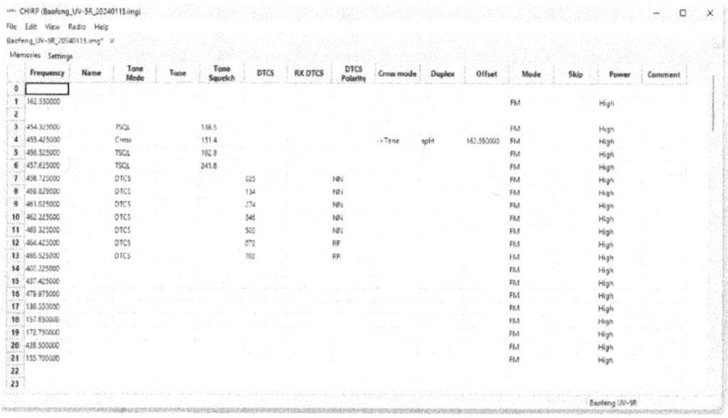

1. Enter your desired frequencies and settings in the Memory channel slots.
2. Once your channels are configured, click the 'Radio' menu and select 'Upload to Radio' to program your entered settings and frequencies into the radio.
3. Remove the cable to test. Toggle through the channel numbers on your radio to verify your programmed frequencies and settings.

Understanding the CHIRP Fields

- Location: The far-left-hand column is filled with numbers. This is the channel number or memory location in your radio, and your radio's capacity defines the number you see.
- Frequency: Sets the channel's receiving frequency in megahertz (MHz). If Duplex is set to None, it also serves as the transmit frequency.

- Name: An alphanumeric label for the channel displayed on the radio's front panel. Your radio's display capabilities define the character limit.
- Tone Mode: Controls transmit/receive squelch tones or selective calling features.
  - Options:
    - None: No tone or code transmitted; receive squelch is open or carrier-triggered.
    - Tone: Transmits a single CTCSS tone; receive squelch is open or carrier-triggered.
    - TSQL: Transmits and requires a matching CTCSS tone for receive squelch.
    - DTCS: Transmits a DTCS/DCS code; receive squelch requires the same code.
    - Cross: A combination of squelch technologies; specifics set in the Cross Mode field.
- Tone: Sets the CTCSS tone for transmission when Tone Mode is set to Tone or Cross.
- Tone Squelch: Sets the CTCSS tone for transmission and receive squelch in TSQL or Cross Mode.
- DTCS: Sets the DTCS code for transmission and receive squelch in DTCS or Cross Mode.
- RX DTCS: Sets the DTCS code for receive squelch when using Cross Mode (if supported by the radio).
- DTCS Polarity: Determines DTCS code polarity for both transmitting and receiving.
  - Options:
    - 'N' for Normal
    - 'R' for Reversed.
- Cross Mode: Configures complex squelch behaviors when Tone Mode is set to Cross. Consists of two technologies separated by an arrow (->). The left side controls the transmit selective-call method, and the right side controls the receive squelch.
  - Possible Values:
    - Tone: Uses CTCSS tones (transmit tone from the Tone column, receive tone from the ToneSql column).
    - DTCS: Uses DTCS/DCS codes (transmit code from the DTCS Code column, receive code from the DTCS Rx Code column).
    - Combination: For example, "Tone->DTCS" means a CTCSS tone is used for transmission (set in Tone), and a DTCS Rx Code is used for receive squelch.
  - Examples of Cross Mode configurations include "Tone->Tone," "DTCS->Tone," and "->DTCS," each combining different technologies for transmit and receive squelch functions.
- Duplex: Sets the channel's duplex mode.
  - Options:
    - None: Transmit and receive frequencies are the same.
    - +/- (Plus/Minus): Transmit frequency is above/below the receive frequency by the Offset value.

    - Split: Uses an absolute transmit frequency, as defined in Offset.
    - Off: Disables transmission on the channel.

- Offset: Defines the frequency difference for transmit in relation to receive frequency (if Duplex is not None or Off). Enter the shift amount (in MHz) or absolute transmit frequency for Split Duplex.
- Mode: Controls the channel's transmit and receive mode.
  - Common Modes:
    - FM: Wide FM (5 kHz deviation) for two-way communications.
    - NFM: Narrow FM (2.5 kHz deviation) for two-way communications.
    - WFM: Wide FM for broadcast (approx. 100 kHz deviation).
    - AM: Used in the aircraft band in the US.
    - DV/DN: Digital modes for specific systems (e.g., D-STAR, SystemFusion).
- Skip: Controls scan behavior for the channel.
  - Options:
    - S: Skip this channel during a scan.
    - P: This channel is a priority.
- Power: Sets the power output for this channel.
- Comment: Space to enter your notes about this channel.

Programming With Software (CHIRP) with CSV Files

CSV stands for "Comma Separated Values," a flexible data encoding format rather than a strict file format.

Creating and Editing a CHIRP-Compatible CSV File

Starting from Scratch:

1. Start CHIRP: Open the CHIRP software.
2. Create a New File: Navigate to File > New to generate an empty CSV file.
3. Add a Memory: Input a frequency (e.g., 155.525) in the first row to create a memory entry.
4. Export the Template: Save your new file with a .csv extension via File > Export.
5. Edit in Spreadsheet Software: Open this CSV file in a spreadsheet application, make your edits, and save.
6. Reopen in CHIRP: To check the format, open the edited file in CHIRP by going to File > Open.
7. Copy Memories: Optionally, you can copy these memories to another radio tab within CHIRP.

Tips for Maintaining Formatting

CHIRP depends on the CSV's format to correctly read and process the data. Follow these guidelines to avoid issues:

- Preserve the Header Row: Keep the original header format so CHIRP knows the value of each column.
- Reorder with Care: You can change the column order if the header row remains consistent.
- Deleting Columns: You can remove columns with default values; CHIRP will assume these defaults when opened.
- Quoting Values: Use double quotes (") around values containing spaces or commas.
- Field Value Accuracy: Ensure field values match the original and accepted formats. Refer to a CHIRP-generated CSV file for examples of correct values.

Transferring Data to Your Radio

Since CSV files are generic and not specific to any radio model, direct uploads to a radio are not possible. Follow these steps to transfer data:

1. Open CSV File: Use File > Open in CHIRP to open your CSV file.
2. Load Radio Image: Open a .img file for your radio, or download one from your radio using Radio > Download From Radio. Ensure the tab indicates the specific radio model.
3. Select and Copy Memories:

- Select the tab with your CSV file.
- Highlight the memories you wish to transfer (use Shift or Control for multiple selections).
- Copy them using Edit > Copy.

1. Paste into Radio's Memory:
    - Switch to the tab for your actual radio.
    - Choose the memory slot to paste into, then use Edit > Paste.
2. Uploading Changes:
    - For clone-mode radios, upload the changes back to your device with Radio > Upload To Radio.
    - For live-mode radios, wait for synchronization to complete (check the status bar).

Handling Missing Columns

If the CSV lacks information needed by your radio (like a "Power" column), CHIRP can still help:

1. Adjust in CHIRP Radio Window: Open the spreadsheet in CHIRP's radio-specific window.
2. Bulk Edit Rows:
    - Highlight relevant rows (click the first row, shift-click the last).
    - Right-click and choose "properties."
    - Check the property ("column") you wish to change and select the new value from the dropdown (e.g., "power"

to "high").

- Click OK to apply changes.

Following these steps and guidelines, you can efficiently use CHIRP to program your radio, ensuring data is formatted and transferred correctly.

I also realize this is a lot of new information. Return to this section until you are 100% comfortable completing these actions.

Overcoming Common Programming Challenges

Despite the step-by-step process, you may face some common challenges when programming your Baofeng radio:

- Compatibility issues if your software version needs to be updated. Check CHIRP's website and upgrade to the latest release.
- Errors in channel numbers or frequencies cause incorrect programming. Double-check your inputs and file contents. Make sure you have all your decimals right!
- Failure to connect radio with computer. Try different USB ports and cables. Enable USB debugging mode on the radio. Here are some steps to debug and troubleshoot if you are unable to connect your Baofeng radio to your computer:
    - Try different USB ports on your computer.
    - Test different USB cables. Cables can fail over time.
    - Restart the CHIRP programming software on your computer and reconnect the cable. Sometimes restarting can clear up USB issues.
    - Update USB drivers for your computer's operating system and Baofeng model. Outdated drivers can disrupt connectivity.
    - As a last resort, try programming software that doesn't require a cable.
- Software freezing or glitching. Reinstall your software and upgrade computer drivers. Restart your PC.
- Overwriting previous program memories accidentally. Always back up your radio memory before programming.
- Loose antenna or battery interrupting the programming process. Ensure all connections are tight.

Remember, encountering a roadblock now and then is part of learning new skills. By troubleshooting these common programming issues, you'll resolve the problem and gain a deeper understanding of how your Baofeng radio works. So, take your time, and don't get frustrated with initial hiccups. Programming gets much easier with practice. **Always back up and verify your inputs**.

Now that you have a programmed Baofeng radio customized to your needs and interests, it's time to focus on achieving the best possible voice communications with your equipment. In the next chapter, we'll explore tips and techniques to make your transmissions crystal clear every time you press that push-to-talk button. From antenna optimization to mic technique and cutting through noise, you'll learn how to make your radio's audio as clean and understandable as possible. Let's move on to ensuring your team can perfectly hear every word coming through loud and clear!

Chapter 5: Crystal Clear – Securing Communications

In the world of radio, where preparedness and clear communication can mean the difference between safety and peril, your Baofeng radio is a vital tool. This chapter is dedicated to honing your skills in achieving crystal-clear radio transmissions, an essential aspect for any prepper.

Here, we dive into the nuances of signal clarity, exploring methods to ensure pristine reception and troubleshooting techniques to eliminate interference. Not only will you learn to banish noise but also to amplify your radio's capabilities through repeater use and optimal antenna selection. This knowledge is invaluable, especially in emergencies when reliable communication can significantly impact your survival strategy.

Get ready to elevate your Baofeng radio from a basic communication device to an exceptional tool in your preparedness arsenal, ensuring it won't disappoint you when you need it the most.

Getting Started: Band Selection

Understanding the nuances of VHF (Very High Frequency) and UHF (Ultra High Frequency) bands is crucial for clear and effective communication. Each band has its unique characteristics and is suited for different environments.

- **VHF (Very High Frequency) Bands:**
  - Remember, 'V' for very long range.
  - Frequency Range: 136–174 MHz.
  - VHF signals excel in open outdoor areas, providing extended range due to line-of-sight propagation. They travel in straight lines, making them suitable for distances without significant physical barriers. A 5-watt VHF signal can optimally reach 1–3 miles (Centers, 2022).
  - Indoors, VHF's performance is limited (around 100–500 feet) because its longer wavelengths struggle to penetrate walls and other obstacles.
- **UHF (Ultra High Frequency) Bands:**
  - Remember, 'U' is for urban, so better at penetrating obstacles.
  - Frequency Range: 400–520 MHz.
  - UHF signals are more effective in environments with obstructions like buildings and foliage. Their shorter wavelengths allow better penetration through barriers, offering improved indoor coverage.

  - In open outdoor settings, UHF has a shorter range (approximately 0.5–1 mile) than VHF, but its ability to navigate obstructions can be advantageous in urban or forested areas.

Understanding these bands helps you select the right one based on your operating environment. Additionally, tweaking other radio settings, such as squelch, bandwidth, and RF gain, and using the correct antenna height and orientation can further enhance signal quality. The frequency band and the radio's output power heavily influence the transmission range.

In summary, grasping the science behind VHF and UHF propagation is essential to realistic expectations regarding range and signal quality (Amateur Radio Service, 2016). Your Baofeng radio supports both bands, allowing you to choose the best frequency for your specific environment, whether communicating in open fields or navigating the challenges of urban or indoor settings. By understanding and leveraging the unique characteristics of VHF and UHF bands, you'll be better equipped to maximize your radio's capabilities and maintain uninterrupted communication even during disasters.

Squelch (SQL)

Mastering the squelch function on your Baofeng radio is a crucial skill for any ham radio enthusiast. Squelch is essentially a gatekeeper for your radio's speaker output, determining when to mute unnecessary background noise and when to let through actual transmissions. A well-adjusted squelch setting is vital for a pleasant and efficient radio experience.

- **Understanding Squelch Levels:**
  - Squelch settings typically range from fully open (low number) to very tight (high number). A lower squelch level (around 3) means the radio will pick up more sounds, including background static. A higher setting (like 8) filters out weaker signals, reducing noise but potentially missing faint transmissions.
- **Balancing Squelch Settings:**
  - The goal is to find a balance where the squelch mutes background static without cutting off legitimate communications. Too loose, and you'll constantly hear static; too tight, and you might miss important signals.
- **Using Carrier Squelch:**
  - If you're plagued by constant static, engage the carrier squelch. Set the squelch level to 0 (fully open) and press the [Moni/Carrier] button. This mode mutes the speaker unless a signal is present, helping to filter out the static hiss. Adjust upwards from level 0 until the noise is muted, but communications still come through clearly.
- **Advanced Squelch with CTCSS and DCS:**
  - For environments with persistent static, consider using tone squelch options like CTCSS or DCS. These settings add a layer of noise filtering, allowing only transmissions with specific tones to be heard.

Using squelch effectively is about tailoring your radio to your environment and needs. A tighter squelch or carrier squelch mode can be invaluable in noisy settings. But remember, the perfect squelch setting is a moving target, varying with your surroundings and the specific signals you're trying to catch.

By understanding and adeptly adjusting the squelch settings, you can significantly enhance your radio communication, making it clearer, more focused, and more enjoyable. Whether chatting with fellow ham operators or scanning during a disaster, a well-adjusted squelch is vital to clear communications.

Troubleshooting Poor Signal Reception

Troubleshooting poor reception is a crucial skill for ensuring clear and effective communication. The difference between spotty and crystal-clear reception often lies in understanding and addressing common issues. Here's a guide to help you diagnose and solve reception problems with your radio.

- **Location and Environmental Factors:**
    - Reception quality can be significantly affected by your physical location, weather conditions, and nearby structures. Moving to a higher elevation or away from large buildings can improve signal strength.
- **Antenna Issues:**
    - The antenna is a common culprit for poor reception. Check the antenna connectors for looseness or damage. Upgrading to a higher-gain antenna can make a substantial difference.
- **Frequency and Offset Settings:**
    - Incorrect frequency or offset settings can prevent your radio from receiving transmissions correctly. Double-check your frequency inputs and ensure they are accurate for your intended communication channel.
- **Squelch and Tone Settings:**
    - An improperly set squelch level can let in too much noise or block out legitimate signals. Adjust the squelch level to find the right balance. Also, verify CTCSS/DCS tone settings if you're using them, as incorrect tones can block desired signals.
- **Battery Power:**
    - Low battery power can lead to reduced transmit power and, thus, poorer reception. Regularly check your battery level, and keep it charged. Carrying spare charged batteries can be a lifesaver in prolonged situations.
- **Adjust the Bandwidth:** The bandwidth setting on your Baofeng radio is like adjusting the focus on a camera lens. It helps you get a clear "picture" by filtering out unwanted "noise." Changing the bandwidth can improve signal clarity, especially in crowded or noisy environments.
    - On a UV-5R, press MENU
    - Press '5' on your keypad (or use the arrows). This will take you to the 'WN' (Wide / Narrow) Bandwidth setting. Press MENU to enter.
    - Using the UP and DOWN keys, select either 'WIDE' or 'NARROW.'
    - Press MENU to save your choice.
    - Press EXIT to leave the menu. Your bandwidth is now set.
    - Try both settings to determine what works best in your current situation.

Poor reception typically involves checking equipment settings, ensuring optimal physical setup, and maintaining your gear. By methodically checking these areas – antenna, location, frequency, squelch and tone settings, and battery power – you can often pinpoint and resolve the root cause of reception issues. Remember, even minor adjustments can lead to significant improvements in reception quality.

This troubleshooting process will improve your experience and deepen your understanding of radio operations, making you a more adept and versatile operator.

Advanced Tactic: Find a Repeater

Utilizing local repeaters is a crucial strategy for extending the range of your communications. Repeaters, essentially radio relay stations, capture your transmission and rebroadcast it at a higher power over a wider area. Knowing how to find and use them effectively can significantly enhance your ham radio experience.

- **Finding Repeaters:**
  - Online Resources: Websites like RepeaterBook (www.repeaterbook.com) offer comprehensive listings of local repeaters. Use GPS coordinates for precise locations.
  - Mobile Apps: Apps such as RepeaterBook provide convenient access to repeater information on the go, allowing you to find nearby repeaters easily.
  - Local Radio Clubs: Joining local ham radio clubs or their social media groups can provide valuable information on area repeaters, updates, and advice from experienced operators.
  - Field Experience: You'll naturally learn about local repeaters as you engage more in ham radio activities and converse with fellow operators.
- **Using Repeaters:**
  - Etiquette: Always monitor a repeater before transmitting to ensure it's not currently in use and avoid interrupting ongoing conversations.
  - Programming Your Radio: Input the repeater's transmit and receive frequencies with the correct offset into your radio.

Understanding and utilizing local repeaters is a valuable skill to extend your range. Repeaters extend the range of your communications, which can mean life or death in a crisis; remember our opening story about Alden Jones. Learning to find and use repeaters correctly opens up a world of enhanced communication possibilities.

Advanced Tactic: Upgrade Antennas

Enhancing your Baofeng radio's performance through antenna upgrades is vital to achieving better reception and a broader communication range. The right antenna can significantly amplify your radio's capabilities, whether you're on the move or stationed at home.

- **Handheld Antenna Upgrades:**
  - For portable handheld units, consider a high-gain rubber duck antenna. Options like the Nagoya NA-320 or Abbree folding antenna can offer 3–4 times the range of standard antennas, making them ideal for on-the-go use.
  - High-gain antennas focus their energy in a particular direction instead of omnidirectional antennas that radiate or receive signals in all directions equally. By concentrating the signal in a specific direction, high-gain antennas effectively increase the signal's strength in that direction.
- **Mobile Antennas:**
  - For mobile operation, magnetic mount antennas are a great choice. They are easy to install and remove, with no permanent installation required. Dual-band 2m/70cm antennas from brands like Comet or Diamond are popular and effective.
- **Base Station Antennas:**
  - At a fixed location, roof-mounted base station antennas can provide excellent range in all directions. Opt for durable materials like fiberglass or steel, and choose dual-band antennas for versatility.

- **Important Considerations:**
  - Ensure your chosen antenna is well-matched to your radio's frequency range and output power.
  - Use high-quality, low-loss feed line cables like RG-8X to minimize signal loss.

Upgrading Antennas:

- Yagi Antennas: Yagi antennas are directional antennas with multiple elements, including a driven element, reflector, and directors. They offer high gain and are widely used in VHF/UHF applications for focused, long-distance signal reception and transmission.
- J-Pole Antennas: The J-pole antenna is a simple, efficient, vertically polarized antenna offering modest gain. It is commonly used in VHF/UHF amateur radio for its omnidirectional pattern and ease of construction. Roll-up J-Pole antennas provide a portable option, which is excellent for field use.
- Discone Antennas: The discone antenna is a wideband, omnidirectional antenna with a unique disc-cone structure. It is primarily used for VHF/UHF frequencies and is valued for its broad bandwidth and non-directional radiation pattern.
- Ladder Antenna: A ladder line antenna, often used in amateur radio, is a balanced feeder with two parallel conductors connected by rungs, known for its low-loss transmission, especially effective for high-frequency bands.

Upgrading your Baofeng radio antenna is a strategic move towards better, more reliable communication. A better antenna can transform your communications, whether in the field, on the road, or at your base station. By carefully selecting and correctly installing an appropriate antenna, you can significantly extend your reach and achieve clearer transmissions.

In the next chapter, we will navigate the important legal aspects of radio operations, including obtaining licenses and call signs.

Chapter 6: Licenses - Staying Legal with Baofeng

ADHERENCE TO LEGAL GUIDELINES is paramount in the radio realm, so it's crucial to understand the regulatory framework governing Baofeng radios. These radios offer affordable and efficient two-way communication, essential in emergency and survival scenarios. However, they operate within a legal framework set by federal regulations.

This chapter is designed to navigate the complex legal landscape surrounding the use of Baofeng radios. It provides clear guidance on using these powerful communication tools responsibly, ensuring you stay within legal boundaries and avoid penalties.

For preppers, this knowledge is not just about compliance; it's about ensuring uninterrupted legal communication capabilities in times of crisis. Understanding these regulations enhances your preparedness strategy, keeping you informed and compliant and ensuring your communication lines remain open and legal when you need them most.

Making Sense of Licensing Requirements

In the United States, the Federal Communications Commission (FCC) regulates the airwaves and designates which frequency bands require licenses for transmission. As a Baofeng radio owner, you must understand if a license is needed based on your intended use. Transmitting without a license on restricted frequencies can potentially lead to fines, equipment seizure, loss of privileges, and even criminal prosecution in egregious cases. Let's make sure you stay legal.

Unlicensed Frequency Bands

Some frequency bands are available for public use without any license required. These are generally allocated for short-range personal and business communications. Common unlicensed bands used by Baofeng radios include (*Family Radio Service* [FRS], 2011; *General Mobile Radio Service [GMRS]*, n.d.):

- **Family Radio Service (FRS)**: Uses the UHF 462–467 MHz band and is license-free but requires FCC-approved radios transmitting at 2 watts or less.
- **Multi-Use Radio Service (MURS)**: This service uses 5 VHF frequency channels near 151–154 MHz. No license is needed, but power is limited to 2 watts.

As long as you operate on these unlicensed bands within the power and equipment rules, no FCC license is required to transmit with your Baofeng radio.

Licensed Radio Services

Legal radio communication involves understanding the licensing requirements for different frequency bands. Here's a breakdown of the licenses needed for specific radio services (General Mobile Radio Service [GMRS], n.d.; Ham Radio

Licenses, n.d.):

- **General Mobile Radio Service (GMRS) Licensing:**
  - A GMRS license allows you to operate high-power two-way voice communication on the 462/467 MHz band.
  - This license, which costs $35 and is valid for five years, covers you and your immediate family members.
  - Unlike ham radio, there's no exam required to obtain a GMRS license.
- **Amateur "Ham" Radio Licensing:**
  - Ham radio operators have access to a wide range of frequency bands, each with its own characteristics and uses.
  - You must pass an FCC exam to obtain a valid operator's license.
  - There are three license classes - Technician, General, and Extra - each offering increasing access and privileges.
  - The difficulty of the exam and the privileges granted ascend with each class.

It's crucial to note that operating on restricted frequencies without a valid FCC license is illegal and subject to enforcement actions. This includes public safety, aircraft, maritime, and other regulated bands. Always ensure you have the appropriate license for your intended radio use to avoid legal complications and to operate within the bounds of FCC regulations. Understanding and adhering to these licensing requirements is critical to a responsible and lawful radio communication experience.

'The Command Packet' includes a printable version of the FRS and GMRS frequencies, which you should keep close.

Baofeng-Specific Guidance

Operating a Baofeng radio involves understanding and adhering to specific guidelines to ensure legal and responsible use. Here's a concise guide on what you need to know:

- **Licensing for Ham Radio Bands:**
  - To transmit on amateur (ham) radio bands (access to all amateur radio frequencies above 30 MHz), you must have a technician-class or higher amateur radio license.
- **GMRS Band Usage:**
  - If you wish to use the GMRS band, especially at higher power levels, you need a GMRS license. This license is more straightforward to obtain as it doesn't require an exam.
- **Adhering to Power Limits:**
  - Be mindful of power limits on certain bands like FRS (Family Radio Service) and MURS (Multi-Use Radio Service). On your Baofeng radio, ensure the power output setting does not exceed the allowed limits, such as 2 watts on the FRS band.
- **Restrictions on Certain Bands:**
  - Transmitting on commercial, public safety, emergency services, or critical infrastructure bands is illegal without proper authorization. Always check and comply with the regulations for each band.

- **Monitoring vs. Transmitting:**
  - While it's legal to listen to any frequency band for monitoring purposes, transmitting on restricted bands without a license is prohibited.

Remember, while Baofeng radios can transmit across a broad spectrum of frequencies, you must restrict your usage to the bands you're legally allowed to operate on. An FCC license grants legal access to certain bands and often allows for higher power output than unlicensed bands. Consider obtaining an appropriate license to enhance your radio operating experience and be fully prepared.

FCC Radio Operator's License for Ultimate Preparedness

Expanding your radio communication capabilities often means obtaining an FCC license, which unlocks access to additional frequencies and increased power limits. Here's a detailed guide to the different types of FCC licenses available:

General Mobile Radio Service (GMRS) License:

- Cost: $35 for a 5-year term.
- Family Coverage: The license covers immediate family members.
- No Exam Required: Easier to obtain as no exam is necessary.
- Frequency Access: Grants access to the 462/467 MHz band with higher power limits up to 50W.
- Repeater Use: Permits the use of repeaters and duplex operations.
- Ideal for short-range personal and family communication needs.
- You will receive a call sign specific to GMRS. Unlike amateur radio, where operators often choose call signs, the FCC automatically assigns GMRS call signs primarily for identification and regulatory purposes.

Amateur "Ham" Radio License:

- License Classes: Three classes - Technician, General, Extra, each with increasing privileges.
  - Technician License: Entry-level; basic access to specific HF, VHF, and UHF bands.
  - General License: Offers more extensive HF band privileges.
  - Extra License: Provides full access to all HF bands and privileges.
- Exams Required: Each class requires passing an exam, with difficulty increasing for each subsequent level.
- Study and Exam Process: Materials and exams are administered by Volunteer Examiner Coordinators.
- Call Sign Issuance: Upon licensing, you receive a call sign to operate on amateur bands.
- Perfect for radio hobbyists and those interested in technical learning.
- A user can apply for a new call sign after reaching each new level to indicate their new status or to keep their existing call sign.

Whether looking for short-range family communication with a GMRS license or diving deeper into amateur radio with a Ham license, these options provide greater freedom and capabilities in your radio use. The choice depends on your specific needs and interest in the hobby, with the Ham radio license offering a path for technical growth and a more comprehensive range of communication experiences.

Ham Radio Licenses Expanded:

In the United States, the Federal Communications Commission (FCC) oversees the licensing of amateur (ham) radio operators, offering three main types of licenses. Each license type caters to different levels of expertise and access within the ham radio spectrum:

1. **Technician License**:

- Level: Entry-level, ideal for beginners in ham radio.
- Exam: A 35-question test covering radio theory, regulations, and operating practices.
- Privileges: Access to all amateur radio frequencies above 30 MHz, including the popular 2-meter and 70-centimeter bands. Limited privileges on certain HF bands (10 meters), mainly for Morse code.
- Why You Want It: The entry-level license is ideal for beginners. It is excellent for local and regional communication via VHF/UHF and basic HF privileges. It is popular with those interested in local communication, emergency communication, and getting started with amateur radio.

1. **General License**:

- Level: Intermediate for those seeking broader communication capabilities.
- Exam: Another 35-question examination, building upon the Technician level.
- Privileges: Expanded HF band privileges and access to most amateur bands, including most of the HF spectrum, allowing for international and long-distance communication.
- Why You Want It: This class is for those who want to explore the full range of amateur radio, including long-distance (DX) contacts. The General class opens up a world of global communication on the HF bands, where much of amateur radio's magic lies.

1. **Amateur Extra License**:

- Level: Advanced for those pursuing the fullest extent of amateur radio privileges.
- Exam: A comprehensive 50-question examination.
- Privileges: All amateur band privileges. Access to exclusive frequency allocations in the HF bands.
- Why You Want It: It's the highest level of licensing, appealing to those deeply involved in the hobby. It grants maximum operational flexibility and access to all available frequencies and modes. It is ideal for the most serious hobbyists who want to explore every aspect of amateur radio, including those interested in the hobby's advanced technical and experimental aspects.

Each progressive level of licensing requires passing a more challenging examination, which tests knowledge of radio theory, regulations, and operating practices. The choice of license depends on your interests in the hobby:

- Local Communication & Basic HF: Technician.
- Global HF Communication & Advanced Modes: General.
- Full Spectrum Access & Technical Experimentation: Amateur Extra.

All Amateur Radio operators in the U.S. must be licensed, with licenses valid for ten years before renewal. While the FCC regulates these licenses, the exams are administered by volunteer groups of Amateur Radio operators. Fun Fact: in 2007, the FCC stopped requiring Morse code proficiency as part of the licensing process.

I recommend starting with your licensing on the National Association of Amateur Radio's website. They will walk you through registering with the FCC and finding an exam location and date.

https://www.arrl.org/find-an-amateur-radio-license-exam-session

Navigating Power Restrictions

Navigating the power limits of your Baofeng radio is crucial for legal and efficient operation. Understanding these limits, set by the Federal Communications Commission (FCC), helps avoid interference with other services and prevents potential penalties. Here's a guide to help you manage power limits responsibly:

Understanding Power Limits: Power limits, expressed in watts (e.g., 1W, 5W, 50W), dictate the maximum transmitting power allowed on specific radio frequency bands. The FCC establishes these limits based on band propagation characteristics and the need to minimize interference between radio services (*Ham Radio Licenses*, n.d.).

Adjusting Transmit Power on Baofeng Radios: To change your Baofeng radios transmit power, typically on models like the UV-5R, follow these steps:

1. Press "MENU" to access settings.

2. Navigate to the "TXP" (Transmit Power) sub-menu using the arrow keys.

3. Enter the "Power" settings by pressing "MENU" again.

4. Choose between "High" (4-5 watts), "Mid" (2-2.5 watts, model dependent), or "Low" (1-0.5 watts) power settings.

5. Confirm and save your choice by pressing "MENU."

Consequences of Exceeding Power Limits: Exceeding FCC power limits is considered harmful interference and can lead to (*Enforcement Overview*, 2020):

- Substantial fines.
- Equipment confiscation.
- License revocation.
- In severe cases, imprisonment.

Best Practices for Power Management:

- Use the minimum power necessary for effective communication.
- On unlicensed bands like FRS/MURS, adhere to the 2W limit.
- For repeater use, higher power like 8W may be necessary.
- For short distances, 1-4W is sufficient and conserves battery life.

Typical Power Limit Examples:

- FRS: 2 watts maximum ERP.
- MURS: 2 watts is the maximum conducted power.

- GMRS: Up to 50 watts ERP, often limited to 5 watts on specific channels.
- Amateur Radio: Up to 1500 watts ERP on some HF bands for Extra Class licensees.

The Importance of Understanding Power Limits: Baofeng radios, capable of transmitting at higher power levels, require careful management to comply with FCC regulations. For instance, the popular UV-5R model can broadcast up to 5 watts, which exceeds the limits for unlicensed bands like FRS and MURS. Always verify FCC regulations for the intended frequency band and adjust your radio's power settings accordingly. This approach ensures compliance and helps you operate your radio within the legal framework, maintaining orderly and efficient spectrum usage.

Legality Before Operating

Before using your Baofeng radio, ensure you're up to date with FCC regulations, as these can change. Quick checks safeguard against illegal transmissions and promote responsible use. Remember, staying compliant protects you and respects the wider radio community.

The amateur radio community is an excellent resource for guidance, especially when in doubt about regulations. This collaborative learning fosters a compliant and interference-free radio environment.

Practicing diligence in adhering to rules helps you enjoy your Baofeng radio hobby without fear of regulatory issues. Be cautious and stay informed to avoid infractions.

As you master your Baofeng radio's basics, our next chapter will introduce advanced features, shifting you from beginner to proficient. Get ready to fully engage with your radio and delve deeper into the exciting realm of amateur radio, an invaluable skill for any prepper looking to enhance communication readiness.

The upcoming chapter will elevate your skills, transitioning you from a novice to a proficient Baofeng user. Prepare to unlock your radio's full potential.

Chapter 7: Advanced Operations – Beyond the Basics

For the adept prepper, mastering advanced features of the Baofeng radio is not just about enhancing technical skills; it's about ensuring readiness and versatility in unpredictable situations. This chapter takes you beyond the basics, equipping you with the knowledge to utilize your Baofeng radio's full potential in scenarios where efficient communication can be a game-changer.

You'll explore the nuances of dual watch for simultaneous channel monitoring, understand the criticality of emergency transmission in crisis situations, and learn effective memory channel management for swift access to vital frequencies. Additionally, insights into battery management and advanced troubleshooting will prepare you to maintain reliable communication under various conditions. These skills are indispensable for any prepper looking to stay one step ahead in emergency preparedness and effective response.

Keeping an Ear on Two Channels: Exploring Dual Watch

Imagine being at an airport, listening in on the ground crew's chatter and the pilot's communication, or in the woods and listening as your two teams and basecamp communicate.

The Dual Watch/Dual Reception feature of your Baofeng radio offers a similar functionality. It lets you monitor two frequencies or channels, alternating between them, so you don't miss any important communication. This advanced capability is handy when staying informed on multiple channels is critical during an emergency.

How Dual Watch Works on a Baofeng Radio:

- First, set your radio to Frequency Mode (VFO/MR button)
- Next, input the first frequency or channel you want to monitor.
- Press the A/B key to switch to the other display line.
- Input the second frequency or channel.
- Press the "MENU" button on your Baofeng UV-5R.
- Use the arrow keys to navigate to the "TDR" (Dual Watch/Dual Reception) setting or directly press "7".
- Press "MENU" again to enter the TDR setting.
- Use the arrow keys to select "ON" for Dual Watch.
- Press "MENU" to activate Dual Watch.

- Press "EXIT" to return to the main screen.
- You'll see both frequencies displayed, and an 'S' will confirm Dual Watch is active. You are now monitoring multiple frequencies.

Understanding the Display:

- A steady arrow or pyramid symbol indicates the frequency you're set to transmit.
- A flashing arrow means you're receiving a transmission on that frequency.

Transmitting on Dual Watch:

- While Dual Watch allows you to receive on two frequencies, you can only transmit on one at a time (GMRS, n.d.).
- To switch the transmitting frequency, press the "A/B" button.

Practical Applications of Dual Watch:

- Emergency Monitoring: Keep an emergency channel on standby while using another for regular communication.
- Repeater Input Monitoring: Listen to a repeater's input and output frequencies.
- Team Coordination: Monitor both command and field team frequencies simultaneously.
- Enhanced Situational Awareness: Stay informed on two vital channels in dynamic environments.

Limitations and Considerations:

- Single Transmission: You cannot transmit on both frequencies simultaneously.
- Model Support: Not all Baofeng models support genuine Dual Watch. Verify your model's capabilities.
- Battery Life: Monitoring two channels can drain the battery faster.
- Distraction Potential: Continuous dual monitoring might be distracting, so use it judiciously.
- Practice Required: Effectively using Dual Watch takes practice to master rapid channel switching and timely transmissions.

In summary, Dual Watch is a versatile tool that, when used effectively, can significantly enhance your communication capabilities and situational awareness. Whether for emergency preparedness, team coordination, or just staying in the loop on multiple fronts, mastering Dual Watch can elevate your experience with your Baofeng radio. Remember to practice and familiarize yourself with this feature to maximize its benefits.

Keep Your Hands Free by Unlocking VOX

"Voice-Operated Transmission," or VOX, is a dynamic feature on many Baofeng radios that brings the convenience of operating hands-free. When VOX is enabled, your radio can transmit automatically in response to your voice, eliminating the need to press the Push-to-Talk (PTT) button for each communication. This feature is paramount when your hands are occupied, or a quick response is critical.

Activating VOX on Your Baofeng Radio: To enable VOX on a Baofeng radio, such as the UV-5R:

- Press the "MENU" button to access the radio's settings.

- Navigate to the "VOX" option (press '4').
- Enter the VOX settings ranging from Off, 1 (Low) to 10 (High) sensitivity.
- Start with Medium sensitivity (5) and adjust as needed to suit your environment.

Setting the Right Sensitivity Level: Getting the VOX sensitivity setting right is crucial:

- Too high, and the radio might transmit unintended background noise.
- If it is too low, your voice may not trigger the transmission effectively. Test the settings in your usual environment to find the sweet spot where your voice activates transmission reliably without false triggers.

Enhancing VOX with External Microphones: For improved VOX functionality, consider using an external headset, lapel, or throat microphone. These accessories are designed for clearer voice capture, reducing the chance of unintended transmissions caused by ambient noise.

The Benefits of Using VOX:

- Convenience: Frees up your hands for other tasks, as the radio transmits automatically when you speak.
- Natural Conversations: Provides a more fluid communication experience, removing the need for button pressing.
- Quick Response: Enables faster communication since it transmits as soon as you start speaking.
- Multitasking: Ideal for scenarios where you must communicate while your hands are busy.
- Accessibility: Beneficial for individuals who find it challenging to use PTT buttons.

Practical Scenarios for VOX Use:

- Climbing: Maintain communication without having to let go of your gear.
- Medical or First-Aid Situations: Communicate with others while attending to injuries or administering first aid, where your hands are occupied with medical supplies or patient care.
- Hiking or Driving: Keep your hands free for safety while staying connected.
- Wilderness Survival: Communication is essential when building shelters, starting fires, or preparing food, as both hands are needed for these survival tasks.
- Search and Rescue Operations: Stay in constant contact during search missions in challenging terrains, where hands-free communication is essential for coordinating efforts and ensuring safety.
- Fishing or Hunting: Maintain communication with partners while handling fishing rods, setting up baits, or aiming during a hunt.

Using VOX Responsibly: To maximize the benefits of VOX while minimizing drawbacks:

- Utilize a headset or lapel mic for focused voice pickup.
- Test and set the appropriate sensitivity level.
- Be aware that your radio is a "hot mic" when VOX is active.
- Disable extra tones like Roger Beep (a beep you hear after you release the PTT button) to prevent feedback loops.

- Monitor battery usage, as VOX can increase power consumption.

Mastering the VOX feature on your Baofeng radio can significantly enhance your communication experience. It offers a hands-free, responsive, and convenient way to stay connected. With some practice and the correct settings, VOX can transform how you use your radio, making it an invaluable tool in various scenarios.

Raising the Alarm With Your Radio: Emergency Alert

The emergency alert feature on Baofeng radios is a potentially life-saving tool designed for critical situations. When activated, it signals a state of distress, alerting nearby radio users who can offer assistance. This feature is powerful but must be used responsibly and exclusively in emergencies.

Activating the Emergency Alarm: On most Baofeng models, such as the UV-5R, activating the emergency alarm is straightforward:

- Press the "CALL" button for about three seconds.
- This action triggers a loud, continuous beeping alarm tone and rapid flashing of the LED indicator, making the distress signal hard to miss.
- The alarm typically sounds for 30 to 60 seconds. To deactivate it manually before it auto-stops, press and hold the "CALL" button again for about two seconds.

Transmitting a Distress Signal: In addition to emitting an audible alarm, this feature also transmits a unique distress signal over the currently tuned channel. Other radios within range can pick up this transmission if tuned to the same frequency. Most Baofeng radios override volume or mute settings to ensure the distress call is heard (*Baofeng - Baofeng Troubleshooting*, 2022).

Practical Applications of the Emergency Alert: This emergency alarm can be crucial in various real-life scenarios:

- Hiking Accidents: Injured hikers can alert their group or nearby rescuers.
- Structure Fires: Notifying occupants of immediate evacuation needs.
- Medical Emergencies: Quickly signaling for help during health crises.
- SOS Signaling: Attracting attention when stranded in remote areas.

Responsible Use of the Emergency Alarm:

- Avoid False Alarms: Be cautious not to activate the alarm accidentally.
- No Pranks: Misusing the alarm for non-emergency purposes is unethical and potentially illegal.
- Confirm Alarms: Verify the legitimacy of any received emergency alarms before responding.
- Legal Compliance: Be aware of local laws regarding using emergency signals and adhere strictly to them.

When used correctly, the emergency alarm on your Baofeng radio is a critical feature that can significantly enhance safety in dangerous situations. Understanding and respecting its function is crucial, ensuring it's used appropriately and only in true emergencies. Doing so contributes to maintaining the alarm's integrity and effectiveness as a vital distress signal.

Personalizing Your Radio: Mastering Memory Channels

Memory channels on Baofeng radios are essential for personalizing and streamlining your radio experience. By saving specific frequencies into these channels, you gain quick, one-touch access to your most frequently used frequencies, bypassing the need for manual entry each time. This functionality tailors your radio to your unique needs, making its operation far more efficient.

Manually Programming Memory Channels on a Baofeng UV-5R: To manually save a frequency into a memory channel. Go back to Chapter 4 for detailed instructions.

Software Programming for Convenience: While manual programming is effective, software like CHIRP is a faster, more user-friendly way to program multiple channels. This method, detailed earlier in Chapter 4, allows bulk frequency input.

Benefits of Using Memory Channels:

- Quick Frequency Access: Easily switch to saved channels faster than manual input.
- Organized Frequencies: Group and organize channels logically, such as sequential numbering for local repeaters.
- Efficient Scanning: Scan through programmed channels to detect active transmissions.
- Customizable Setup: Tailor the channel lineup to your specific needs.

Tips for Optimizing Memory Channel Use:

- Logical Grouping: Organize channels in a way that makes sense for your usage, like grouping all local repeaters together.
- Naming Channels: Name channels for easy identification if your programming software allows.
- Separate Transmit/Receive Storage: Use pairs of memory channels to store separate transmit and receive frequencies for repeaters.
- Leave Space for Expansion: Skip some numbers between channel groups for future additions.
- Routine Scanning: Regularly scan through channels to stay informed of active transmissions.
- Back-Up Your Setup: Save your configuration externally for easy recovery.

Understanding Memory Channel Limitations:

- Limited Number: Baofeng radios typically offer 128–250 memory channels, which may require thoughtful management.
- Temporary Storage: Memory channels hold settings until overwritten or reset, necessitating periodic backups.
- Manual Selection Speed: Selecting channels manually can be slower than automatic scanning.
- Organizational Discipline: Effective grouping and naming are essential for quick and easy channel identification.

In conclusion, mastering memory channels is vital to becoming proficient with your Baofeng radio. These channels not only provide a fast way to access your preferred frequencies but also allow you to organize and customize your radio for optimized use. With practice and thoughtful setup, you'll find memory channels indispensable in your everyday radio operations.

Your Radio's Lifeline: Battery Life

Batteries are the lifeline of your Baofeng radio, providing the power needed for all its functions. Understanding how to manage and maintain your radio's batteries effectively ensures long-lasting performance and reliability.

Types of Batteries for Baofeng Radios: Most Baofeng radios use rechargeable lithium-ion (Li-ion) batteries. These batteries are known for their high energy density and long life span. They are typically lightweight, making them ideal for handheld radios like the Baofeng models.

Maximizing Battery Performance:

- Initial Charge: When you first get your Baofeng radio, charge the battery fully before using it. This ensures optimal battery performance.
- Regular Charging: Li-ion batteries do not suffer from memory effects, so it's best to charge them regularly and keep them topped up.
- Deep Discharge: Avoid completely draining the battery, which can shorten its life. Recharge the battery when you notice a decrease in performance.

Understanding Battery Capacity:

- Baofeng batteries are often rated in milliampere-hours (mAh), indicating how much charge they can hold. Higher mAh ratings mean longer battery life per charge.
- For instance, an 1800 mAh battery will last longer than a 1500 mAh battery under the same usage conditions.

Charging Your Baofeng Battery:

- Use the charger provided with your radio or a compatible replacement.
- Avoid overcharging the battery; remove it from the charger once fully charged.
- Some Baofeng models come with charging cradles for convenient charging.

Storage and Care:

- If you're not using your radio for an extended period, store the battery separately in a cool, dry place.
- Avoid exposing the battery to extreme temperatures, as heat can degrade its performance and lifespan.

Replacing the Battery:

- Over time, you may notice a reduction in battery life. This is normal for rechargeable batteries and indicates it's time for a replacement.
- Use only Baofeng-approved or reputable third-party batteries to ensure compatibility and safety.

In conclusion, properly managing your Baofeng radio's battery ensures that it remains a reliable communication tool. Regular charging, appropriate storage, and timely replacement are vital for maintaining battery health. Understanding and caring for your radio's battery allows you to enjoy extended use and avoid interruptions in communication.

Advanced Troubleshooting

As you delve deeper into your Baofeng radio's use and explore its advanced features, you might encounter complex issues that call for more sophisticated troubleshooting techniques. Here's a guide to help you navigate some common advanced problems and their potential solutions:

Troubleshooting Tone Squelch Activation Failures: If you're using CTCSS or DCS tone squelch settings and suddenly find yourself unable to receive signals, the issue often lies in mismatched codes between your radio and the transmitting radio. For successful communication:

- Ensure both radios use the exact same CTCSS subaudible tone frequency or DCS code values.
- Double-check the squelch settings to confirm that your radio's tone frequency or code value precisely matches the transmitting radio's settings.

Addressing Repeater Connection Issues: If you're experiencing difficulties connecting to repeaters or if transmissions are broken or distorted, consider these checks:

- Confirm that the transmit offset in your radio matches the repeater's input frequency.
- Check that any required tone encode (usually CTCSS) for the repeater is correctly set on your radio.

Managing Shorter Battery Life: A rapidly depleting battery can be a sign of various issues:

- Avoid exposing the radio to extreme temperatures.
- Regularly inspect the battery for swelling, which indicates damage.
- Ensure consistent and complete charging after use.
- Use original Baofeng batteries and chargers for optimal compatibility and performance.

Solving Bluetooth Connectivity Issues: For problems pairing with Bluetooth devices, these steps can help:

- Engage pairing mode on your radio and the Bluetooth device while ensuring they are close.
- Minimize physical obstructions between the radio and the Bluetooth device.
- Update your radio with the latest firmware to enhance Bluetooth stability.
- Consider resetting both the radio and the Bluetooth accessory to factory settings before attempting to pair again.

Leveraging Community Expertise: Don't hesitate to seek help from the Baofeng user community for advanced troubleshooting:

- Online Communities: Platforms like Reddit's Baofeng subreddit and Facebook groups like "Baofeng Radio Club" are valuable places to share troubleshooting advice.
- Manufacturer Forums: Official forums like BaofengTech can provide specific guidance for advanced issues.
- Local Hamfest Meets: Meeting experienced users in person can offer direct, hands-on assistance.
- YouTube Tutorials: Visual guides from experienced users can be beneficial for complex troubleshooting.

Remember, while learning to solve advanced issues independently is rewarding, tapping into the collective wisdom of the Baofeng community can save you time and frustration. Over time, this shared knowledge will significantly enhance your understanding and proficiency in handling your Baofeng radio.

Empowered Through Continuous Learning

Mastering your Baofeng radio is an ongoing journey of learning and adaptation. As technology evolves with new models, enhanced features, and expanded capabilities, staying informed and engaged with the Baofeng community is key to honing your skills.

Stay Informed Through Official Sources:

- Regularly visit Baofeng manufacturers' websites, such as BaofengTech and BaoFengRadio, to stay up-to-date on the latest products, updates, and accessories.
- Subscribe to their newsletters and follow their social media channels for the latest news.

Enhance Your Operator Skills:

- Consider obtaining an amateur radio license. Earning your license and call sign is a badge of commitment and expertise.
- Even if you choose not to get licensed, continuously improve your operator skills. Practice navigating your radio's menu and settings daily. Familiarize yourself with programming memory channels and experiment with different accessories to optimize your setup.

Connect with Local Clubs:

- Local amateur radio clubs are invaluable resources. Regular meetings offer a chance to learn from experienced hams. These clubs can provide practical advice, legal best practices, equipment recommendations, and hands-on training.
- Participating in club activities and events also helps you build a network of fellow radio enthusiasts.
- In these groups, you will meet and become friends and build networks that will be valuable assets during a crisis.

Stay Proactively Engaged:

- Actively participate in online forums and platforms dedicated to Baofeng radios. Engaging with fellow users allows you to tap into a wealth of collective experience and knowledge.
- Share your own experiences and learnings. The more you contribute, the more you gain from the community.

After mastering the advanced aspects of your Baofeng radio, it's time to explore one of its most vital uses: emergency and disaster communications. When traditional communication infrastructures fail, your Baofeng radio becomes a crucial link to the outside world. The next chapter will delve into effectively using your radio in times of crisis, ensuring you're prepared to communicate when it matters most.

Chapter 8: Staying Connected – Baofeng Radios During Disasters

In times of disaster, when traditional communication networks fail, Baofeng radios emerge as crucial tools for maintaining vital connections. This chapter delves into the indispensable role of Baofeng radios during emergencies, offering guidance on preparation, effective communication, and maintenance.

Baofeng radios stand out in emergencies due to their resilience. These battery-powered devices enable direct radio wave transmissions over considerable distances, providing a reliable alternative when standard communication infrastructures collapse. Their ease of network setup allows for quick and efficient coordination during crises.

Understanding the tactical advantage of Baofeng radios is essential. We will explore their functionality, adaptability, and decentralized operation, which are crucial for effective crisis management. This chapter will equip you with practical knowledge for using Baofeng radios during emergencies – preparing your device, maintaining it, and best practices for communication.

We aim to empower you with the skills to use Baofeng radios confidently in emergencies. You'll gain insights into transforming these devices from simple communication tools into strategic assets for emergency response. Whether you're a first responder or a concerned citizen, this knowledge is vital for ensuring preparedness and effective response in times of need.

Natural Disasters

Mother Nature is beautiful yet unpredictable. A sunny day can quickly turn into a stormy night, a calm sea into a raging torrent. Natural disasters such as hurricanes, earthquakes, floods, and wildfires can strike without warning, disrupting normal communication channels and isolating communities. In such situations, your Baofeng radio can serve as a critical lifeline.

With its ability to operate independently of mobile networks and internet connections, your Baofeng radio can transmit and receive crucial information when conventional communication methods fail. Whether receiving updates on the disaster situation, requesting assistance, or coordinating relief efforts, your Baofeng radio can help you stay connected and informed even in the heart of a natural disaster.

Hurricane Katrina

Picture the scene: a city submerged, homes wrecked, and communication networks crippled. This was the grim reality of New Orleans in 2005, in the aftermath of Hurricane Katrina. Amid the chaos and destruction, Baofeng radios emerged as an unexpected hero. As cell towers and internet lines succumbed to the storm, these sturdy radios kept their signal, providing a lifeline to stranded residents and first responders alike.

Emergency teams used their Baofeng radios to coordinate rescue operations, communicate with relief centers, and provide real-time updates on the situation. Citizens used them to request help, share information, and stay connected with their loved ones. In a city silenced by the storm, the airwaves buzzed with the static and chatter of Baofeng radios, echoing the resilience of the human spirit.

California Wildfires

Now, let's shift our gaze to the golden state of California, known for its sunny beaches, Hollywood glamour, and, sadly, devastating wildfires. In 2018, the state faced one of its deadliest wildfire seasons. As the flames spread, engulfing homes and forests alike, Baofeng radios again stepped into the fray.

Firefighters used their Baofeng radios to communicate with each other, coordinate their efforts, and navigate the dangerous terrain. Evacuees used them to receive updates on the fire's progress, find safe evacuation routes, and stay connected with the outside world. Amid the smoke and flames, the humble Baofeng radio emerged as a beacon of hope, its signal cutting through the haze of uncertainty.

Wildfires can be extremely deadly and fast. They can overtake an area with only a minute warning based on wind patterns. This is when you need to activate your sudden evacuation plan. A bug-out bag is crucial for disaster preparedness, and alongside your Baofeng radio, it should include essentials for at least 72 hours of survival. Ensure it contains water purification methods, non-perishable foods, clothes, a basic first-aid kit, and tools like a multi-tool and flashlight. Include important personal documents, cash, and maps for navigation. Tailor it to your needs, check it regularly, and keep it ready for immediate departure.

Search and Rescue Operations

Picture a game of hide-and-seek in a vast forest. The seeker calls out, and the hider responds, their voices guiding each other through the dense undergrowth. In search and rescue operations, your Baofeng radio plays a similar role. It's a voice in the wilderness, a beacon in the dark, and a critical tool in locating and rescuing lost or injured individuals.

Search and rescue teams often operate in remote and challenging terrains where regular communication systems may not be reliable. Baofeng radios, with their robust performance, long battery life, and ability to communicate over large distances, can provide a reliable means of communication. Coordinating search efforts, sharing information about the missing person, or calling for medical help every transmission can be a step towards a successful rescue.

Configure a "Grab and Go" Kit

Assemble a self-contained Go Bag for your handheld with essential accessories for emergency use. Include spare charged batteries, a 12V cigarette lighter charger, a foldable external antenna, a printed reference sheet of programmed channels, headsets, and a small notepad and pen for written messages. Keep this kit prepared and easily accessible for rapid deployment when needed. Your radio and all its essentials.

Nepal Earthquake

Lastly, we journey to the foothills of the Himalayas, to Nepal, a country known for its stunning landscapes and rich culture. In 2015, a massive earthquake struck Nepal, causing widespread destruction. As the ground shook and buildings crumbled, communication networks collapsed, leaving millions cut off from the rest of the world.

Once again, Baofeng radios rose to the occasion. Rescue teams, relief workers, and even ordinary citizens turned to these radios as their primary means of communication. They used them to coordinate relief efforts, call for medical assistance, and provide comfort and reassurance to those affected by the disaster. In the face of unimaginable devastation, Baofeng radios served as a lifeline, connecting hearts, hands, and hopes.

Earthquakes destroy city infrastructures. Roads, bridges, and ways out of the impacted area are blocked. Supply chains and emergency services may be unavailable for days. Electricity might be out for weeks. Your stocked home, base station, and Baofeng will be invaluable here.

Powering Your Radio with a Car Inverter

When the power grid fails, as it most likely will after an earthquake, your car can become a valuable asset for transportation and a power source for your Baofeng radio and other small electronics. Here's how to turn your vehicle into a makeshift power plant with the help of an inverter.

Utilizing a Car Inverter

- For Light Loads: A cigarette lighter socket in your car typically supports up to 150 watts, sufficient for charging phones and radios, powering small LED lights, or running a compact fan.
- For Heavier Loads: If you need to power devices that require more energy, you'll need to clamp an inverter directly to the car battery. Mounting the inverter on a block of wood is advisable to prevent it from moving and ensure safety.
    - Remember that your car's alternator generates about 400-600 watts. Therefore, if you operate appliances requiring more power than this, they will drain the battery after the alternator's output is exceeded.
    - Ensure you are idling your vehicle during these types of loads.
- Efficient Use: Your car, equipped with an inverter, is ideal for powering household items within the lower power range. It's impractical to start a generator to charge your phone or radio; use your car's battery instead.
- Fuel Efficiency: A car or truck with a full gas tank can idle and provide power for approximately 24 hours, making it an efficient emergency power source.

Safe Fuel Storage for Extended Power Supply

- Fuel Storage: Invest in several 5-gallon fuel cans for gasoline storage. This fuel can be used for your car or a backup generator.
    - Monthly Rotation: Fill up one can each month and mark it with the month's number.
    - Yearly Cycle: After storing the fuel for a year, rotate it into your vehicle's tank. This ensures the fuel doesn't age too much and keeps your stock fresh. Pour the can with this month's number into your car. Then go fill up your gas can with new gas.

Following these guidelines, you can ensure you're prepared for power outages with a reliable and efficient system for powering essential devices like your Baofeng.

Power Outages

Imagine a sudden blackout, an unexpected power outage. The lights go out, computers shut down, and mobile networks go offline, plunging the world into a communication blackout. In such scenarios, your Baofeng radio can be a beacon of light, enabling you to stay connected even when the power grid fails.

During power outages, Baofeng radios can operate on their battery power, allowing you to receive critical updates and communicate with others. Whether contacting family members or local authorities or coordinating community response efforts, your Baofeng radio can ensure you're never truly in the dark.

Power outages are unpredictable, and service restoration depends on what happened and whether the electric company has workers willing to leave their families to reconnect services. This is when your extra batteries, battery bank, car inverter, generator, and fully stocked fuel rotation come into play with your Baofeng.

A home battery bank can be invaluable in a power outage, especially for keeping critical communication lines open, such as charging your Baofeng radios. Here's how to create and utilize a battery bank tailored for small-scale energy needs.

Creating A Simple Home Battery Bank

Begin by purchasing the following components:

- **Deep Cycle Marine Battery**: Choose a Group 31 or 29 Flooded Lead Acid Battery, with a capacity of around 75 Amp-hours, to ensure enough storage for your needs.
- **Intelligent 3-Stage Computerized Battery Charger**: This charger will optimize your battery's health and longevity, adjusting the charge as needed. Opt for a 10 AMP charger to balance speed and battery care.
- **Voltmeter**: Essential for monitoring your battery's charge level and overall health. The voltmeter needs to be the type that can be plugged into an outlet.
- **Inverter**: A simple 150-to-400-watt converter is more than enough for our purposes. The inverter, which will have clamps that connect directly to the battery, converts the stored DC power in your battery to AC power for your devices.
- **Compact Table Lamp with Low Wattage Bulb**: This low-energy lighting solution can run off your battery bank, so you won't need a flashlight when you go to your battery bank.
- **Orange T-shape 1 Plug to 3 Outlet Splitter Adapter:** Buy several. They allow you to turn your one outlet plug into three outlet plugs.

Setting Up Your Bank

- Mount the battery onto a table, shelf, or workbench in a dry, temperature-controlled location.
- Place the battery charger next to the battery and plug it into an AC outlet. Connect the charger to the battery.
- Place the inverter next to the battery and connect it directly to the battery. You now have power!
- Connect the voltmeter to the inverter. This will provide you with a visual display of the health of your batter.
- Connect your orange T-shape splitters so that you have more outlets to use.
- Connect your table lamp for light.
- In one of your orange splits, connect your Baofeng Radio charger.

Your bank is now set up and ready!

Using Your Battery Bank

With a 120-watt load, your battery bank should last 4-6 hours of continuous use. However, you don't need to use it continuously. That's not their purpose. The purpose is to charge your devices and disconnect from the bank once fully charged. Doing this will extend its functional use.

Remember, it's designed for small devices such as:

- **Yes**: Radio chargers, lights, phones, and tablets. To charge them to full health.
- **No**: Refrigerators, freezers, or coffee machines. Don't even think about running these devices.

Recharging Your Battery Bank

To recharge:

- Jump Start the Battery from Your Car: Jump the battery as usual. This method is suitable if you need a quick charge to keep essential devices running.
- From a Car Inverter: Connect your battery charger to an inverter attached to an idling car using an extension cord.
- From a Generator or Car Inverter: Connect your battery charger to a generator and let it recharge the battery.

Keep in mind that the battery bank is primarily for communications and lighting.

By following these steps, you'll have a reliable power source for your most critical devices, ensuring you can stay informed and connected even when the grid is down. Remember, the proper preparation can make all the difference in an emergency.

Civil Unrest

Communication plays a crucial role in times of civil unrest. It's a thread that holds communities together, a channel that disseminates critical information, and a platform that gives voice to the voiceless. Whether it's a peaceful protest, a political rally, or a sudden outbreak of violence, your Baofeng radio can be an indispensable communication tool.

Baofeng radios can help you stay updated on the situation, receive instructions from authorities, or coordinate with fellow community members. Operating independently of cellular networks and the internet ensures you can maintain communication even when these regular channels are disrupted.

Remember your time at your local ham club and setting up your neighborhood net? This is when your investment in time pays off.

Outdoor Adventures

Venturing into the great outdoors is exhilarating but comes with risks. A simple hike can become a survival situation, a fun camping trip into an unexpected night in the wilderness. Your Baofeng radio can be valuable for your survival kit in such situations.

Whether lost on a trail, injured during a climb, or stranded due to bad weather, your Baofeng radio can help you call for help, receive weather updates, or stay connected with your group. It's not just a radio; it's your lifeline in the wild.

In the face of an emergency, your Baofeng radio is more than just a device; it's a lifeline, a beacon of hope. It's a testament to our resilience and capacity to adapt, innovate, and turn a simple communication tool into a lifeline. So, as you hold your Baofeng radio, remember: it's not just about the frequencies it can tune into or the signals it can transmit. It's about the lives it can touch, the hope it can inspire, and the difference it can make in times of need.

Independent Local Radio Nets

Baofeng's flexibility, portability, and ease of use allow setting up independent local "radio nets" when needed (Cribbs, 2019a). A radio net is a network of stations communicating on a shared channel during an emergency.

Nets can quickly share urgent news and weather, safety alerts, request medical assistance, coordinate logistics like equipment and volunteers, and check on vulnerable community members. Having access to such information can be lifesaving.

Your chance to barter begins now! With this group of like-minded individuals.

Preparing Your Baofeng Radio for Emergencies

With some planning and preparation, your Baofeng radio can provide a resilient communication lifeline when standard infrastructure fails. Follow these tips to ensure your handheld is ready for disasters:

Program Emergency Frequencies

Use PC software like CHIRP to pre-program your region's radio with critical emergency frequencies. Good sources are NOAA weather radio, FEMA, Red Cross, local fire/police, and Amateur Radio emergency networks. Having these ready to select instantly saves precious time when a crisis strikes.

National Weather Service (NWS) Channels

The National Weather Service (NWS) operates a network of radio stations known as NOAA Weather Radio (NWR), which broadcast continuous weather information 24/7. These channels are real-time weather updates, keeping you informed about changing weather conditions, forecasts, and severe storm warnings.

Your Baofeng radio can tune into these NWS channels, ensuring you have access to this vital information. This can be particularly useful in emergencies, enabling you to stay ahead of the weather and make informed decisions.

The Emergency Alert System (EAS)

EAS plays a similar role in the modern world. It's a national public warning system that allows the President and local authorities to communicate emergency information to the public via television, radio, and other communication devices.

Your Baofeng radio can receive these EAS alerts when tuned to an appropriate frequency. This ensures you're promptly informed of local or national emergencies, enabling you to take appropriate action.

Install a Base Station

Baofeng radios, known for their versatility and affordability, can be an excellent choice for establishing a base station radio setup. Unlike their portable counterparts, base station radios are stationary units, typically placed on a desktop, and often include an integrated power supply within a frame alongside an external antenna. While you sacrifice portability, you gain significant advantages in power and range. Typically, these radios boast transmit power levels between 25 to 50 watts. Installing an outdoor antenna as high as possible is advisable to maximize your communication reach. However, if your range needs are modest, a simpler solution, like a mag-mount mobile antenna on a metal surface, such as a file cabinet, can suffice.

Installing a Baofeng base station at your home or bug-out location provides a steadfast communication hub, particularly critical if the power grid fails. You can effectively connect with aid and resources by operating on simplex or repeater frequencies covering 20–50 miles. Moreover, ensuring your base station is equipped with battery backup power is crucial for maintaining communication during power outages. By understanding these essentials, beginners can set up a reliable and efficient base station, leveraging Baofeng's capabilities for robust local communication.

Waterproof Your Handheld

Consider a waterproof carry case or plastic sealing bag for your Baofeng to protect it from water damage if caught out during flooding or storms. Ensure any accessories like batteries and microphones are likewise protected from moisture ingress.

The last chapter on maintenance includes what to do and how to care for your wet radio.

One of the benefits of Baofengs is their price. Being very affordable, you can, and should, often buy several and keep some in reserve as backups.

Hold Practice Drills

Schedule periodic emergency drills to rehearse your family's, neighborhood's, or team's disaster communication plan. Identify any equipment, training, or procedural knowledge gaps to address before an actual crisis. Consistent practice cements skills and readiness.

A question to ask yourself: Does your family know how to use your radio? Or are you the bottleneck and gatekeeper? If you get hurt, can anyone at your home use the radio to call for help or establish contact with others?

Learn and Grow

Expand your knowledge through local ham radio club workshops, online training programs, and FEMA emergency communication courses. Consider training for volunteer emergency roles like Community Emergency Response Teams (CERT), which provide crucial radio support.

- CERT (Community Emergency Response Team): CERT is a program that educates volunteers about disaster preparedness for hazards that may impact their area. It trains them in basic disaster response skills, such as fire safety, light search and rescue, team organization, and disaster medical operations. CERT aims to enable volunteers to assist others in their neighborhood or workplace following an event when professional responders are not immediately available to help.

The easiest way to find a local CERT team is to type 'Find Local CERT' into Google. Then click on the FEMA website link, which will bring you to the exact page you need.

Hardening Your Station

Professional-grade station surge suppressors protect your radio equipment against lightning strikes and power surges. Consider shielding from solar flares and EMP (Electromagnetic Pulse - a burst of electromagnetic radiation from certain natural occurrences like lightning or man-made events such as a nuclear explosion) with simple Faraday solutions like conductive mesh bags. Redundancy is key. Here are some more detailed steps on hardening your amateur radio station:

- Install whole-house surge suppressors or lightning arresters to protect all equipment. Look for units with high joule ratings.

- Use coax surge protectors rated for DC-1500 MHz on antenna feedlines (also known as transmission lines) to bleed static charges. These play a crucial role in connecting the radio transmitter and receiver to the antenna; feedlines carry radio frequency (RF) signals between the radio equipment and the antenna with minimal energy loss. Replace them periodically.

It's time for a quick deep dive on feedlines and bleeding static charges.

- **Static Charges on Antennas**: Antennas, especially those outdoors, are prone to accumulating static electricity. This can be due to various environmental factors like wind, rain, or even the natural build-up of static in the atmosphere. Static charges can become problematic because they can create noise in the receiver, potentially damaging the radio's sensitive electronic components.

- **Bleeding Static Charges**: To protect the radio equipment, it's crucial to have a method of safely discharging these static charges. This process is often referred to as "bleeding" static charges. It provides a path for the static electricity to discharge safely to the ground without passing through the radio's circuits.

- **Implementation**: This can be achieved by grounding the antenna system. Grounding involves connecting the metallic parts of the antenna and feedline system to the earth, creating a direct path for static charges to flow into the ground. Proper grounding not only helps in bleeding off static electricity but also enhances the overall performance of the antenna and provides a safety mechanism against potential lightning strikes.

    - Plug all station accessories like tuners, amplifiers, and computers into power strips with built-in surge protection.

    - Unplug antennas and power when not in use to minimize the chances of damage—disconnect feedlines during storms.

    - Store radios in conductive mesh Faraday bags when not in use to shield them from electromagnetic pulses.

    - Build a low-cost Faraday cage from wire mesh and ground it appropriately to store backup gear.

It's time for a quick deep dive into Faraday cages.

- A Faraday cage is a protective enclosure made from conductive material, such as metal, which shields its contents from external electromagnetic fields. Named after Michael Faraday, who invented it in 1836, this cage works by redistributing electromagnetic charges around its exterior, effectively neutralizing electromagnetic radiation like static, radio waves, and EMPs (Electromagnetic Pulses) inside the cage. It is commonly constructed from a metal mesh, the size of which is smaller than the wavelength of the blocked radiation.

Other Tactics:

- Keep spare radios in Pelican cases (Pelican is a brand of rugged, durable cases designed to protect sensitive equipment from various kinds of damage, including impacts, dust, and water. These cases are widely known for their strength and reliability) lined with copper tape to create portable shielded boxes.
- Use surge-protected power strips and 12V DC power options for off-grid flexibility.
- Install gas tubes or spark gap arresters on antennas tuned to shortwave bands for added protection.
    - Gas tubes and spark gap arresters are surge protectors that safeguard electronic equipment from high-voltage surges. Gas tube arresters have a gas-filled tube that becomes conductive under high voltage, diverting the surge to ground, while spark gap arresters use two electrodes with an air gap, where a surge creates a spark to discharge excess voltage. Both types quickly respond to voltage spikes.
- Follow proper grounding procedures for all station equipment and consider multiple grounding rods.
- Check your homeowner's insurance policy regarding amateur radio equipment coverage. If a disaster hits and society doesn't collapse, you at least get your damaged equipment reimbursed by the insurance company.

Analog Backup

While convenient, strictly digital modes may be disrupted in disasters. Maintain analog capabilities like AM/FM transmission/reception for redundancy. You may be using analog radio without even realizing it. If you like to listen to the radio in your car, any FM station transmits audio to you via the analog system. Broadband antennas and older transceivers provide backup utility. Here are some more detailed steps for maintaining analog backup capabilities:

- Keep spare antenna parts like wire, insulators, and baluns (BALanced to UNbalanced: Baluns are devices that convert between balanced and unbalanced signals in radio; they are crucial for minimizing interference and matching impedance. They are used in connecting antennas to radios) to construct emergency broadband antennas. Ladder lines, long wires, and dipole antennas can cover large frequency ranges.
- Set aside basic analog equipment for backup, such as AM/FM portable radios and transceivers. Look for self-powered and weather-resistant models.
- Learn to tune and operate vintage analog transmitters like tube-based AM (Amplitude Modulation) and SSB (Single Sideband) equipment. These can transmit voice with just power and a wire antenna.

AM and SSB are two different methods of modulating radio signals for voice communication in ham radio. AM is simpler but less efficient, while SSB, by transmitting only a single sideband, offers greater efficiency and is better suited for long-range communication.

- Stock spare tubes, crystals, dynamotors, and other analog parts that may be hard to source later. Learn how to service and repair old gear.
- Practice using CW (Continuous Wave) and Morse code skills on HF bands. In the worst case, CW may work when voice modes fail.

CW is a mode of communication that uses a continuous wave signal to transmit information, typically in the form of Morse code.

Morse code is a system of encoding textual characters as standardized sequences of two different signal durations, known as dots and dashes.

In summary, CW is the method or mode used in radio communication, often employing Morse code as the language for transmitting messages.

- Establish an emergency cache of batteries, generators, fuel, and solar power options to keep analog stations transmitting if the grid goes down.
- Consider getting licensed for shortwave bands to transmit international AM/SSB signals during crises.
- Drill with local radio groups to practice passing voice and Morse code messages using vintage analog equipment. Build skills and experience.
- Become an Elmer, an affectionate term for an experienced amateur radio operator who mentors, guides, or assists newcomers to the hobby. Document your expertise to act as a resource for others.

Maintain Your Gear

Check emergency kit supplies for expiration or deterioration twice annually. Inspect antennas and cables for damage. Cycle batteries monthly. Keep all equipment ready for immediate use when required.

That was a lot! Don't worry about completing all of it right now. Focus on the basics and the easy things first. As you learn and gain experience, you can advance to the higher-level preparations.

Establishing Effective Communication During Emergencies

When normal communication channels fail in a crisis, your Baofeng radio becomes invaluable for reliable updates and coordination. However, ineffective use can hinder response, so follow emergency communication best practices.

Listen More Than Transmit

During disasters, designated radio frequencies carry critical traffic from responders and agencies. Keep transmissions brief and focused only on vital information. Avoid tying up the channel with casual chatter. Listen attentively to stay continuously updated on the evolving situation.

Remain Calm and Clear

Speak slowly, calmly, and clearly. Take your time with your transmission despite the urgency. Identify yourself and your location or situation. Repeat critical information to confirm receipt. In high-stress situations, a measured delivery ensures your message is understood.

Conserve Battery Power

In prolonged emergencies, conserve handheld radio battery power through judicious transmit times, minimal standby monitoring, and low audio volume. Have spare charged battery packs and emergency chargers readily available. Manually turn off the radio to avoid draining batteries in standby mode. But periodically check emergency channels for critical updates.

Avoid Overusing Alert Tones

The emergency alert tone on Baofeng radios should only be used for immediate, life-threatening crises requiring urgent response. Do not use this alert for non-critical issues, leading to alert fatigue. Always cancel the alert once the escalation call is acknowledged.

Maintaining Your Baofeng Radio in Severe Conditions

Emergencies often entail operating radios in harsh environments very different from everyday conditions. Situations like storms, floods, fires, or extreme cold present unique challenges for maintaining and protecting your equipment. Let's explore valuable tips for keeping Baofeng handhelds functioning in demanding conditions.

Water and Moisture Resistance

Floods, storms, and heavy rainfall can expose radios to soakings that risk internal corrosion and electrical shorts. There are several protective measures you can take:

- Use a waterproof case or dry bag to shield the radio, securing it well with o-ring gaskets. Some purpose-built radio cases even allow the operation of the controls through a transparent membrane (*Baofeng UV-5R+ Plus Review*, 2022).
- For lightweight protection, seal the radio in a plastic bag or food wrap, securing the open end with a rubber band. However, this won't permit access to the controls (*Baofeng UV-82HP vs UV-5R*, 2022).
- Apply corrosion-inhibiting gel or grease on any exposed metal plugs and contacts if water intrusion is expected. Avoid greasing plastic parts (*Storage of a Baofeng UV-5R*, n.d.).
- Use a speaker microphone accessory to keep the radio in a bag or case while allowing push-to-talk operation. Just ensure the accessory jack is well sealed (Baofener, 2023a).
- If your radio does get wet, remove the battery immediately and allow the unit to dry thoroughly for at least 24 hours before reconnecting the power (Baofener, 2023b).

Dust and Dirt Resistance

Mud, dust, and sand storms can wreak havoc on radio joints, buttons, and exposed jacks. To protect from gritty grime:

- Start with a waterproof case, sealed plastic bag, or quality holster as your first line of defense (Twowayradiogear, 2014).
- Periodically blow out any particles from crevices using short bursts of compressed air (*How to Preserve Your Two Way Radios*, n.d.).
- Use rubbing alcohol and cotton swabs to remove dirt from tight spots, avoiding harsh scrubbing (*How to Clean and Care for Your Two-Way Radio*, n.d.).
- After severe grime exposure, disassemble the radio following the manufacturer's steps and thoroughly clean all components (Twowayradiogear, 2014).

Temperature Extremes

Both hot and cold temperature extremes can impact handheld radio function. To mitigate the damage:

- In hot weather, store the radio in the shade and use passive cooling methods, such as placing it near cool running water. Actively cooling an overheated radio can cause further problems from condensation.
- In cold, keep the radio insulated as much as possible and ideally close to your body heat. Avoid charging batteries in extreme cold.
- Whether hot or cold, keep the radio off when not in actual use to prevent compounding any temperature damage. The unit will assume ambient conditions.

- In vehicles, avoid direct sunlight exposure on the dash and use shade, tinting, or reflective barriers to prevent excessive heat buildup.Select lithium-ion replacement batteries when available, as they handle temperature fluctuations better than NiMH batteries.

Care During Power Outages

When normal electrical service is disrupted, take precautions when charging handheld radios to avoid voltage spikes or surges when power is restored:

- Avoid using uninterruptible power supplies (UPS) or standalone generators for charging. Voltage regulation may be poor.
- If you are charging the radio using a vehicle, disconnect the charger when the engine starts to avoid peak alternator output.
- When power is restored, wait 5–10 minutes before reconnecting any charger, allowing grid voltage to stabilize.
- Select power strips or chargers with built-in surge suppression specifically rated for electronics.

Handling Corrosion Issues

Remove the batteries immediately if your radio is exposed to moisture during use, and dry the radio thoroughly with a microfiber cloth. Do not use hair dryers or other heat sources, as they can damage the internals. If humidity is an issue, place moisture-absorbing packs in the storage box.

Check for any signs of corrosion like discoloration, pitting, or deposits around metal components and solder joints. Corrosion inhibitor gels and contact sprays can protect susceptible components. If corrosion is severe, replace affected parts.

Storage Best Practices

- Store handheld radios in impact-resistant, waterproof cases with padding to protect them from physical damage. Avoid fabric bags that allow moisture ingress.
- Remove batteries from the radio and charger before storage to prevent battery drain or acid leaks.
- Place several silica gel desiccant packs in the storage container to actively absorb moisture from the air and prevent humidity damage.
- Avoid storing radios in non-climate-controlled environments with extreme temperature and humidity fluctuations like attics or sheds.

Anticipating the harsh conditions likely present during emergencies allows you to properly prepare your Baofeng radio to withstand the elements. Protecting the unit from moisture, grit, and temperature extremes keeps it functioning when needed. Follow preventative maintenance steps; your hearty handheld will provide reliable communication through any challenges ahead.

From the submerged streets of New Orleans to the fire-ravaged landscapes of California to the earthquake-shattered towns of Nepal, Baofeng radios have repeatedly proven their mettle. They've shown that in times of crisis, when all else fails, communication endures. And as you hold your Baofeng radio, know that you own more than just a device. You have a lifeline, a beacon, a voice in the wilderness, ready to spring into action when duty calls.

Now, let's continue to delve deeper into the world of Baofeng radios. In the upcoming chapter, we'll explore how you can use your Baofeng radio in your everyday life. You'll discover how your Baofeng radio can enhance daily communication, from neighborhood watch programs to local ham radio clubs. So, stay tuned, and let's continue this together.

'The Command Packet' includes a Morse code cheat sheet and the 3-3-3 Radio Plan; activated during SHTF events.

Chapter 9: Baofeng Unleashed - Everyday Real-World Communication

THE ACTUAL TEST OF your Baofeng radio mastery is more than just understanding its functions; it's about applying those skills in real-world scenarios where effective communication can be a lifesaver. This chapter focuses on practical applications of your Baofeng radio, especially in situations that preppers often encounter.

You'll learn how to use your radio for outdoor activities like hiking and camping, where reliable communication can keep you safe and connected. The chapter also covers using Baofeng radios for event coordination and off-grid communication, essential skills for managing community events, or navigating through areas without cellular coverage. These real-life applications of Baofeng radios enhance your outdoor experiences and prepare you for handling emergencies with confidence and expertise.

Communicating in the Great Outdoors

Baofeng radios are ideal for communicating during outdoor activities where cell phone coverage is unreliable or nonexistent. Their simplicity, durability, and flexible power options make them well-suited for treks into the wilderness. Let's explore typical usage scenarios where a Baofeng can keep you connected.

'The Command Packet,' which includes 10 radio operational checklists for real-life scenarios to ensure your radio is always ready and functional.

Hiking and Backpacking

For group hiking trips, especially in remote locations, a Baofeng radio is invaluable for keeping team members coordinated over potentially vast distances along the trail. Assign a communications lead to monitor critical frequencies. Group members can provide periodic check-ins regarding their status and location. If anyone strays off course, the radio allows them to be quickly redirected. In case of injuries, it summons help rapidly. The radios also enable coordinating meetups, water/food drops, and pickups. If caught in storms, getting real-time weather alerts from NOAA frequencies can help the group take appropriate precautions (*Best Baofeng Radios*, 2022). In the quiet of the wilderness, the crackle of their Baofeng radios is a comforting reminder of the connection with your group.

Camping

When camping off-the-grid, beyond cell phone service, Baofengs are ideal for keeping fellow campers coordinated, especially after dark when moving around the site. They operate like supercharged walkie-talkies amongst family or friends. Quick communications allow for checking that the perimeter is secured, finding companions who've wandered off, and requesting supplies from others at camp. Monitoring NOAA weather channels provides advance alerts for any severe storms approaching, giving time to batten down gear and take shelter. The radio can summon emergency help quickly should someone face a medical event or threatening wildlife emerge. Or they can be used for just good old entertainment. Tune into local FM stations to enjoy music, news, or radio shows during your outdoor adventures. The versatility and range of a Baofeng radio transforms camping connectivity and safety.

Fishing

On remote fishing trips, a handheld radio provides vital benefits. Frequently scanning weather frequencies keeps the anglers updated on any systems moving in that might cut the excursion short. Voice-activated transmission (VOX) allows quickly radioing others hands-free when reeling in a catch. The emergency tone feature rapidly signals the group if a participant is injured or overboard. And in the worst case of an engine failure at sea, the radio may become a lifeline to call for help when miles from shore and cell coverage. The importance of a Baofeng radio on open water cannot be overstated.

Hunting

For hunters venturing into the wild, a radio is equally crucial. Keeping an ear on NOAA weather channels provides advance notice to exit the field before storms strike. The emergency tone capability rapidly summons aid should a hunter become lost or injured in rough terrain. Voice activation allows quietly transmitting to companions if game is spotted without fumbling for a handset. Some radios even have silent vibration modes to avoid startling wildlife. And if recovering a downed deer or hauling equipment, quick coordination with others via the radio brings essential assistance. For safe and successful hunts, a radio is indispensable gear.

Motorsports

In the context of motorsports, Baofeng portables genuinely shine. Quick radio communication can save lives at remote race events like desert rallies, or motocross held miles from cellular service. An accident that immobilizes a vehicle can leave racers stranded and vulnerable. A Baofeng allows contacting event coordinators immediately to dispatch medical responders before a situation becomes dire. Or if a race vehicle breaks down out of sight, the radio can guide support crews to the disabled driver. Two-way radios are long-standing saviors in motorsports, and Baofeng models deliver exceptional range and reliability.

In summary, whether deep in the mountains, miles offshore, or racing across the desert, exploring the great outdoors poses risks beyond cell phone service. However, a properly utilized Baofengs provides a flexible emergency lifeline. With their durability, excellent range, and ease of use, these radios empower users with confidence-inspiring communications when adventuring off-grid. So, make a Baofeng mandatory gear for your next foray beyond the pavement. Their importance is immeasurable if an emergency arises.

Using Your Radio for Event Coordination

For events large and small, Baofeng radios provide an indispensable tool for communication and coordination between staff spread over a wide area. Let's explore typical use cases where these versatile radios shine for events.

Neighborhood Watch Program

In a quiet neighborhood where the most exciting event is the weekly book club meeting, the arrival of a Baofeng radio sparked a small revolution. When a spate of burglaries shook the peace of this tranquil community, the residents decided to take action. They formed a neighborhood watch program, and their Baofeng radios became their most trusted tool.

The radios facilitated real-time communication among residents, from coordinating patrols to reporting suspicious activities. The local frequency became a hub of community coordination, a testament to the spirit of camaraderie and vigilance. The burglaries dropped, and the peace was restored. The residents continued to use their Baofeng radios, not just for the neighborhood watch but for coordinating community events, sharing local news, and sometimes, just for a friendly chat.

Local Ham Radio Club

In a bustling city, a group of radio enthusiasts found solace in the airwaves. They formed a local ham radio club, and their Baofeng radios were their tickets to this exclusive fellowship. The club members used their radios to learn from each other, share their experiences, and revel in the joy of radio communication.

They organized 'fox hunts,' in which a transmitter is hidden, and members use their Baofeng radios to locate it. They also conduct workshops to help new members get their licenses and master their radios. They even set up a station to communicate with astronauts aboard the International Space Station! For this radio club, their Baofeng radios are more than just devices; they are keys to a world of endless learning and exploration.

Local Community Events

For small local events like school fundraisers, fun runs, and street fairs, handhelds allow rapid communication between event organizers and volunteers. Channels can be pre-programmed for key roles—ticket sales, parking coordination, setup/teardown crews, etc. Requests can be made over the radios if extra staffing is needed in the ticket booth or parking. For widely dispersed events, radio contact keeps everything running smoothly.

Festivals and Concerts

At large music festivals spanning extensive grounds, reliable communication is vital yet challenging. Cell networks quickly become overloaded with thousands of attendees, so radio is essential. Baofengs allow staff to remain connected across the site. If the water station needs supplies, the medical tent requires an ambulance, or gates require more security—urgent assistance can be summoned quickly over the radio. Channels help sort important communications, like coordinating artists and VIPs discreetly via dedicated security channels.

Marathons and Races

Marathons and races pose similar challenges, with thousands of participants spread over a lengthy course. Spotters and medical personnel along the route can radio in any developing situation. Race officials communicate progress, lap counts, and pack positions. Baofengs prove perfect for the mobile nature of races, keeping coordinators connected across long distances. Reliable communication is crucial if any emergencies arise.

Rural Events

Cell coverage is often nonexistent in remote rural locations. However, Baofengs provide clear communication regardless of terrain or cell congestion. Radio is the only way staff can efficiently coordinate and respond to issues for outdoor events like cattle shows, air shows, or motocross rallies in the sparse country. Emergency services also utilize the radio frequencies to provide urgent medical care when required.

Corporate Events

Seamless coordination ensures perfect execution at highly choreographed corporate events and conferences with key executives in attendance. Baofengs enable staff and technical directors to flawlessly sync speeches, videos, and lighting cues. Security personnel can also discreetly communicate threats if issues arise. These robust radios operate reliably even in large buildings and exhibition halls.

Transportation Support

Radios are indispensable at transportation hubs like bus depots and airports. Staff can use them to redirect passengers, dispatch buses efficiently during peak loads, report mechanical problems, and request maintenance crews rapidly. Radio contact between drivers and dispatchers keeps operations moving smoothly for widely dispersed parking lots and transit centers. Reliable communication is essential for airports and transit systems.

Emergency/Contingency Planning

While festivals and events are joyous, emergencies can arise from weather, accidents, or other unforeseen issues. Event planners prepare contingency protocols using radio to coordinate urgent responses. Channels are pre-allocated for emergency services, medical teams, security, and evacuation decision-makers. Radios allow swift deployment of contingency plans even if normal communications channels become overloaded or fail.

In summary, smooth coordination underlies every successful event, whether an intimate gathering or a massive concert.

Baofeng radios enable robust and reliable contact between planners and staff, even in remote locales or congested environments. For rapid response to changing needs, contingencies, and safety, professional two-way radios prove indispensable tools behind the scenes of events worldwide.

Off-Grid and Survival Situations

Baofeng radios truly shine during crises when standard infrastructure fails. Being able to communicate off-grid can save lives. Let's explore some scenarios where dependable handheld radios prove invaluable for resilience.

Wilderness Expeditions & Survival

Envision a scenario where you're hiking, injured, and without a cell signal. In such a moment, a Baofeng radio enables you to reach Search and Rescue teams or connect with your group for guidance. For outdoor adventurers, a reliable handheld radio like the Baofeng is an essential safety tool, providing communication where cell networks don't.

While satellite phones are helpful in remote areas, affordable, portable radios like Baofengs are vital for team communication and emergency coordination. They offer advanced communication capabilities, ensuring you're prepared for any situation. Regular training with your Baofeng makes you a valuable asset to your community in emergencies, turning you from a mere user into a crucial part of a larger support network.

Severe Weather Events

When violent storms, flooding, or fires disrupt everyday communications, amateur radio fills the void. Hams band together to relay status reports and call for assistance. During Hurricane Katrina and many other disasters, ham radio was the only consistently functioning communication system. Baofengs and other handhelds are perfect for rapidly deploying ad-hoc networks. Your skills could save lives by summoning rescues for stranded victims.

Homesteading and Off-Grid Living

For modern homesteaders living sustainably off-grid, especially in remote regions, Baofeng radios have become a crucial tool, offering a vital link to surrounding communities. These radios enable the sharing of news and organizing assistance. Still, they are also pivotal in monitoring threats and facilitating bartering, providing a reliable form of communication that is indispensable in areas where landline phones are vulnerable to weather disruptions.

Baofeng handhelds and base stations offer an affordable, hardened redundancy that off-grid groups like preppers widely recognize as essential gear for self-reliance. When traditional communication networks are down in emergencies, Baofeng radios serve as a critical lifeline, allowing homesteaders to receive updates on natural disasters, seek help, and coordinate community responses. They also facilitate regular check-ins with neighbors, enhancing local networking and community building, essential for remote living.

One of these radios' most crucial uses is monitoring weather and environmental conditions. Access to real-time weather updates and alerts is vital for planning and preparedness, especially in areas prone to severe weather conditions. For those involved in agriculture, Baofeng radios enable efficient coordination across large regions, which is essential for coordinating harvests and managing livestock.

Additionally, these radios play a significant role in security and surveillance, providing a means for homesteaders to communicate instantly in case of suspicious activities or threats. They are also used as a tool for educational purposes, facilitating the exchange of information and skills crucial for sustainable living.

The technical advantages of Baofeng radios are particularly suited to the needs of homesteaders. Known for their durability and reliability, they are well-equipped to handle the demanding conditions of off-grid living. With extended battery options, these radios ensure communication lines stay open, even during power outages. Users can also program specific frequencies for private group communication, enhancing privacy and reducing interference.

Furthermore, the range of these radios can be extended with external antennas and repeaters, a vital feature for use in sprawling rural areas. Their affordability makes them a cost-effective communication solution, lowering the entry barrier for homesteaders and preppers. Additionally, Baofeng radios are user-friendly and require minimal technical knowledge for basic operations, ensuring they are accessible to many users.

Baofeng radios are more than just communication tools for the modern homesteader and prepper; they are a gateway to a self-reliant, resilient lifestyle. Offering versatility, reliability, and accessibility, they are a wise investment for anyone seeking to establish a sustainable, connected homestead. Their practical applications in emergency, agricultural, and community settings and their technical benefits make them indispensable to off-grid living.

Societal Collapse

In a societal collapse scenario, Baofeng radios emerge as invaluable tools for survival and rebuilding communities. Traditional communication infrastructures will likely be disrupted or utterly nonfunctional during such tumultuous times. In this context, Baofeng radios offer a reliable means for maintaining communication, which is crucial for safety, coordination, and community resilience.

Baofeng handhelds and base stations have become pivotal in establishing emergency communication networks. They enable individuals and groups to share critical information, such as safety alerts and resource locations, and coordinate group efforts in response to ongoing challenges. When timely information can mean the difference between safety and danger, these radios provide a lifeline.

The radios' versatility in accessing a wide range of frequencies is particularly beneficial. Users can tap into various channels to gather news, connect with others, or set up localized networks for community defense and mutual aid. Their ability to operate on VHF and UHF frequencies allows for local and extended-range communications, which is essential in a fragmented society.

Regarding logistics and resource management, Baofeng radios facilitate organizing essential services such as food distribution, medical aid, and shelter management. They enable groups to coordinate scavenging missions, communicate about safe and dangerous areas, and organize trading or bartering systems necessary for survival.

From a security standpoint, these radios are crucial for setting up surveillance and alert systems. In a societal collapse, safeguarding communities against various threats is paramount. Baofeng radios provide a means for constant vigilance and quick response, enhancing collective security.

The rugged and durable nature of Baofeng radios ensures their reliability in harsh conditions. Coupled with their long battery life and the option for rechargeable batteries or alternative charging methods (like solar chargers), they are well-suited for prolonged use in scenarios where conventional power sources are unavailable.

Moreover, their affordability and ease of use make them accessible to a broader population. This accessibility is crucial in a societal collapse, where complex or expensive communication solutions would be impractical or unattainable for most people.

In the face of societal collapse, Baofeng radios are essential for survival, organization, and community resilience. Their ability to facilitate reliable communication, coordination, and security in a world where traditional infrastructures have failed makes them indispensable in navigating and adapting to the new challenges of a drastically changed world. Their technical robustness, user-friendliness, and affordability ensure they can be widely adopted to rebuild and reconnect fragmented communities.

Emergency Radio Groups

Should catastrophes ever disrupt normal society, resilient communication will prove critical. Amateur radio groups like ARES, RACES, and CERT drill extensively for these emergency scenarios.

- **ARES (Amateur Radio Emergency Service)**: ARES is a program run by the American Radio Relay League (ARRL). It consists of licensed amateur radio operators who volunteer to support emergency communications during disasters and public service events. ARES members train to offer their skills in radio communications to assist local authorities and emergency response agencies when regular communication systems fail or are overloaded.

- **RACES (Radio Amateur Civil Emergency Service)**: RACES is a public service provided by amateur radio operators, established by the Federal Emergency Management Agency (FEMA) and the Federal Communications Commission (FCC). In a national emergency, RACES volunteers provide essential communications and are activated by local, county, or state emergency management agencies. Unlike ARES, which can be activated for a broader range of public service events, RACES is strictly limited to emergency or disaster situations where everyday communications are unavailable.

Rapidly deploying radio nets using optimized equipment like Baofeng portables enables community coordination for safety, news, and resources. Hams become the communication backbone, recovering order from chaos. Should the unthinkable occur, dedicated preppers equip themselves with radios, knowledge, and training to assist others.

Your Baofeng radio is more than just a radio; it's a versatile tool, ready to assist you in various aspects of your daily life. So, whether you're chatting with your neighbor, tracking a storm, exploring the great outdoors, or volunteering at a local event, your Baofeng radio will be right there with you, enhancing your experiences and ensuring you're never truly alone.

Now that you've mastered operation, features, and real-world applications for your Baofeng radio, let's focus on keeping it running smoothly long-term. The key to reliability is proper maintenance and knowing how to troubleshoot issues when they arise.

In our final chapter, we'll cover essential topics like care, cleaning, accessories, and troubleshooting common problems with your radio. Proper upkeep will ensure your Baofeng provides years of flawless performance when needed.

Chapter 10: Field Guide - Maintenance, Care, and Problem Solving

In the world of prepping and preparedness, your Baofeng radio is not just a tool but a critical asset for communication in times of crisis. Understanding proper maintenance and troubleshooting is essential to ensure its reliability when you need it most. This chapter is dedicated to empowering you with the knowledge to keep your Baofeng radio in top working condition. You'll learn the ins and outs of regular upkeep so your radio remains operational in various conditions. We'll also delve into common issues and their solutions, equipping you with the skills to diagnose and fix problems quickly.

Maintaining Your Baofeng Radio

Proper maintenance and care ensure your Baofeng radio delivers years of flawless performance. With the appropriate upkeep, these radios can serve reliably even in harsh conditions. We will cover essential maintenance best practices, including cleaning, storage, accessories care, and periodic refresh procedures.

Exterior Cleaning

Maintaining the exterior of your Baofeng radio is crucial for preserving its appearance and functionality. Over time, exposure to dirt, dust, and grime can lead to various issues, including:

- Blocked speaker holes, which can diminish audio quality.
- Buttons becoming stiff or sticky due to grime accumulation.
- Potential abrasion damage to the display if dust is left uncleaned.

To ensure your radio remains in top condition, follow these cleaning guidelines:

- Always turn off the radio and remove the battery before cleaning.
- Use a lightly dampened microfiber cloth with a bit of mild detergent. Be sure to wring out any excess water from the fabric.
- Gently clean all external surfaces, focusing on crevices, buttons, and the display area.
- Do not introduce moisture into sensitive areas such as the battery compartment, speaker holes, and microphone holes.
- After cleaning, use a dry cloth to remove moisture and let the battery air dry completely before reassembling.

- For tougher dirt, a small amount of isopropyl alcohol can be effective, but ensure it doesn't seep into the radio's internal components.
- Do not use harsh chemicals, solvents, bleach, or abrasive cleaners, as these can damage the radio.

Regular external cleaning will keep your radio looking new and prevent performance issues related to dirt and dust accumulation.

Antenna Care

Proper care of your radio's antenna is essential, as it is crucial in signal transmission and reception. Regular antenna maintenance ensures it functions at its best, providing optimal range and clear communication. When inspecting your antenna, particularly for Baofeng models like the UV-5R, pay attention to the following:

- Check the SMA male connector for any signs of oxidation or carbon buildup, which can occur due to exposure or aging. If present, gently clean the connector using isopropyl alcohol.
- Examine the antenna's rubber gasket. If it's cracked or deformed, replace it to maintain the antenna's water resistance.
- Check the metal antenna shaft for any physical damage, such as cuts, nicks, or dents. Damage like this can degrade the antenna's performance, so it's best to replace it in these cases.
- Inspect the antenna tip for carbon deposits, which might result from arcing. Carefully remove these using fine sandpaper.

Battery Maintenance

Proper battery care practices can significantly extend the battery's runtime and overall lifecycle. Here's how to get the most out of your Baofeng radio's battery:

- Always use manufacturer-approved Li-ion battery packs. Refrain from using cheaper, non-approved clones to ensure safety and efficiency.
- Regularly inspect the battery for any signs of damage, such as swelling, and safely dispose of any compromised batteries.
- Keep the contact terminals clean to prevent corrosion. Gently wipe them with alcohol periodically.
- To prolong battery life, avoid completely draining the battery before recharging.
- When not using the battery, remove it from the radio and store it separately.
- Charge the battery in moderate temperatures. Avoid charging in extremely hot or cold environments like an uninsulated garage or shed.
- Occasionally, perform a complete discharge and recharge cycle. This helps to calibrate the battery and can maintain its capacity.

Storage Best Practices

Storing your Baofeng radio correctly when it's not in use is crucial for preserving its condition and extending its service life. Proper storage methods can prevent damage and maintain the radio and its batteries in optimal condition (*Baofeng Cheat Sheet,* n.d.). Here are essential practices to follow:

- Store the radio in a cool, dry place, away from extreme temperatures. High-heat areas, like vehicles, should be avoided.
- Prefer indoor storage to protect the radio from dust and moisture, which can be harmful if exposed to temperature

or moisture changes, for example, in a shed or non-temperature control garage.

- Avoid prolonged exposure to direct sunlight, as it can damage the radio's surfaces and compromise its seals.
- Store the radio upright, in its intended orientation, rather than laying it flat. This helps prevent potential damage to its buttons and other components.
- For long-term storage, remove the batteries from the radio. This practice helps maintain the battery's health and prevents potential leakage or damage.

Waterproofing Considerations

Ensuring your Baofeng radio's protection against water is essential, especially since most models are not inherently designed to be waterproof. Even though some high-end models may offer water resistance, taking additional measures to safeguard your device in wet conditions is crucial. Here are some effective waterproofing strategies:

- Utilize a waterproof carry case or bag when operating in moist environments. Many cases are designed to allow access to the radio's buttons and display while providing protection.
- Consider applying waterproofing sprays to the radio's exterior for temporary defense against light rain or splashes. Remember to reapply the spray following any exposure to water.
- After using the radio near or in saltwater, such as during boating or coastal activities, rinse it with fresh water to prevent corrosion damage.
- Avoid submerging the radio in water unless enclosed in a case specifically designed for immersion.

By taking these precautions, you can significantly extend the life and maintain the functionality of your Baofeng radio, even in challenging environmental conditions.

Programming Backups

Maintaining the safety and integrity of your radio's programmed frequencies and settings is crucial, as they are stored in internal memory that could be susceptible to corruption or loss. Regular backups are an intelligent way to safeguard this vital data. Here's how you can efficiently back up and maintain the settings of your radio:

- **Backup Process:**
  - Connect your radio to a computer using the appropriate cable.
  - Utilize software like CHIRP to read and save the current configuration of memory channels and settings.
  - Consider storing backups in multiple locations, such as cloud storage or an external hard drive, for extra security.
- **Verifying Backup Data:**
  - After backing up the data, carefully review the saved channel information to ensure no errors or missing details.
  - Regular checks are essential to confirm the integrity and completeness of your backup.
- **Maintaining Backup Data:**
  - It's advisable to refresh your backup data annually or whenever you experience issues like dropped channels or other anomalies.
  - Keeping your backup current ensures you have saved the latest settings and configurations.

- **Restoration:**
    - In the event of data corruption or loss, having a recent backup allows you to swiftly restore your radio to its previous, optimal configuration.
    - This quick restoration is precious in field operations where reliability is paramount.

Adhering to these best practices for backup and maintenance will ensure that your radio's settings and frequencies are well-protected and easily recoverable. This proactive approach will save time in troubleshooting and ensure your radio is always ready for use with your preferred configurations.

Accessory Maintenance and Care

Maintaining and caring for your chargers ensures ongoing performance and reliability. Here's how you can effectively maintain these critical components:

Charger Maintenance:

- Regularly inspect the charger contacts for any build-up of grime or oxidation. If necessary, clean them using alcohol to ensure efficient charging.
- Avoid leaving the radio on the charger continuously after fully charging. Overcharging can significantly reduce the battery's lifespan.
- If the charger is damaged or its efficiency is reduced, replace it with an original equipment manufacturer (OEM) product to ensure compatibility and safety.

Drop-in Desktop Charging Station: A drop-in desktop multi-bay charging station can be incredibly convenient for those who frequently use their radios. Here are its key benefits:

- Allows for simultaneous charging of multiple radios and spare batteries, enhancing efficiency.
- Eliminates the need for constantly connecting and disconnecting individual chargers.
- The cradles in the station ensure secure docking and maintain proper contact for consistent charging.
- Indicator lights provide a quick and easy visual of the charging status of each bay. Such a charging station is a time-saver for heavy users, keeping batteries fully charged and ready for use. Common models include 6 or 12 bays catering to different usage needs.

Advanced Troubleshooting

Software Diagnosis

Advanced troubleshooting of your Baofeng radio can be conducted with programming software like CHIRP, especially for users comfortable with more technical aspects. This software offers a range of diagnostic tools and capabilities that can aid in identifying and resolving complex issues:

- Reading Radio Data: By accessing the radio's internal data, CHIRP can help identify issues like memory corruption or incorrectly stored frequencies, which can be crucial in pinpointing problems that aren't immediately obvious.
- Error Logs and Debug Modes: CHIRP often provides access to error logs and debug modes, offering more profound insight into any faults or anomalies your radio might be experiencing.
- Monitoring Signals: The software can monitor transmitting and receiving signals in real-time, which is invaluable

for on-the-spot debugging and ensuring optimal radio performance.

- Factory Reset: In cases where issues arise from incorrect custom programming, CHIRP can facilitate resetting the radio to its factory settings, often resolving such glitches.

For most users, basic troubleshooting methods are sufficient. However, programming software unlocks a higher diagnostic and analysis potential, allowing advanced users to delve deeper into their radio's functionality and address more complex issues. This advanced approach can be incredibly beneficial for maintaining the reliability and performance of your Baofeng.

Full Factory Reset

A full factory reset on your Baofeng radio can be necessary if you encounter persistent glitches or anomalies that standard troubleshooting doesn't fix. Here's a step-by-step guide to ensure a successful reset:

1. **Backup Radio Memory:**
    - If you want to retain your channel programming, back it up using computer software. This step ensures that your channel information is preserved for later restoration.
2. **Initiating Full Factory Reset:**
    - Press the MENU button on your Baofeng radio.
    - Use the arrow keys to navigate to the 'RESET' menu option. On most versions, it will be the very last Menu option.
    - Select the option for a full reset and confirm the action when prompted.
3. **Reprogramming the Radio:**
    - After the reset, reprogram your radio from scratch. Re-enter any saved channels and settings if necessary.
4. **Testing Post-Reset:**
    - Thoroughly test the radio to ensure that all issues have been resolved. Monitor its performance across all features before putting it back into regular use.

A full factory reset is often a last resort but can effectively resolve severe misconfigurations or programming bugs. This careful approach will help restore your radio to its optimal functioning state.

Key Preventative Maintenance Schedule

Maintaining your radio in top condition requires a proactive approach. Here's an easy-to-follow preventative maintenance schedule to help you stay ahead of potential issues:

Daily Checks Before Use:

- Conduct a visual inspection of your radio for any signs of damage.
- Ensure the battery is adequately charged and the charger is ready.

Monthly Care After Heavy Field Use:

- Give the radio a thorough external clean, focusing on the body, buttons, display, and connectors.
- Inspect the antenna connection and its gasket for any wear or damage.

- Clean the battery terminals if you notice any buildup.

Annual Maintenance:

- Perform a complete functionality check and a range test to verify the radio's performance.
- Back up and refresh your memory channel data to safeguard your settings.
- Go through a complete battery discharge and recharge cycle to optimize battery health.

As Required:

- Apply waterproofing treatments if you frequently use the radio in wet conditions.
- Update your programming software to the latest version for enhanced functionality.

By dedicating time to these primary care routines, you can identify and address minor issues before they escalate. Developing expertise in these maintenance procedures is a surefire way to prevent more significant, more complex problems down the line. Remember, even the most reliable equipment requires consistent care—don't overlook the importance of routine maintenance for seamless operation.

Understanding & Troubleshooting Common Issues

Understanding and troubleshooting common issues with your Baofeng can enhance your user experience and ensure longevity and preparedness. While technical glitches can be frustrating, most problems have straightforward solutions. This section will guide you through the most common issues users face with their radios and provide practical fixes.

Radio Not Powering On The inability to power on can be highly disruptive, but common causes include:

- Dead Battery: Verify the battery's charge with a spare pack. New batteries may need charging.
- Faulty Power Button: Inspect for dirty or damaged contacts if the battery is functional.
- Blown Fuse: Use a multimeter to check for continuity.
- Charging Circuit Issues: Faulty components can inhibit charging and power from the battery.
- Mainboard Failure: Internal regulator or power control issues may require professional repair.

Poor or No Reception: Reception problems often stem from antenna issues:

- Loose or Detached Antenna: Ensure it's fully screwed on for proper contact.
- Damaged Antenna: Inspect for physical damage and replace if necessary.
- Damaged Center Pin: Check for damage or corrosion in the antenna connector.
- Defective Receiver: The receiver module may need repair if the antenna is OK.

Battery Issues: Common battery-related problems include:

- Rapid Drainage: Regularly discharge and recharge to maintain battery health.
- Dirty or Faulty Contacts: Clean contacts to prevent power issues.
- Physical Damage: Check for swelling or leaks and replace damaged batteries.

- Counterfeit Batteries: Use only branded batteries for reliability.

Keypad and Button Issues: Wear and tear can lead to unresponsive buttons:

- Contamination: Clean around buttons to remove dirt.
- Stuck Contacts: Inspect and gently adjust the interior rubber membrane.
- Physical Damage: Repair or replace damaged components.

Sudden Loss of Transmit Ability: If transmit capability is suddenly lost, consider the following:

- Antenna Mismatch: Ensure antenna impedance compatibility.
- Faulty Power Amplifier: Test and replace if necessary.
- Internal Wiring Issues: Inspect coaxial cables for damage.
- Mainboard Problems: Check transmit-related components for faults.

Distorted or Reduced Volume: Audio issues can be caused by:

- Obstructed Speaker: Clean any blockages carefully.
- Low Battery Voltage: Replace with a fully charged battery.
- Damaged Speaker: Replace speakers with physical damage.
- Bad Solder Joints: Professional repair may be required for board issues.

Programming Difficulties: Challenges in programming can arise from:

- Outdated Software: Use the latest version of programming software.
- Cable Issues: Check for faulty connections and use quality cables.
- User Error: Review programming steps and data.
- Memory Corruption: Reset the radio to factory settings if necessary.

Repair Options: Consider various repair options before replacing your radio:

- Manufacturer Warranty: Utilize the warranty for eligible repairs.
- Local Repair Shops: Obtain an estimate for economical fixes.
- DIY Repairs: Simple replacements like antennas or batteries can be user-performed.
- Mail-In Service: For complex issues, consider specialized repair services.

Avoiding Future Problems: Preventive measures can help avoid common issues:

- Careful Handling: Use cases and straps to prevent physical damage.
- Moisture Prevention: Keep the radio dry, especially if it's not water-resistant.
- Proper Storage: Store in protective cases away from heavy objects.

- Regular Inspection: Check for wear and tear and replace parts proactively.
- Accessory Maintenance: Use only compatible, branded accessories.

By understanding these common issues and their solutions, you can swiftly address problems and ensure your Baofeng radio remains a reliable communication tool. Regular care and preventive measures will also contribute to its durability and functionality.

Seeking Troubleshooting Help

When facing complex technical issues with your Baofeng that surpasses your expertise or when special diagnostic tools are required, seeking external help can be invaluable. Here's how you can find assistance to address your radio problems effectively:

- **User Forums:** Online communities like the Baofeng subreddit or RadioReference forums are great resources. Here, fellow radio enthusiasts share troubleshooting tips and solutions. You can search for existing threads addressing similar issues or post your queries for personalized advice.
- **Local Repair Shops:** For intricate component-level problems, local electronics repair shops can be a lifesaver. They have specialized tools, like oscilloscopes, to diagnose issues precisely. Often, the cost of repairs is more economical than replacing your radio.
- **Technical Manuals:** Detailed service manuals designed for technicians offer comprehensive information on your radio. These manuals include schematics, standard measurement values, and guidance on disassembling your device for advanced troubleshooting.
- **Manufacturer Support:** If your radio is under warranty, contacting Baofeng's support can be beneficial. Be prepared to provide details like the model, a summary of the issue, and any troubleshooting steps you've already attempted.
- **YouTube Tutorials:** Numerous video tutorials cover common repairs and troubleshooting for specific Baofeng models. These visual guides can walk you through the process step-by-step, making complex fixes more approachable.

If you struggle with persistent radio issues, remember that a wealth of knowledge and assistance is available. From the collective wisdom of online forums and instructional videos to the expertise of professional repair services, these resources can help you effectively resolve even the most challenging problems. Don't hesitate to tap into these options to get your Baofeng radio back in optimal working condition.

Case Studies: Troubleshooting in Action

Overcoming Reception Issues in a Dense Forest

Consider a group of outdoor enthusiasts exploring the dense, verdant woods of the Pacific Northwest. Their Baofengs, strapped to their backpacks, are their lifelines, keeping them connected as they wander through the wilderness. But as they delve deeper into the forest, they notice a problem. The lush canopy overhead, though beautifully serene, is causing reception issues on their radios. Instead of succumbing to frustration, they take a pragmatic approach. They recall the troubleshooting tips they've learned and put them into action. They check their antennas, ensuring they are properly connected and not damaged. They tweak the squelch settings on their radios, filtering out background noise and improving signal clarity. Yet, the dense foliage remains a formidable adversary.

One of the group members, an experienced radio operator, suggests a simple yet effective solution - elevation. They find a clearing with a tall tree. One of them, an adept climber, volunteers to climb up, radio in hand. As she ascends, the reception starts to clear. The higher she climbs, the clearer the signal becomes. From her elevated perch, she is able to relay information to the rest of the group and successfully guide them through their woodland adventure. Alternatively, this group could have packed a portable antenna system. This could be a lightweight, portable antenna specifically designed for field operations. An example would be a collapsible or roll-up J-pole antenna made from lightweight materials. This type of antenna can be easily deployed, and offers improved reception over the standard rubber duck antennas typically found on handheld radios.

In addition, knowing if there are repeaters in the area is a viable option that should be tried.

Solving Battery Drain During a Power Outage

In a small town struck by a severe storm, the power grid has failed, plunging the townsfolk into darkness. Their Baofeng radios are their only means of connecting with the outside world. But as the hours turn into days, they notice their radio batteries draining rapidly. One resident, a seasoned radio operator, steps up to tackle the issue. He advises the townsfolk to turn off unnecessary features on their radios, such as the flashlight and the backlight on the display. He then teaches them how to set their radios to Battery Save Mode, which conserves power when the radio isn't actively receiving a signal.

Despite these measures, the battery drain continues. That's when the radio operator pulls out his secret weapon - a car charger adapter for Baofeng radios. Using the power from a car battery, he is able to recharge the radio batteries, ensuring they stay operational until power is restored. His quick thinking and practical knowledge save the day, keeping the lines of communication open during this crucial time. This is another perfect example of where your battery bank will keep you safe!

Fixing Programming Errors Before a Community Event

In a vibrant community preparing for an annual fair, Baofeng radios are buzzing with activity. The event organizers use the radios to coordinate preparations, assign tasks, and keep everyone updated. But as the event day approaches, they encounter a problem. Their attempts to program new channels into their radios for different event teams are met with errors and frustrations. Noticing their struggle, a member of the community steps forward. She's a radio enthusiast with a knack for troubleshooting. She looks at the radios and identifies the problem - a simple programming error. The organizers were trying to save the new channels in already occupied slots.

She guides them through deleting the existing channels before programming the new ones. She also shows them how to name the new channels according to their respective teams, making it easier for everyone. Thanks to her, the programming errors are quickly resolved, and the community event is a resounding success.

Addressing Audio Problems in a Search and Rescue Operation

A search and rescue team is on a mission high in the mountains. Their Baofeng radios are their lifelines, connecting them and their base camp. But as they ascend higher, they start experiencing audio problems on their radios. The volume is too low, and the sound is distorted, making communication difficult. One of the team members, an experienced radio operator, decides to troubleshoot the issue. He checks the volume settings on their radios, ensuring they are turned up. He inspects the speakers for any blockage or damage but finds none.

Then, he remembers a trick he learned. He suggests using an external speaker/mic. The team has a few in their gear meant to facilitate hands-free communication. They connect the speaker/mics to their radios, and instantly, the audio is louder and clearer. With the audio problems resolved, the team can communicate effectively, coordinate their efforts, and complete their rescue mission.

Navigating the airwaves with your Baofeng is like embarking on an exciting adventure. Each transmission is a step forward, each frequency a new path to explore. And like any adventure, there might be challenges along the way. But with every challenge comes an opportunity to learn, grow, and become a better radio operator. So, whether you're scaling the heights of a mountain, navigating the depths of a forest, powering through a blackout, or bringing a community together, your Baofeng radio will be right there with you, ready to conquer any challenge that comes your way.

Conclusion

Wrapping Up: The Journey So Far

As our exploration of the Baofeng radio universe draws to a close, it's time to reflect on the incredible journey we've embarked upon together. Our voyage has been enriching and enlightening, from the fundamental understanding of radio anatomy to mastering programming techniques and recognizing these devices' indispensable role in everyday and emergency scenarios. We began this adventure with a simple goal: to break through the complex world of Baofeng radios.

What we uncovered was far more profound-a realization of these radios' immense impact on our lives and safety. Beyond their technical capabilities, Baofeng radios have emerged as powerful instruments of connection and change, fostering community ties and serving as critical lifelines in times of need and disaster. These devices remind us that despite the many barriers in our world, the airwaves offer a universal platform for connection and understanding. They empower us as radio operators and active participants in a global dialogue, even when our phones fail us.

Keep Learning: Join the Baofeng Community

Our adventure doesn't conclude here. The Baofeng radio world is vast and continuously evolving. I encourage you to delve deeper, seek new knowledge, and continually expand your understanding. A wealth of resources is available, online communities, user groups, and advanced literature, that can further refine your proficiency and enjoyment of Baofeng radios. I warmly invite you to become an active member of the vibrant Baofeng community. Share your experiences, learn from fellow enthusiasts, and contribute to a global network of radio aficionados. Every transmission sent and received is a step towards thoroughly preparing your communications.

Signing Off

In these pages, we aimed to transform beginners into informed, skilled guerilla Baofeng users. From selecting the right model to mastering initial setups and advanced operations, our goal was to simplify and clarify, turning complexity into confidence. We've tackled key aspects like effective programming, achieving clear communication, navigating legalities, and delving into advanced features, equipping you for a fulfilling long-term engagement with your radio and the preparedness to survive.

Remember, this book is but an introduction to a lifelong journey. Continue to engage with the community, participate in forums, drill, attend events, and train with your radio regularly. Challenges are growth opportunities, and proficiency comes with practice. It has been an honor and a pleasure to introduce you to the fascinating world of Baofeng radios. I hope you feel empowered and curious enough to embark on or continue your journey. The world of two-way radio communication is vast and waiting for you.

Your real adventure begins now as a guerilla Baofeng user who has broken through the complexity secured their communications, and become prepared for disaster utilizing prepper tactics!

- QRV?

Bonus Section: A Beginner's Essential Guide - Your Next Steps

Congratulations! You made it through the book, but I'm guessing you still have questions. You still aren't 100% sure what to do next. That's exactly how I felt at this same stage. I wanted just a little more direction. So here you go! This cheat sheet provides a simplified roadmap to get you started on the right foot. Follow these steps to start moving from the beginning guerilla to Baofeng master.

Step 1: Purchase the Baofeng UV-5R Radio

The UV-5R is the classic starter model, renowned for its affordability, versatility, and tremendous user community support. I recommend buying it from an authorized dealer like BaoFengTech.com or Amazon to guarantee authenticity.

Some key benefits of the UV-5R for beginners (*Best Baofeng Radios*, 2022; Tobias, 2015):

- Dual-band support for VHF and UHF frequencies right out of the box. This allows you to access a wide range of amateur radio bands as you get started.
- Highly customizable via easy PC programming with CHIRP software. The UV-5R was one of the first accessible handhelds that opened programming for beginners through a computer.
- There is a vast online knowledge base and active user forums. As the most popular starter radio, every aspect of the UV-5R has been discussed and documented by the global community. This community provides troubleshooting support and hacks/mods to customize the radio as your skills progress.
- Key Benefit: Affordably priced between $25–50 depending on accessories bundled. This low cost allows anyone to get started without a significant investment. Which is a very important and often overlooked aspect for a beginner.

While imperfect, the UV-5R is built to last and, with proper care, will serve you well for years. Overall, it's the ideal starter radio.

Step 2: Obtain Your GMRS License

While the UV-5R can operate license-free on FRS frequencies, I highly recommend getting the GMRS license. This unlocks higher transmit power on GMRS frequencies for enhanced range.

Key features of this license:

- No test needed to obtain.
- Minimal fee of $35, which is valid for ten years.

- You will receive a GMRS call sign to use in your communications.

The minimal investment is well worth it. Your UV-5R can transmit up to 5 watts on GMRS repeaters and frequencies for better reach. This opens up your communication versatility—you can still use FRS for short-range neighborhood watch coordination, but switch to GMRS when you need more distance, like on hikes or remote events.

Follow this Process to Obtain Your GMRS License

- Obtain your 10-digit FCC Registration Number (FRN) by clicking on 'REGISTER' here: https://apps.fcc.gov/cores/userLogin.do
- Once you have your FRN, login to the FCC License Manager website: https://wireless2.fcc.gov/UlsEntry/licManager/login.jsp
- Click on the 'Apply for a New License' button.
- Then, select 'ZA – General Mobile Radio Service (GMRS).' It begins with a 'z,' so it will be near the bottom of the dropdown list.
- Complete the application and then submit your payment.
- Your new call sign will be assigned to you.

Step 3: Find Frequencies Relevant to You

Start with the basic FRS/GMRS national calling channels for interoperability with other users - program in all the 22 designated frequencies in the UHF band.

With your GMRS Callsign, register with *myGMRS.com* and find local repeaters in your area:

Finally, if you haven't done it already, find and program your local NOAA weather radio (refer to Chapter 4). This frequency provides weather and emergency alerts for your local area.

This gives you a basic group of frequencies, from which you will expand over time.

Step 4: Download CHIRP

Download It Here:

Use CHIRP to program the frequencies from step #3. Refer to Chapter 4 for step-by-step instructions.

Step 5: The Essential Accessories

Start with an extra battery for all-day operation, an aftermarket antenna upgrade for increased range, and a lapel microphone for convenient hands-free use. Quality accessories dramatically improve your experience (*Best Baofeng Radios*, 2022).

Specific recommendations

- Programming Cable: If your radio bundle didn't include a programming cable, I recommend buying one. Ensure it works with your model type.
- Extra battery: An 1800+ mAh high-capacity Li-ion replacement battery provides longer runtime. Avoid cheap clones. With batteries, always remember '2-is-1 and 1-is-none'. If you lose your battery, you also lose your radio. That's why I always say '3-is-for-me.'

- Aftermarket antenna: A compact Nagoya NA-771 15.6-inch upgraded antenna gives 2–3x the range over the stock antenna.
- Lapel microphone: Clip the mic near your mouth for clear voice pickup and hands-free operation. Look for models with an earphone.
- Carry case: A sturdy belt clip case protects your radio and keeps it handy while out and about.

These affordable add-ons truly elevate the usefulness of your UV-5R for outdoor activities and emergencies.

Step 6: Join a Baofeng and Ham Group

Connect with the passionate Baofeng user community. Stick with the basics. Find one local in-person group and then join a few Facebook groups. They provide troubleshooting support, inspiration for expanded learning, accessory recommendations, and a group of like-minded individuals.

Suggestions

- In-Person Local Clubs: Type "ham groups in my area" into Google.
    - Everyone has a website these days; this is a straightforward way to find local groups.
- Facebook: Type "Baofeng Radio" into Facebook groups and look at the hundreds that pop up. Some good ones that I'm involved in include:
    - Radio Preppers
    - USA Baofeng Owners
    - Baofeng Outlaws
    - Baofeng Owner's Club

Step 7: Practice & Drill Regularly

Nothing beats hands-on experience for developing skills and confidence operating your Baofeng. I recommend:

- Take your radio when hiking and camping for actual field use.
- Practice scanning all frequencies to determine active ones in your area. Try this at different times of the day and on different days of the week. Write down the frequencies.
- Use an online resource like www.repeaterbook.com to find local repeaters and practice hitting them from different locations.
- Develop a cheat sheet of frequently used settings like squelch, power, and bandwidth optimized for your common locations.
- Drill with emergency simulations—e.g., set up a roadside assistance scenario communicating within a separate partner vehicle. Note where you can and can't hear them. Try adjusting the settings and note how that changes the connection.

Regular practical use makes quickly configuring and operating the UV-5R second nature. Applied experience provides the most significant learning.

Final Advanced Step

After following this initial checklist, you now have an optimized foundation to enjoy your new skill. Next, continue expanding your knowledge by:

- Study for the amateur radio license exam and get your license.
- Consider getting your friends and family started in the hobby by gifting Baofeng handhelds. Pre-program their radios to your local channels. Start building your neighborhood net.

The world of amateur radio is rich, and this guide only scratches the surface. But by taking it step-by-step, you now have an optimal foundation. I wish you the best.

Jared@MorseCodePublishing.com

References

Admin. (2023, January 3). *The best car radio antennas for 2023.* Hamtronics. https://www.hamtronics.com/best-car-radio-antennas/

Baofeng - Baofeng troubleshooting. (2022, January 6). *RadioReference.com* forums. https://forums.radioreference.com/threads/baofeng-troubleshooting.436574/

Baofeng cheat sheet. (n.d.). W7APK. Retrieved October 20, 2023, from https://w7apk.com/baofeng

Baofeng UV-5R+ plus review, a better UV-5R? (2022). *Besthamradio.* https://www.besthamradio.com/baofeng-uv-5r-plus/

Baofeng UV-82HP vs UV-5R - Which one should I buy? (2022, October 18). Walkie-Talkie-Guide. https://www.walkie-talkie-guide.com/baofeng-uv-82hp-vs-uv-5r/

Centers, J. (2020, December 4). *Baofengs are fine radios, but they don't have hundreds of miles of range.* The Prepared. https://theprepared.com/blog/baofengs-are-fine-radios-but-they-dont-have-hundreds-of-miles-of-range/

Chambers, R. (2021, August 13). *10 Best Baofeng radios (2022 update) buyer's guide.* Best Survival. https://bestsurvival.org/best-baofeng-radio/

Chen, J. (2021, November 30). The facts about distance: What's the range for radio? *Baofeng.* https://www.baofengradio.com/blogs/news/the-facts-about-distance

Draper, J. (2023, January 19). Best Baofeng radio in 2023 - Reviews & buying guide. *Savenetradio.* https://www.savenetradio.org/best-baofeng-radio/

Encyclopædia Britannica, Inc. (2022). *How radio works using radio waves.* Britannica. https://www.britannica.com/video/214986/How-radio-works-overview-radio-waves-frequency-amplitude-modulation

Enforcement overview. (2020). Federal Communications Commission. https://www.fcc.gov/sites/default/files/public_enforcement_overview.pdf

Family radio service (FRS). (2011, August). Federal Communications Commission. https://www.fcc.gov/wireless/bureau-divisions/mobility-division/family-radio-service-frs

Ham radio licenses. (n.d.). *ARRL.* Retrieved October 20, 2023, from https://www.arrl.org/ham-radio-licenses

Ho, J. (2023, October). 5 Best car radio antennas - Oct. 2023. *BestReviews.* Retrieved October 20, 2023, from https://bestreviews.com/automotive/stereos/best-car-radio-antennas

How far can I talk? (n.d.). Quality2wayradios. https://quality2wayradios.com/store/radio-range-distance

How long will a Baofeng UV-5R portable radio last? (2019) Quora. https://www.quora.com/How-long-will-a-Baofeng-UV-5R-portable-radio-last

Martens, M. (2014, November 15). Get a better signal out of your handheld radio. *KB9VBR Antennas.* https://www.jpol

e-antenna.com/2014/11/15/get-a-better-signal-out-of-your-handheld-radio/

Matt. (2023, January 24). Resource review: The guerilla's guide to the Baofeng radio. *Everyday Marksman*. https://www.everydaymarksman.co/resources/guerillas-guide-to-the-baofeng/

Mclaren, S. (2023, June 23). Ham radio emergency frequencies and common uses [PDF]. *Stryker Radios*. https://strykerradios.com/ham-radios/ham-radio-emergency-frequencies-common-uses/#:~:text=34.90%20MHz:%20Nationwide%20emergency%20channel

Pekelny, J. (2021, July 14). Review: Yaesu FT-60R vs BaoFeng BF-F8HP for new ham radio operators. *The Prepared*. https://theprepared.com/blog/review-yaesu-ft-60r-vs-baofeng-bf-f8hp-for-new-ham-radio-operators/

Problem with Boafenf BF-888 radios. (n.d.). *Radio Stack Exchange*. Retrieved October 20, 2023, from https://ham.stackexchange.com/questions/17596/problem-with-baofeng-bf-888-radios

Quinton. (2023, June 2). Baofeng BF-F8HP vs UV-82HP- Ham radio planet. *Ham Radio Planet*. https://hamradioplanet.com/baofeng-bf-f8hp-vs-uv-82hp/

RatedRadarDetector. (2023, October 3). What is the best Baofeng radio?. https://www.ratedradardetector.org/best-baofeng-radio/

Reed, A. (2020, March 8). 7 best FM antennas of 2020 to have a crystal clear reception - AW2K. *A Web Not to Miss*. https://www.awebtoknow.com/reviews/best-fm-antennas/

Retevis. (2017, February 19). *Squelch and squelch level*. Two Way Radio Community. https://www.twowayradiocommunity.com/squelch-squelch-level/

Richard. (2020a, June 7). 7 Best Baofeng radios of 2023 - Reviews & buying guides. *Amateur Radio Wiki*. https://www.amateur-radio-wiki.net/best-baofeng-radio/

Rick. (2016, July 6). What to do if the radio does not power on. *Buy Two Way Radios*. https://www.buytwowayradios.com/blog/2016/07/what_to_do_if_the_radio_does_not_power_on.html

Ricky. (2022, January 11). How to choosing the right radio antenna for your car. Sound Audio Guru. https://soundaudioguru.com/chossing-radio-antenna-for-car#:~:text=How%20to%20Choose%20the%20Right%20Radio%20Antenna%20for

Six tips to maintain your two way radio. (2014). *Two Way Radio Gear*. https://twowayradiogear.com/blogs/news/six-tips-to-maintain-your-two-way-radio

Smith, D. (n.d.). Beginners guide - CHIRP. *Chirp.danplanet*. https://chirp.danplanet.com/projects/chirp/wiki/Beginners_Guide

Tate, A. (2021, August 28). Real life survival story: Baofeng, the little ham radio that could -. *The Organic Prepper*. https://www.theorganicprepper.com/survival-story-baofeng-ham-radio/

Thomas, T. (2022). How to clean and care for your two-way radio (+ 2 printable lists). *Waveband Communications*. https://www.wvbandcoms.com/blogs/blog/cleaning-two-way-radio

Tobias. (2015, February 22). How program a Baofeng UV-5R with CHIRP software. *GeekPrepper*. https://geekprepper.com/program-a-baofeng-radio-with-chirp/

Twowaydirect. (n.d.). How to preserve your two way radios. *Two Way Direct*. https://www.twowaydirect.com/two-way-direct-blog/how-to-preserve-your-two-way-radios/

Walkie-talkie-guide. (2022, May 10). Tutorial: Baofeng UV-5R programming. *Www.walkie-Talkie-Guide.com*. https://www.walkie-talkie-guide.com/baofeng-uv-5r-programming/

GMRS Radio Field Manual

The Visual Command System for FCC Licensing, Programming CTCSS/DCS Privacy Tones, Using Repeaters, and Building an Off-Grid Family Emergency Network

MorseCode Publishing

Introduction

The Communications Gap

Objective: Shift your mindset from a disappointed consumer to a capable "Citizen Operator" by exposing marketing lies and rejecting the gatekeeping of hobbyists.

Key Deliverable: A rejection of the "35-Mile Lie" and adoption of the "Visual Command" mindset that values utility over theory.

You are standing in the aisle of a sporting goods store, or maybe scrolling through Amazon late at night. You're looking at a blister pack of two camo-print handheld radios. The packaging features a rugged-looking guy scaling a mountain, and in bold, jagged yellow font, it promises: **"UP TO 35-MILE RANGE."**

You buy them. You charge them. You hand one to your spouse or your kid, and you drive down the road to test this military-grade technology.

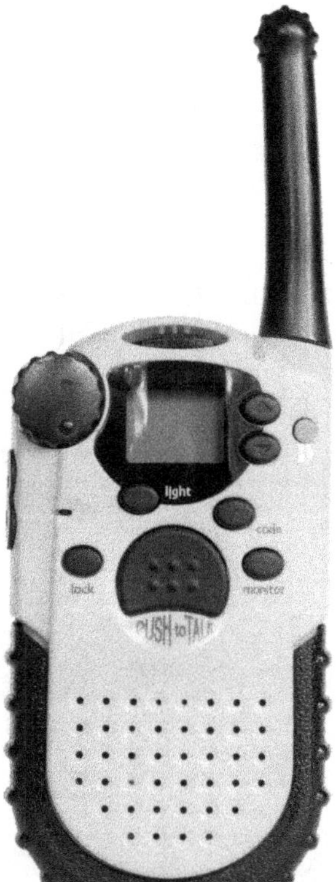

At half a mile, the signal gets fuzzy. At one mile, it sounds like bacon frying in a jet engine. At two miles - nothing. Silence. The confidence you felt in the store evaporates, replaced by the sinking realization that if this were a real emergency, you would be alone.

You didn't get 35 miles. You barely got to the grocery store.

This is the **Communications Gap**. It is the chasm between what the marketing department promised you and what physics actually allows. If you are reading this book, chances are you've felt that specific sting of disappointment. You might have tossed those radios in a junk drawer, assuming you bought "toys," or perhaps you assumed you did something wrong.

You didn't do anything wrong. You just fell for the "35-Mile Lie."

Manufacturers are technically allowed to claim that range based on a "perfect scenario" - usually involving two mountaintops with zero atmosphere, zero trees, zero buildings, and the curvature of the earth conveniently flattened out. But you don't live in a vacuum on a flat earth. You live in the suburbs, or you drive through dense pine forests, or you explore canyons.

In the real world, radio is messy. It bounces off skyscrapers, gets absorbed by wet pine needles, and dies instantly inside a metal parking garage.

This manual exists to close that gap. We are going to take the radio from a magical mystery box that disappoints you, and turn it into a reliable tool that secures your tribe.

The Identity Crisis: You Are Not a "Sad Ham"

Before we fix your radio, we need to fix your mindset.

If you have ever tried to research radio online, you likely stumbled into a forum or a Facebook group populated by what we affectionately call "Sad Hams."

The Sad Ham is a gatekeeper. He has been into Amateur Radio (Ham) since 1978. He speaks entirely in acronyms. He is obsessed with electrical theory, soldering irons, and obscure regulations. If you ask a simple question like, *"Which radio is best for my Jeep?"* the Sad Ham will lecture you. He will tell you that you're too stupid to operate a radio, that your equipment is "Chinese junk," and that you need to study for a difficult technical exam to earn the privilege of pushing a button.

The Sad Ham treats communication as a **hobby**. He wants to build the radio. He wants to talk to a stranger in Slovenia about the weather using a wire strung up in his backyard.

That is not you.

You are a Citizen Operator.

For him, the radio is the destination. For you, the radio is just the vehicle. It is a tool that supports your actual hobbies and responsibilities.

- You are the **Tactical Dad** ensuring your family can coordinate during a power outage or natural disaster.
- You are the **Overland Adventurer** who needs to tell the truck behind you that there is a washout around the blind curve.
- You are the **Community Protector** organizing a neighborhood watch.

You don't care about the physics of wave propagation in the ionosphere. You care about: *If I press this button, will my wife hear me?*

You want utility, not theory. You want results, not a science project. This book is written specifically for the Citizen Operator. We are explicitly rejecting the gatekeeping culture of the radio world. We don't care if your radio cost $30 or $300; we care if you know how to use it when the cell towers go down.

The Solution: The Visual Command System

Why is radio so confusing? Because it is invisible.

You can't see radio waves. You can't see why a signal hits your friend standing on a hill but misses your friend standing in the valley. When things are invisible, they feel like magic. And when magic stops working, we feel helpless.

To become a competent operator, we have to make the invisible **visible**.

This book utilizes the **Visual Command System**. We are going to bypass the complex electrical engineering formulas (which you will never use in the field) and replace them with physical, visual analogies that your brain can instantly grasp.

- We won't talk about "electromotive force"; we will talk about **Water Pressure**.
- We won't talk about "RF propagation shadowing"; we will talk about **Flashlight Beams**.
- We won't talk about "squelch hysteresis"; we will talk about a **Gatekeeper**.

If you can understand how a garden hose nozzle focuses water, you can understand how to double your range with a better antenna. If you can understand how a flashlight casts a shadow behind a wall, you can understand why your radio doesn't work in the city.

This system relies on "If-This-Then-That" logic. We are stripping away the fluff to give you a decision tree.

- *Static on the line?* Do this.
- *Can't hear your buddy?* Move here.
- *Radio beeping loudly?* Turn this off.

By the end of this manual, you won't just be a person holding a plastic box. You will be a Commander of your immediate environment. You will look at the terrain - the hills, the buildings, the trees - and you will visually "see" where your signal is going. You will operate with confidence because you understand the mechanical reality of what is happening in the air around you.

How To Use This Manual

This is not a textbook. Do not put this on your bookshelf between your college history books and that novel you haven't read yet.

This is a **Field Manual**.

It is designed to live in your glovebox, your center console, or the top pouch of your Go-Bag. It is designed to get dog-eared, stained with coffee, and marked up with a sharpie. The chapters are short for a reason. The diagrams are bold for a reason. When you are stuck on a trail in the rain and your comms go down, you don't have time to read a dissertation. You need an answer.

The Go-Bag Size

You will notice we have kept the formatting tight. We have stripped out the history of the FCC and the biography of Marconi. We focus entirely on GMRS (General Mobile Radio Service) because it is the "Goldilocks" of radio for the regular person - more powerful than the toy blister-pack radios (FRS), but requiring no tests or technical exams, unlike Ham radio.

The Icons

To help you scan this book quickly when you are in the field, we use a system of visual icons:

THE SAFETY ALERT When you see this icon, pay attention. This indicates a warning. Usually, this means "Don't do this or you will fry your radio" or "Don't do this or you will interfere with emergency services." These are the guardrails. Stay between them, and you can't mess up.

FIELD TACTICS This icon marks a force multiplier. These are the tips and tricks that instantly boost your range or clarity. These are the hacks that separate the rookies from the pros. When you see this, you're about to learn a shortcut that actually works.

MENTAL MODEL This icon introduces a visualization tool. This is where we turn a complex invisible concept into a simple picture. Read these sections carefully; they are the foundation of the Visual Command System.

You have already taken the first step. You realized that a cell phone is a fragile leash. You realized that in a real emergency, or even just deep in the woods, the grid does not care about you.

You bought the radio. Now, let's learn how to actually use it. Turn the page, and let's get to work.

Jared (WSLC230) – Jared@morsecodepublishing.com

Chapter 1

The Physics of Reality

Visualizing Radio Waves

Objective: Replace invisible magic with physical intuition regarding how terrain, obstacles, and antenna height dictate signal success.

Key Deliverable: The "Range Expectation Checklist"—a tool to instantly predict signal failure based on your environment (Urban Canyon vs. Open Desert).

Most people try to solve their communication problems with their credit card.

They buy a more expensive radio. They buy a "high-gain" antenna that looks like a tactical whip. They buy a 50-watt mobile unit to replace their 5-watt handheld, assuming that ten times the power equals ten times the range.

Then they go out into the field, press the button, and fail.

They fail because they are fighting physics, and physics always wins. You cannot buy your way out of the laws of nature. However, you can *outsmart* them.

Welcome to the foundation of the **Visual Command System**. In this chapter, we are going to stop thinking about radio waves as invisible, magical forces. We are going to visualize them as physical objects. Once you can "see" the radio waves leaving your antenna, you will intuitively know where to stand, where to drive, and when to give up on a bad position.

We are not going to do math. We are going to do geometry.

1.1 The Invisible Terrain

To understand GMRS (General Mobile Radio Service), you have to understand where it lives on the spectrum. GMRS operates in the UHF (Ultra High Frequency) band, specifically around 462 MHz and 467 MHz.

Unless you are an electrical engineer, those numbers mean nothing to you. So let's replace them with a mental model.

Mental Model: The Flashlight vs. The Bouncing Ball

Imagine you are holding two different tools in a dark room full of furniture.

1. The CB Radio (Low Frequency): Think of a CB radio (27 MHz) like a **super-bouncy rubber ball**. You throw it (transmit). It hits a wall, it bounces. It hits a ceiling, it bounces. It hits the ground, it bounces. In the right conditions (atmospheric skip), you can throw that ball, and it might bounce off the sky (ionosphere) and land in another state. It's chaotic, it's long-range, but it's unreliable. It goes around corners easily, but it has no precision.

2. The GMRS Radio (UHF Frequency): Think of your GMRS radio like a **tactical flashlight**. When you press the PTT (Push-to-Talk) button, you are turning on a beam of light.

- **Does light go through a mountain?** No. It hits the rock and stops.
- **Does light go around the curve of the earth?** No. It travels in a straight line until it hits the horizon.
- **Does light go through a thick forest?** Sort of. Some beams get through the gaps in the leaves, but a lot of it gets absorbed or scattered.

This is the most critical rule of GMRS: **Line of Sight.**

If your antenna cannot "see" the other antenna (optically speaking), you probably won't talk to them. The radio waves travel in straight lines, just like light. If there is a massive granite ridge between you and your convoy, you are in the dark. No amount of wattage will blast through that rock.

The "Urban Canyon": Why GMRS Beats CB

If GMRS is just a "flashlight" that gets blocked by obstacles, why do we use it? Why not use the "bouncy ball" CB radio?

Because we don't live on a flat desert. We live in the **Urban Canyon**.

The Urban Canyon is any environment with man-made structures - cities, suburbs, or even dense industrial parks. In this environment, the "bouncy ball" of the CB radio is too big. The wavelength is 11 meters long (about 36 feet). Trying to get a 36-foot wave into a parking garage or through a window is like trying to drive a semi-truck through a doggy door. It doesn't fit.

GMRS waves (UHF) are short - about 70 centimeters (2 feet).

This is small enough to fit through windows. It's small enough to bounce down a hallway. It's small enough to punch through drywall and standard suburban housing materials (with one major exception: **Stucco**, which often contains a metal wire mesh that kills signals).

Field Tactic: The Window Breach If you are inside a building and can't reach your team, move to a window. Glass is transparent to UHF radio waves, just like it is to light. But **watch out for insect screens**. If your window has a metal screen, it acts as a Faraday cage and blocks the signal. You must open the screen to let the 'beam' escape. If you are in a basement? You are dead. The earth eats radio waves. Get high, get to a window.

Diagramming Line of Sight (LOS)

Picture two people standing on a flat prairie.

- **Person A** is 6 feet tall.
- **Person B** is 6 feet tall.

Due to the curvature of the earth, the horizon is only about 3 miles away. If Person A and Person B stand 7 miles apart on perfectly flat ground, they cannot see each other. The earth creates a hump between them.

Since GMRS is a "flashlight," the beam from Person A's radio hits the dirt hump of the earth before it reaches Person B.

This is why the "35-Mile Range" on the box is a lie. That 35-mile claim assumes Person A is on top of Mount Everest and Person B is in a valley, with absolutely nothing in between. In the real world, the earth curves.

The Solution? You have to elevate the flashlight. If Person A climbs a water tower, he can now "see" over the curvature of the earth. His radio beam clears the obstacles. This leads us directly to the single most misunderstood concept in radio: Power vs. Height.

1.2 The Wattage Myth

If you hang out in truck stops or 4x4 forums, you will hear guys bragging about their wattage. *"I've got a 50-watt mobile unit in the dash." "I hacked my Baofeng to push 8 watts."*

They treat wattage like horsepower in a truck. They think 50 watts is "stronger" and will therefore push the signal further, like a battering ram smashing through obstacles.

This is the Wattage Myth.

Mental Model: The Water Hose

Imagine your radio is a water hose.

- **Wattage** is the **Water Pressure** (PSI).
- **The Antenna** is the **Nozzle**.

If you have a hose pointing straight at a brick wall (a mountain), you can turn the pressure up to 1,000 PSI. What happens? You just splash more water back on yourself. You don't punch through the wall.

Adding wattage does not help you penetrate obstacles. It only helps you push the signal further *if the path is already clear.*

The Inverse Square Law (The 'Napkin Math' Version)

Here is the hard truth that makes grown men cry: **To double your distance, you have to quadruple your power.**

Actually, it's often worse than that. In real-world conditions, the returns diminish incredibly fast.

Let's look at the difference between a standard handheld and a high-power mobile unit:

- **5 Watts (Handheld):** This is your baseline. Let's say this gets you 2 miles in the suburbs.
- **50 Watts (Mobile Unit):** This is 10 times the power. Do you get 20 miles? **No.**

You might get 3 miles. Maybe 4.

You have burned 10x the battery power for a measly 20% gain in distance.

Why? Because you are still fighting the curvature of the earth and the trees. You are just "yelling louder" at the trees. The trees don't care.

When DOES wattage matter? Wattage matters for **clarity**, not necessarily distance. Think of a crowded room (static/interference).

- A 5-watt radio is speaking in a normal voice.
- A 50-watt radio is shouting through a megaphone.

If you are within range, the 50-watt radio will sound crisp, loud, and authoritative. It will burn through the background static ("noise floor"). But it will not magically bend around a mountain.

The "Height is Might" Principle

If you have $200 to spend on improving your comms, and you have to choose between a more powerful radio or a better antenna location, choose the antenna every single time.

Height is the only thing that defeats geometry.

Remember the flashlight.

- **Scenario A:** You stand on the ground with a 10-million-candlepower searchlight (High Wattage). You shine it forward. It hits the trees 100 yards away. Range: 100 yards.
- **Scenario B:** You climb a 50-foot tower with a cheap disposable penlight (Low Wattage). You shine it forward. You can see for miles.

Field Tactic: The "Roof Rack" Rule Never use a handheld radio inside a vehicle if you can avoid it. The metal body of your car is a "Faraday Cage" - it blocks radio waves. It keeps the signal *in*.

- **Bad:** Holding a 5-watt radio in the driver's seat. (Effective power escaping the car: maybe 1 watt).
- **Good:** Connecting that handheld to a magnetic mount antenna on the roof.
- **Best:** A dedicated mobile unit with a hard-mounted antenna on the roof rack.

By moving the antenna from your lap to the roof, you gain about 4 to 6 feet of elevation. In the radio world, that is huge. You have cleared the engine block, the dashboard, and the heads of your passengers. You have given your "flashlight" a clear view of the horizon.

The Golden Rule: A 5-watt radio on a mountaintop beats a 50-watt radio in a valley. Every. Single. Time.

1.3 FRS vs. GMRS: The Hybrid Confusion

Now that we understand the physics, we have to clear up the legal and technical mess of the channels.

You will often hear terms like FRS (Family Radio Service) and GMRS (General Mobile Radio Service) used interchangeably. This is because they share the same "driveway."

Years ago, the FCC made a decision that confused everyone. They allowed FRS (the cheap, blister-pack radios you buy at Walmart) to operate on the exact same frequencies as GMRS (the prosumer, licensed gear).

This means your high-speed, 50-watt, waterproof, overland GMRS mobile unit can talk to your nephew's $15 Spider-Man walkie-talkie.

But there are rules.

The "Bubble Pack" Chart

Think of the radio spectrum like a highway with 22 lanes (Channels 1 through 22).

Channels 1–7 (The "Low" Shared Lanes)

- **Who is here?** Everyone. FRS and GMRS.
- **Power Limit:** limited to 5 watts for GMRS (formerly less, but rules modernized).
- **Vibe:** Crowded. This is where you hear kids playing tag and people parking RVs.

Channels 8–14 (The "Weak" FRS-Only Lanes)

- **Who is here?** technically FRS only, with very strict power limits (0.5 watts).
- **GMRS Restriction:** Most proper GMRS radios are hard-locked to transmit at very low power here, or they skip

these channels entirely.

- **Vibe:** Silence or static. Because the power is so low (0.5 watts is like a whisper), these channels are useless for anything beyond "across the campsite" comms. **Avoid these for tactical use.**

Channels 15–22 (The "High" GMRS Power Lanes)

- **Who is here?** The Big Dogs.
- **Power Limit:** 50 Watts.
- **Vibe:** This is where we live. This is where you get the range. If you are running a convoy, a neighborhood watch, or a farm operation, you park your comms here.

SAFETY ALERT: The Wide vs. Narrow Band Trap High-quality GMRS radios use "Wide Band" FM audio. Cheap FRS radios use "Narrow Band." If you talk from a GMRS radio (Wide) to an FRS radio (Narrow), you will sound incredibly loud and distorted to them - like you are swallowing the microphone. If they talk to you, they will sound quiet and thin.

- **Action:** If you know your group has a mix of cheap and expensive radios, go into your GMRS radio settings and switch your bandwidth to **NARROW (NFM)**. This levels the playing field. Remember: The cheap FRS radio is **locked** to Narrow band. The user cannot change it. You are the 'Command' station; it is your job to adapt your equipment to hear them, not the other way around.

Why "Repeater Channels" Are Different

You might see channels 15-22 listed, and then another set of channels called "RPTR 15" or "Rep 15."

These are the **Repeater Inputs**. We will cover repeaters in depth in Chapter 5, but for now, visualize a repeater as a "Range Extender Tower."

- **Channels 1-22:** You talk directly to another radio (Simplex). Like tossing a ball to a friend.
- **Repeater Channels:** You talk to a Machine on a mountain, and the Machine screams your message down to the valley (Duplex). Like throwing a ball at a backboard.

Only GMRS radios can access repeaters. FRS radios cannot. This is the main reason you paid for the GMRS license. It unlocks the "Cheat Mode" of using someone else's tower to get 50 miles of range.

Visual Command Action Step: The "Range Expectation" Checklist

Stop guessing. Stop hoping. Use this checklist to set realistic expectations for your team before you step out the door.

This table assumes you are using a standard **5-Watt Handheld Radio** (the most common scenario).

THE TERRAIN	VISUAL ANALOGY	REALISTIC RANGE	WHY?
Inside a Vehicle	The Faraday Cage	0.5 – 1 Mile	The metal roof blocks the signal "flashlight" from escaping.
Suburban Neighborhood	The Urban Canyon	1 – 2 Miles	Houses, fences, and trees scatter the beam.
Dense Forest (Flat)	The Sponge	0.5 – 1.5 Miles	Pine needles and water-filled trunks absorb UHF energy.
Open Water / Desert	The Billiard Table	4 – 6 Miles	No obstacles to block the beam. (Earth curvature eventually cuts it).
Hilltop to Valley	The Sniper's Nest	5 – 25+ Miles	You have defeated the earth's curvature. Pure Line of Sight.

Your Homework: Go outside. Look at your environment. Don't look at it like a landscape; look at it like a geometry problem. Where are the lines of sight? Where are the shadows?

If you are in a valley, **move**. If you are behind a building, **move**. If you are frustrated, **get your antenna higher**.

That is the physics of reality. Now that you understand the playing field, let's look at the equipment you need to dominate it.

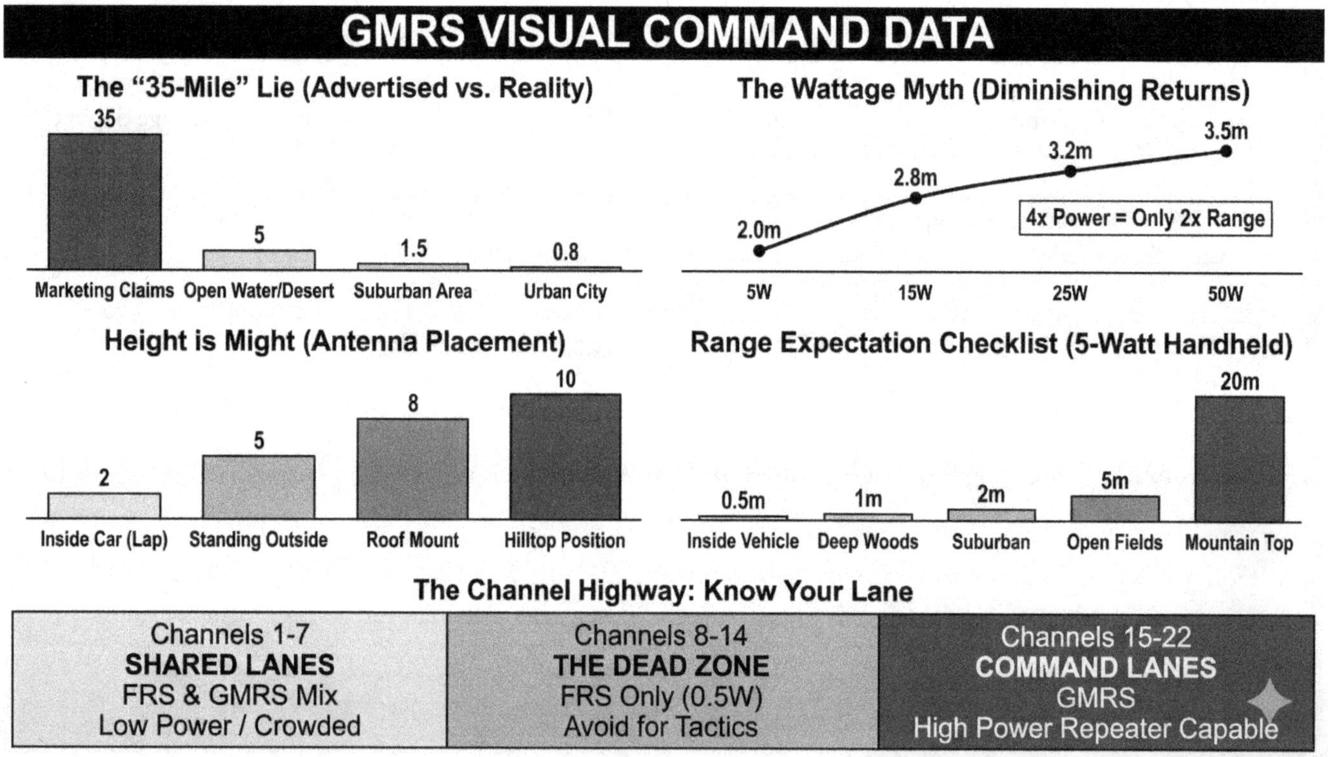

Chapter 2

Licensed to Operate

The 15-Minute Legal Fix

Objective: Demystify the legal bureaucracy to legitimize the entire family unit without fear of exams, Morse code, or "Sad Ham" judgment.

Key Deliverable: The "Family Umbrella" strategy that covers your whole tribe with a single $35 fee, plus a printable "Call Sign Card" for the glovebox.

There is a dirty secret in the off-road and prepping communities: **Most people are operating illegally.**

You see it all the time. A guy buys a $30 radio on Amazon, unboxes it, and starts blasting out transmissions on GMRS frequencies without a license. He assumes that because he bought the radio at a store, he is allowed to use it. Or, he assumes that getting a license involves sitting in a high school cafeteria on a Saturday morning, taking a calculus exam proctored by three guys in ham radio vests who are judging his lack of knowledge about Ohm's Law.

This fear - the fear of the "Test" - is the single biggest barrier to entry for the Citizen Operator.

Let's kill that fear right now.

There is no test. There is no exam. There is no Morse code. There is no interview.

Getting a GMRS license is exactly like getting a fishing license. You go to a government website, you fill out a form that says you aren't a foreign spy, you pay a fee (currently $35), and they email you a PDF.

That's it.

The entire process takes about 15 minutes. It covers your entire family for 10 years. It gives you legal access to powerful repeaters that can triple your range.

If you can order a pizza online, you can get a GMRS license. This chapter is your walkthrough. We are going to navigate the government bureaucracy together, dodge the confusing acronyms, and get you legitimate.

2.1 The "No-Test" License

Why does the "Sad Ham" (Amateur Radio Operator) have to take a test, but you don't?

It comes down to **responsibility**.

- **Ham Radio:** The operator is allowed to build their own radios, modify electronics, and use massive power levels on frequencies that can bounce around the world. Because they can build dangerous things, they need to prove they understand the physics.

- **GMRS:** The radios are certified appliances. You cannot build a GMRS radio; you can only buy one that has been FCC-approved. Since the equipment is "idiot-proofed" by the manufacturer, you don't need to know the physics. You just need to know the rules.

Therefore, the GMRS license is not a "Certificate of Competence." It is simply a tax. It is a "Pay-to-Play" fee that registers you in the database.

The "Family Umbrella" Rule

This is the killer feature of GMRS.

In Ham radio, the license belongs to the *individual*. If you have a license, your wife cannot talk on the radio unless she gets her own license (or you stand right next to her as a "control operator").

In GMRS, the license belongs to the **System**.

When you pay your $35, you receive a Call Sign (e.g., WRXP-555). The FCC regulations (47 CFR § 95.1705) state that this single license covers you **and your immediate family members.**

This means you do not need to buy a $35 license for your wife. You do not need one for your son. You do not need one for your daughter. One license covers the tribe.

Mental Model: The Family Flowchart

Who exactly is "family"? The FCC definition is surprisingly generous.

Picture a family tree. **YOU (The License Holder)** are the trunk.

Who is covered?

- **Spouse:** Yes.
- **Children / Step-children:** Yes.
- **Grandchildren:** Yes.
- **Parents / Step-parents:** Yes.
- **Grandparents:** Yes.
- **Siblings:** Yes.
- **Aunts/Uncles/Nieces/Nephews:** Yes.
- **In-Laws:** Yes (including Father/Mother-in-law, Brother/Sister-in-law, and Son/Daughter-in-law).

Who is NOT covered?

- **Cousins:** No. (This is the most common mistake. Cousin Eddie needs his own license).
- **Employees:** No. If you run a landscaping business, you cannot use a personal GMRS license for your crew. That requires a Commercial Business License (Part 90), which is a totally different beast.
- **Friends/Neighbors:** No. Your buddy in the truck behind you needs his own license.

Field Tactic: The "Guest Operator" Loophole "But wait," you ask. "I'm taking a friend camping. I hand him a spare radio. Is he operating illegally?"

Technically? Yes. However, the rules allow for emergency messages. If safety is at risk, anyone can use any radio. But for general chit-chat ("Hey, grab me a beer"), your friend is technically in violation.

The Solution: Tell your friend to spend the $35. If your group has 3 heads of households (you, your brother, your best friend), you need 3 licenses. That covers all the wives, kids, and grandparents in the convoy. It is a small price to pay for a 10-year permit.

2.2 Navigating the FCC Nightmare

The hardest part of GMRS isn't the radio; it's the website.

The Federal Communications Commission (FCC) uses a system called **CORES** (Commission Registration System) and **ULS** (Universal Licensing System). These websites appear to have been built in 1998 and never updated. They are clunky, they time out, and they use confusing terminology.

Do not let the bad UI (User Interface) defeat you. Follow this "If-This-Then-That" walkthrough.

Phase 1: Get Your FRN (FCC Registration Number)

Before you can apply for a license, you need a customer ID number. This is your FRN.

1. Go to the **FCC CORES** website.
2. Click **"Register"** to create a username (use your email).
3. Once logged in, select **"Register New FRN"**.
4. **Select "An Individual"** (unless you are a corporation, which you aren't).
5. **Contact Address:** Yes, you must use your real address. Yes, this becomes public record.
    - *Privacy Note:* If you are concerned about OPSEC (Operational Security) and don't want your home address searchable in the FCC database, use a P.O. Box. The FCC does not care if you live there, but they must be able to mail you enforcement letters.
6. **Social Security Number:** You must provide it. This is to ensure you aren't a felon avoiding federal fees.
7. **Submit.** You will instantly get a 10-digit number. **Write this down.** This is your FRN.

Phase 2: The Application (The "ZA" Code)

Now that you have an FRN, you can buy the license.

1. Log in to the **FCC ULS License Manager**.
2. On the left menu, click **"Apply for a New License"**.
3. **The Dropdown Menu Trap:** You will see a list of radio services with confusing acronyms. You are looking for **"ZA - General Mobile Radio (GMRS)"**. It is usually at the very bottom of the list.
    - *Note:* Do not click "Amateur" (Ham). Do not click "Land Mobile". Click **ZA**.
4. **Applicant Questions:**
    - *Has the applicant been convicted of a felony?* Answer honestly. (A felony does not automatically disqualify you, but it triggers a review).

- *Do you agree to the rules?* Yes.

5. **Summary:** Verify your address and name.

6. **"Certify"**: This is your digital signature. Type your name.

Phase 3: The Payment Loop (The Final Boss)

This is where 50% of people fail. After you click "Submit Application," the ULS system will say "Application Submitted." **You are not done.**

You must now switch *back* to the **CORES** system to pay. The two systems do not talk to each other well.

1. Look for a button that says **"Log in to CORES to Pay Fees"**.

2. If you miss it, just log back into CORES separately.

3. You should see a "Red Light Status" or "Pending Payment" notification.

4. Select the pending application.

5. Pay the **$35**. (Credit card is best).

Still having issues finding it? Follow this path on the website: FCC website > FCC Registration page > Manage Existing FRNs > FRN Financial page

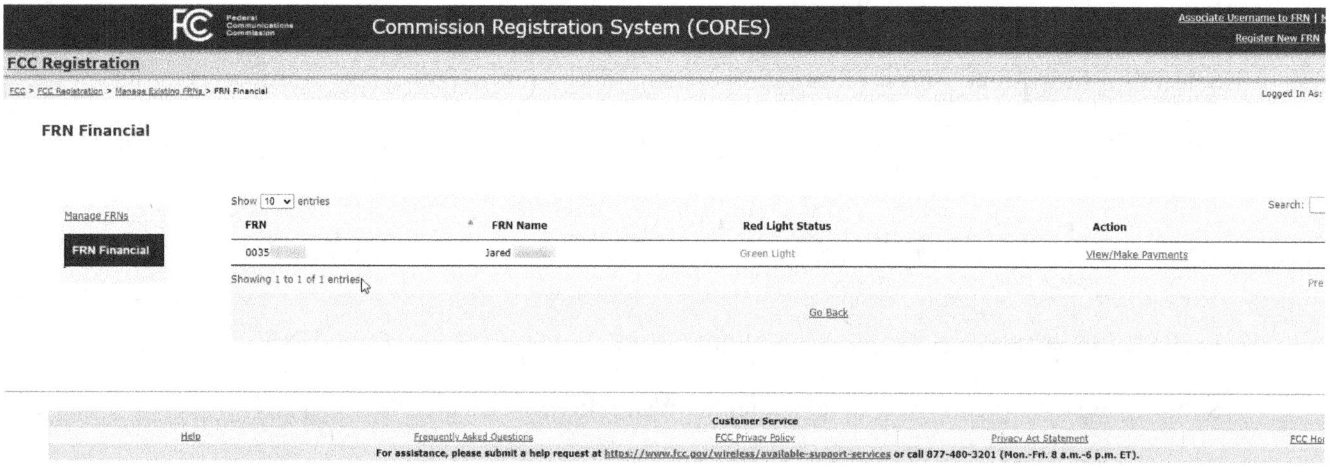

Phase 4: The Waiting Game

You will not get your license instantly. It usually takes **24 to 48 business hours**. Check your email (including spam) for a message from the FCC with a link to "Download Authorization."

SAFETY ALERT: Do Not Transmit Yet Do not pick up that radio and start talking while your application is 'Pending.' You are not legal until the FCC grants the license and issues the Call Sign. Wait for the email.

Field Tactic: The Reference Copy Trap When you finally log in to download your license, you will see two options:

1. "Reference Copy" (Watermarked "Reference Only")

2. "Official Authorization"

Download the **Official Authorization.** Print it out. Put one copy in your glovebox. Put one copy in your Go-Bag. Save a screenshot of it on your phone in a "Important Docs" album. You will likely never be asked for it by law enforcement, but repeater owners (the guys who own the big towers) will often ask for your Call Sign and proof of license before letting you use their system.

2.3 Your Call Sign

Congratulations. You are now **WRXP-555** (or whatever jumble of letters and numbers they assigned you).

This is your **Call Sign**. It is your legal identity on the airwaves.

In the movies, people use "Handles" like *Rubber Duck* or *Eagle Eye*. In the real world of GMRS, we use Call Signs.

When to Identify

The rules (47 CFR § 95.1751) are simple, but people overcomplicate them.

You must state your Call Sign:

1. **Every 15 minutes** during a conversation.
2. **At the end** of a conversation.

You DO NOT need to state it:

- At the very beginning (though it is polite).
- In every single sentence.

Example of a Proper Exchange:

- **Start:** *"Unit 1 to Unit 2, radio check."*
- **Middle:** (You talk about the trail, the weather, the lunch stop for 10 minutes).
- **End:** *"Unit 1 clear. WRXP-555."*

That's it. You just have to tag the end of the chain. If you are talking for an hour, just throw your call sign out whenever there is a lull. *"Hey, just ID-ing, this is WRXP-555."*

The "Unit Number" System (Tactical Clarity)

Your Call Sign is a mouthful. *"Whiskey Romeo X-ray Papa Five Five Five."* Trying to say that every time you want to tell your wife to turn left is annoying.

The Solution: Use **Unit Numbers** for tactical comms, and **Call Signs** for legal ID.

Assign everyone in your family/group a tactical name:

- **Dad:** "Unit 1" (or "Lead")
- **Mom:** "Unit 2"
- **Kids:** "Base" (if at camp) or "Unit 3".

The Conversation: *"Unit 1 to Unit 2, slow down, big rock on the left." "Unit 2, copy."* (Conversation continues...) *"Unit 1, we are stopping for gas." "Unit 2, copy. WRYZ-123 clear."* (Mom drops the legal ID at the end to cover the whole family).

Since the license covers the whole family, **only one person needs to ID** to make the whole transmission legal. If Dad says the Call Sign, Mom is covered.

The Myth of "The FCC Van"

Let's address the paranoia. If you forget to ID? If you say your Call Sign wrong? Will a black van with antennas skid into your driveway and arrest you?

No.

The FCC Enforcement Bureau is underfunded and overworked. They are hunting two things:

1. **Commercial Interference:** A construction company using police frequencies and blocking 911 calls.
2. **Malicious Interference:** Someone setting up a high-power transmitter specifically to jam a repeater or scream profanity for days on end.

They are not hunting "Tactical Dad" who forgot to ID his call sign while hiking.

However... While the *government* might not be listening, the **Sad Hams** are. The GMRS community is self-policing. If you hop on a public repeater and start acting like an idiot - playing music, not using a call sign, burping into the mic - the owner of that repeater can and will block you. They can remotely turn off the repeater or ban your specific radio ID if you are using digital modes.

Respect the Commons. We share these airwaves. Using your Call Sign isn't just a legal requirement; it's a signal to everyone else that you are a "Citizen Operator" who knows the rules, not a "Blister Pack Rookie" who is just making noise. It buys you respect.

Workbook Element: The "Call Sign Card"

Do not rely on your memory. In an emergency, your adrenaline will be high, and you will forget whether you are "WRXP" or "WXRP".

Print this template, fill it out with a Sharpie, and tape it to the back of every radio in your fleet. ***Scan to Download Printable Copies.***

Chapter 3

Hardware Selection

Buying the Right Tool

Objective: Cut through "Gear Paralysis" to select hardware based on Mission Profile (Burner vs. Operator) rather than confusing marketing specs.

Key Deliverable: A "Mission-Ready" hardware loadout that eliminates weak points like bad antennas, heat traps, and incompatible connectors.

If you ask a forum full of radio enthusiasts what radio you should buy, you will get 50 different answers, and 49 of them will be wrong for you.

One guy will tell you to buy a $600 commercial Motorola that requires proprietary software from 1998 to program. Another will tell you to buy a $20 Baofeng because "it's just as good" (it isn't). A third will tell you to build your own radio out of a tuna can and copper wire.

This is **Gear Paralysis**.

In the Tactical Dad and Overlanding world, we love gear. We love researching specs. But with radio, the specs on the box rarely tell the full story. A radio that promises "50 Watts" might have a receiver so sensitive that it goes deaf the moment you drive near a Taco Bell drive-thru.

In this chapter, we are going to cut through the marketing noise. We are going to categorize hardware not by price, but by **Mission Profile**.

Are you equipping a neighborhood watch? Are you outfitting a Jeep for a cross-country trip? Are you handing a radio to a toddler?

We need the right tool for the job.

3.1 The Handheld (HT) Landscape

The "HT" (Handie-Talkie or Handheld Transceiver) is the gateway drug of GMRS. It is portable, battery-powered, and usually affordable.

However, the market is flooded with junk. To navigate this, we divide handhelds into two distinct tiers: The **"Burner" Tier** and the **"Operator" Tier**.

Tier 1: The "Burner" (Baofeng & Clones)

- **Price:** $25 - $40
- **Typical Models:** Baofeng UV-5G, UV-9G, Retevis RA85.
- **The Vibe:** Disposable.

Everyone owns a Baofeng. They are the Harbor Freight tools of radios - cheap, they get the job done for a while, and you won't cry if you drop one in a mud puddle.

The "Glass Jaw" Problem (Front-End Overload) The biggest weakness of cheap radios is their receiver chip. They use a "Direct Conversion" receiver, which is like having ears that cannot filter sound.

- **Scenario:** You are deep in the woods. The radio works great.
- **Scenario:** You drive up to a hilltop near a city to get a better signal. Suddenly, your radio goes silent. Or it starts screeching static. You can't hear your buddy 100 yards away.

Why? Because there is a cell phone tower or a paging transmitter on that same hill. The cheap radio chip gets overwhelmed by *all* the RF energy nearby, not just the GMRS signal. It gets "blinded" by the noise. It's like trying to listen to a whisper while standing next to a jet engine.

The Verdict: Buy these as backups. Buy them to hand out to friends who didn't bring a radio. Buy them for your kids. But do not trust your life to them.

Tier 2: The "Operator" (Wouxun & High-End)

- **Price:** $100 - $250
- **Typical Models:** Wouxun KG-935G, KG-S88G, Rocky Talkie 5 Watt.
- **The Vibe:** Reliable Tool.

These radios often use **Superheterodyne** receivers (or very high-quality filtering chips). This is the "Bouncer" at the door. The radio ignores the cell towers and paging systems and only lets the specific GMRS frequency through. You can stand right next to a radio tower and still hear a weak signal from your convoy.

The Verdict: This is your primary carry. It survives drops, it filters noise, and it works when things get loud.

The Essential Features Checklist

Do not buy a handheld radio in 2025 unless it passes this 3-point inspection.

1. USB-C Charging (Non-Negotiable) Old-school radios require a "cradle" charger that plugs into a wall. This is a logistical nightmare in the field.

- *The Requirement:* You must be able to charge the battery directly via USB-C.
- *Why:* Because you already have a USB-C cable in your truck, in your solar bank, and in your laptop bag. If you lose the charging cradle for an old radio, that radio is a brick. If you have USB-C, power is everywhere.

2. Removable Antenna (SMA Connector) Some "blister pack" radios have fixed antennas glued to the body. Avoid these.

- *The Requirement:* You must be able to unscrew the antenna.
- *Why:* The stock antenna is trash. You need the ability to swap it for a high-gain whip or connect the radio to a roof-mount antenna cable.

3. IP Rating (The "Dunk" Test) Check the "IP" (Ingress Protection) code.

- **IP54:** Splash proof. Okay for light rain. Dust protected. (Minimum Standard).
- **IP67:** Submersible. You can drop it in a creek for 30 minutes.
- *Reality Check:* If you are an overlander, IP54 is fine. If you are a kayaker or a duck hunter, get IP67.

3.2 The Mobile Command Unit

Handhelds are great for spotters and hiking. But if you are in a vehicle, a handheld is the wrong tool.

We discussed the **Faraday Cage** effect in Chapter 1. The metal body of your vehicle blocks radio waves. To get a signal out, you need an external antenna. You *could* connect your handheld to an external antenna, but that leaves you with a cable dangling across your steering wheel and a battery that dies in 8 hours.

Enter the **Mobile Unit**.

These are 20-watt or 50-watt boxes hard-wired to your car battery. They never run out of power. They push 10x the wattage of a handheld.

The Problem: "Where do I mount this brick?"

In a 1995 Ford Bronco, you had plenty of dashboard space to screw in a CB radio. In a modern Toyota Tundra or Jeep Wrangler, every square inch of the dash is covered in touchscreens, airbags, and leather. There is nowhere to mount a radio box.

The Solution: "All-in-the-Handset" Controls

This form factor has taken over the market (e.g., Midland MXT275, MXT575, Wouxun KG-1000G Plus).

How it works:

1. **The Black Box:** The actual radio body is a blind black box. You hide this under the driver's seat, behind the glovebox, or in the center console.
2. **The Mic:** The microphone contains the screen, the volume knob, the channel selector, and the speaker.

3. **The Connection:** You run one ethernet-style cable from the hidden box to the dashboard.

Why this rules the world:

- **Stealth:** When you park at the trailhead, you unplug the mic and lock it in the glovebox. To a thief looking through the window, your truck looks stock. There is no expensive electronics box visible.
- **Ergonomics:** You don't have to reach across the dash to change channels. The controls are in your hand.

SAFETY ALERT: The 50-Watt Heat Problem If you buy a high-power (50W) unit, remember that efficiency is never 100%. A lot of that power turns into **Heat**.

- *Do not* mount a 50W radio body inside a sealed center console wrapped in foam. It will overheat and shut down.
- *Do not* zip-tie it directly to a heater vent.
- Give the heatsink fins room to breathe. Under a seat is fine for daily drivers, but if you ford rivers in a Jeep, mount it high (behind the glovebox) to avoid drowning the radio in deep water crossings.

3.3 Antenna Anatomy

The antenna is 80% of your performance. The radio is just the engine; the antenna is the tires. You can have a Ferrari engine (50W radio), but if you have bald tires (bad antenna), you aren't going anywhere.

Visual Guide: The Three Shapes

1. The "Rubber Duck" (Stock Handheld)

- *Shape:* Short, stubby, flexible.
- *Physics:* It is a coiled spring inside rubber. It is physically shorter than the radio wave requires, so it uses a coil to "trick" the radio.
- *Performance:* **Poor.** It acts like a resistor. It absorbs signal. It is designed for convenience, not range.
- *Use Case:* Clipping the radio to your belt while walking around camp.

2. The "Ghost" (NMO Low Profile)

- *Shape:* A cylinder, about 3 inches tall. Looks like a soup can or a salt shaker.
- *Physics:* Unity Gain (no magnification). It radiates the signal in a perfect sphere (like a lantern).
- *Performance:* **Reliable.** It doesn't snag on tree branches. It fits in parking garages. It handles high power.
- *Use Case:* The daily driver. Because it is only 3 inches tall, you never have to take it off to enter your home garage or a parking structure. It is the 'Set it and Forget it' option.

3. The "Whip" (High Gain)

- *Shape:* A thin steel rod, usually 32 inches or longer. Sometimes has a "coil" in the middle.
- *Physics:* High Gain (3dB to 6dB). It squashes the signal sphere into a flat pancake. It takes energy that would have gone up into the sky and pushes it out toward the horizon.
- *Performance:* **Maximum Range.** This is how you hit repeaters 30 miles away.

- *Use Case:* Desert running, open plains, highway travel.
- *Downside:* It hits every... single... tree branch. *Whack. Whack. Whack.*

Mental Model: The Ground Plane (The Mirror)

This is the concept that confuses everyone. Most mobile antennas are **1/4 Wave** antennas. This means they are physically only one-quarter of the length of the radio wave.

Where is the other three-quarters? **Your car is the other half of the antenna.**

Think of a mirror. If you place a candlestick on a mirror, you see two candlesticks: the real one going up, and the reflection going down. Together, they look like one long object.

Your metal roof is the mirror. The antenna sticks up. The radio wave reflects off the metal roof, creating a "virtual" antenna pointing down. This completes the circuit.

The Implications:

- **Center of Roof:** This is the best spot. The "mirror" is equal in all directions. Your signal goes out 360 degrees perfectly.
- **Hood Mount (Edge):** The "mirror" is behind the antenna (the windshield/roof). Your signal will be strong behind you, but weak in front of you. The pattern is distorted.
- **Plastic Roof (Jeep/Bronco): Problem.** A plastic roof is not a mirror. It is invisible to radio waves. If you mount a standard antenna on a fiberglass roof, you have no Ground Plane. The signal has nowhere to push off of.
  - *The Fix:* You must use a special "No Ground Plane" (NGP) antenna kit, OR mount the antenna on a metal bracket (hood lip, tailgate) that is electrically bonded to the chassis.

Field Tactic: The "Cookie Sheet" Hack If you are setting up a base station at a campsite with a magnetic mount antenna, but you only have a wooden picnic table... Put the antenna on a metal baking sheet (or a pizza pan). Congratulations, you just built a Ground Plane. Your range just doubled.

SAFETY ALERT: The Aluminum Trap If you drive a Ford F-150 (2015+) or a Super Duty (2017+), your truck body is Aluminum. **Magnets do not stick to aluminum.** Do not buy a magnetic mount antenna for these trucks. It will slide right off. You must use a drill-in NMO mount or a dedicated 'fender clamp' bracket.

Action Step: The "Gear Audit"

Before you order anything, we need to talk about **Connector Hell.**

The radio industry cannot agree on a standard plug. If you buy a Baofeng, a Midland, and a dedicated antenna, chances are none of them will fit together without an adapter.

You need to perform a physical audit of your "Male" and "Female" parts.

The "Pin and Hole" Check

Look at the connector on the **Radio Body**.

- **SMA-Male:** Has a center PIN sticking up. (Common on Yeasu, Wouxun).
- **SMA-Female:** Has a center HOLE. (Common on Baofeng, BTech).

Now look at the **Antenna**. It must be the opposite.

Rule of Thumb:

- Buying a **Baofeng**? You usually need an antenna with a **Male** pin (SMA-Female Radio requires SMA-Male Antenna).
- Buying a **Wouxun/Midland/Yaesu**? You usually need an antenna with a **Female** hole (SMA-Male Radio requires SMA-Female Antenna).

The "Gender Changer" Danger You can buy little brass adapters (SMA-M to SMA-F) to bridge the gap.

- **Audit Step:** Order 2 or 3 of these adapters specific to your radio combination.
- **Warning:** Every adapter adds a tiny bit of "insertion loss" (signal weakening) and mechanical stress. Do not stack 4 adapters like a Tower of Babel. If you have to stack adapters, you bought the wrong cable.

The Standard Mobile Connector: PL-259 (UHF)

Most mobile units and roof-mount cables use a much larger connector called the **PL-259** (or SO-239). It looks like a thick silver screw-on plug.

- If you buy a handheld radio and want to connect it to your roof antenna, you will need a **"Pigtail Adapter"**.
- *One End:* Tiny SMA (for the handheld).
- *Other End:* Big SO-239 (for the roof cable).
- *The Cable:* Thin, flexible wire in between to relieve stress.

Your Shopping List:

1. **Primary Radio:** (Mobile 20W+ recommended for drivers, HT for hikers).
2. **The Antenna:** Ghost (Woods) or Whip (Desert).
3. **The Mount:** Lip mount, mag mount, or roof rack bracket.
4. **The Pigtail:** If connecting a handheld to the roof antenna.
5. **USB-C Cables:** One for the car, one for the bag.

Stop looking for the "Perfect Radio." It doesn't exist. Buy the radio that fits your dashboard. Buy the antenna that fits your garage height. Install it cleanly. Then, stop worrying about the gear and start worrying about how to speak on it.

MISSION CONFIG

Hardware Selection & Setup

Cut through the gear paralysis. Whether you are equipping a neighborhood watch or an overland convoy, choose the tool that fits the mission. Don't buy specs; buy reliability.

Handhelds: Burner vs. Operator

The market is flooded. We categorize radios into two tiers. **The Burner** (Baofeng) is cheap and disposable but has a "Glass Jaw"—it goes deaf near cities due to poor filtering. **The Operator** (Wouxun/High-End) filters noise and survives the field.

TIER 1: BURNER (Baofeng)
- $25–$40 price point
- Poor noise filtering
- Backup use only

TIER 2: OPERATOR (Wouxun/High-End)
- $100–$250 price point
- Superheterodyne (filters noise)
- IP Rated (Waterproof)

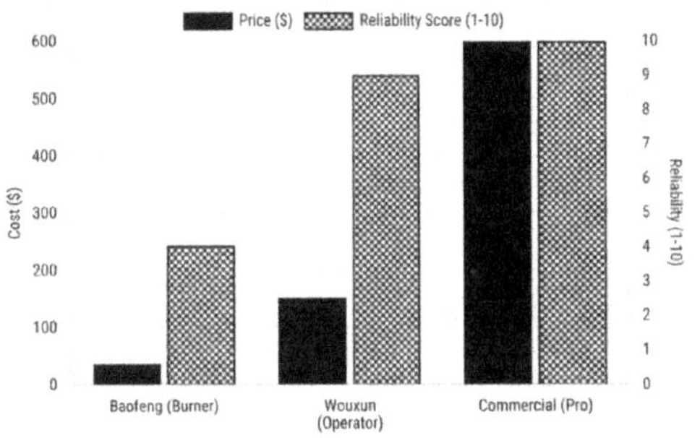

Figure 3.1: Price vs. Signal Reliability

MOBILE COMMAND: THE HEAT TRAP

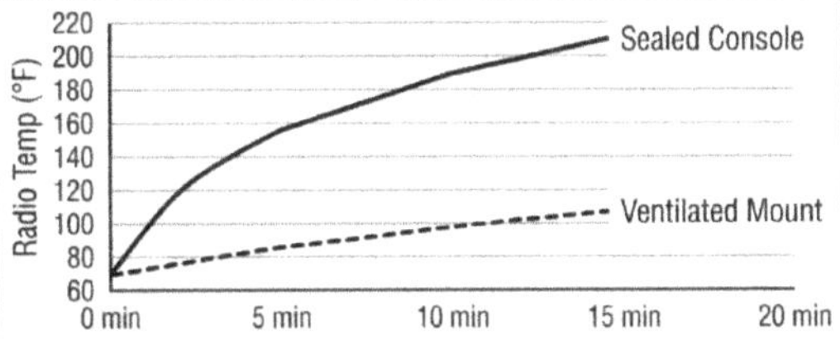

Safety Alert: Never zip-tie a 50W radio directly to a plastic heater vent.

ANTENNA ANATOMY

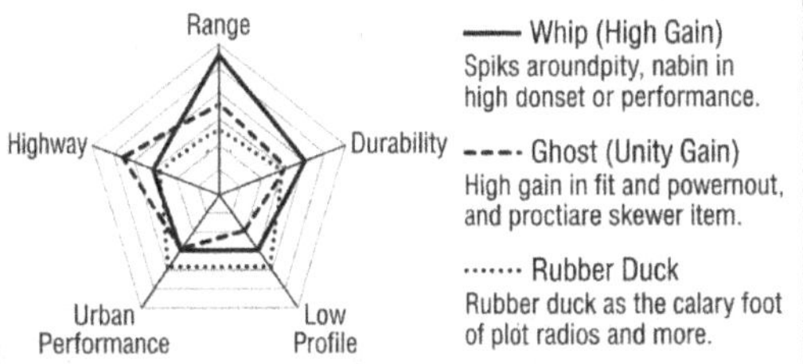

THE GROUND PLANE

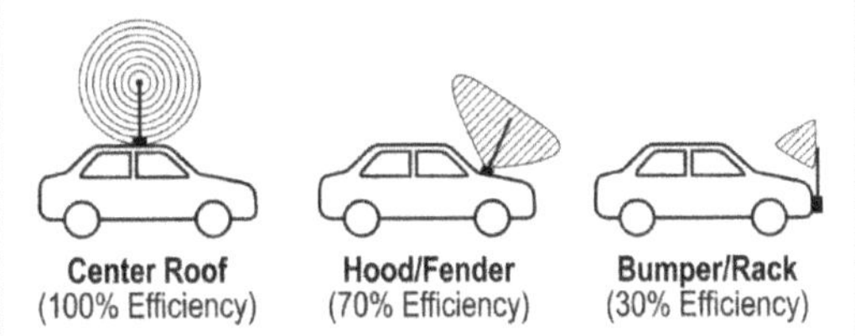

CONNECTOR AUDIT

RADIO BRAND	RADIO CONNECTOR	ANTENNA NEEDED
Baofeng / BTech	SMA-Male (Pin)	SMA-Female (Hole)
Wouxun / Yaesu	SMA-Female (Hole)	SMA-Male (Pin)
Mobile Units	SO-239 (Big Socket)	PL-259 (Big Plug)

The Gender Changer Danger: Every adapter adds signal loss ("Insertion Loss"). Avoid stacking adapters.

Chapter 4

The Visual Command Programming Guide

Conquering CHIRP

Objective: Overcome the "Programming Gap" to transform the radio from a confusing factory brick into a customized, standardized tool.

Key Deliverable: The "Golden Config"—a standardized memory map cloned across every device in the fleet for seamless operation.

If Chapter 2 (Licensing) was the bureaucratic hurdle, this chapter is the technical firewall.

You have bought the radios. You have your license. You have charged the batteries. Now, you want to put in a specific channel - maybe a local repeater, or a specific frequency your 4x4 club uses.

You open the manual that came with your $35 radio. It is written in broken English, translated from Chinese by a robot that clearly hates you. It tells you to press "MENU + 4 + 7 + UP + ARROW + EXIT" to change one setting. You try it. The radio beeps angrily. You throw the manual across the room.

Welcome to the **Programming Gap**.

Most handheld radios, especially the Baofeng and Wouxun models that dominate the market, are terrible to program by hand. The keypads are too small, the menus are nested like Russian dolls, and the logic is backward. If you have to program a radio in the field using the keypad, you have already failed.

The solution is **CHIRP**.

CHIRP is free, open-source software that allows you to plug your radio into your computer and program it using a spreadsheet. It turns a 4-hour nightmare of button mashing into a 5-minute copy-paste job.

But CHIRP has its own demons. It requires drivers. It requires cables. It requires understanding what "Tone Squelch" actually means.

In this chapter, we are going to build your **"Golden Config."** We will troubleshoot the connection issues that make grown men cry, and we will build a programming file that you can blast into every radio in your fleet, ensuring that when you say "Go to Channel 4," everyone actually goes to Channel 4.

4.1 The Driver Disaster

Before we even open the software, we have to talk about **The Cable**.

When you bought your radio, it might have come with a programming cable in the box. Or maybe you bought a cheap one on eBay for $4.

Throw it away.

I am serious. Walk to your trash can and throw it away.

The "Counterfeit Chip" War

Here is the inside baseball of the radio world. The cable that connects your radio to your computer is not just a wire. Inside the USB plug, there is a tiny microchip that translates the radio's language (Serial/TTL) into your computer's language (USB).

There are two main companies that make these chips:

1. **FTDI (The Good Guys):** Reliable, expensive, works instantly on Windows and Mac.
2. **Prolific (The Complicated Guys):** Older, cheaper, finicky.

Chinese factories, trying to save pennies, flooded the market with **counterfeit Prolific chips**. Microsoft (Windows) got tired of these fake chips causing crashes. So, they updated Windows to automatically detect fake chips and *brick* them - disable them so they don't work.

The Result: You plug in your cheap cable. Windows installs the latest driver. The driver sees the fake chip. The driver kills the connection. You spend 6 hours on Google trying to figure out why your computer won't see the radio.

The Solution: Buy an FTDI Cable

You need to buy a programming cable that explicitly says **"Genuine FTDI Chip"** in the listing. It will cost $20 instead of $5. It is worth every penny.

- **BTech PC03** is the gold standard for Baofeng style (Kenwood 2-Pin) radios.
- **Wouxun** often uses their own high-quality cables.

If you ignore this advice and use the cheap cable, you are entering the "Driver Dance."

Troubleshooting Decision Tree: "My Computer Won't See the Radio"

So, you plugged it in, opened CHIRP, clicked "Download from Radio," and got an error message: *"Radio refused to enter programming mode"* or *"COM Port not found."*

Follow this logic path to fix it.

STEP 1: The "Click" Check

- **The Problem:** The Kenwood 2-pin connector (the plug that goes into the radio) requires a frightening amount of force.
- **The Fix:** Push it harder. Seriously. If you didn't hear a distinct *SNAP*, it's not in. On water-resistant radios (like the Baofeng UV-9G or Wouxun KG-935G), the rubber seal fights you. You have to push until your thumbs hurt.
- **Retest:** Try to download again.

STEP 2: The Device Manager Check (Windows)

- Right-click the **Start Button** -> **Device Manager**.
- Look for **"Ports (COM & LPT)"**.

- **Scenario A:** You see nothing new when you plug it in.
  - *Diagnosis:* Dead cable or bad USB port. Try a different port.
- **Scenario B:** You see "Prolific USB-to-Serial" with a **Yellow Triangle (!)** icon.
  - *Diagnosis:* The Driver Disaster. Windows updated your driver and killed the fake chip.
  - *The Fix:* You must "Roll Back" the driver.
    i. Right-click the device -> **Update Driver**.
    ii. Select **"Browse my computer for drivers"**.
    iii. Select **"Let me pick from a list..."**.
    iv. Look for an older version (usually from **2008** or **2011**). Select that one.
    v. Click Next. The Yellow Triangle should disappear.

STEP 3: The Mac "Security" Block

- If you are on a Mac, the OS often blocks the driver installation because it's not from the App Store.
- **The Fix:** Go to System Settings -> Privacy & Security. Look for a message saying "System software from developer was blocked." Click **Allow**. Reboot.

STEP 4: The "Volume Knob" Trick

- Some radios need to be turned ON to program. Some need to be OFF.
- **Standard Rule:** The radio must be **ON**, and the volume should be turned up to about 50%.
- **Exception:** Some older radios require you to hold a button while turning them on to enter "Cloning Mode." Check your manual (or Google).

Once you have conquered the cable, CHIRP will open, and you will see a blank spreadsheet. Now, we build the brain.

4.2 The "Golden Config" Spreadsheet

When you successfully "Download from Radio," CHIRP presents you with a grid. It looks like Excel. This is the **Memory** of your radio.

Rows 1 through 22 are usually your standard GMRS channels. Rows above 22 might be blank.

We are going to explain the columns that matter. Ignore the ones that don't (like "Skip" or "Power" for now). The Big Five are: **Frequency**, **Name**, **Tone Mode**, **Tone**, and **Duplex/Offset**.

Column 1: Loc (Location)

This is just the channel number.

- **Tactical Tip:** Make sure your "Fleet" agrees on the numbers. If "Channel 5" is GMRS 5 on your radio, but "Channel 5" is a local repeater on your friend's radio, you are going to have a bad day.
- **Standardization:** Keep 1-22 as the standard GMRS frequencies. Put your special custom channels (Repeaters) in slots 23 through 50.

Column 2: Name (The Alpha Tag)

Cheap radios just show the frequency (e.g., "462.550"). Good radios show a Name (e.g., "GMRS 15").

This column allows you to name the channel. You usually have 6 to 8 characters.

- **Bad Name:** "RPT1" (Vague).
- **Good Name:** "DEN-RPT" (Denver Repeater).
- **Bad Name:** "EMERG" (Vague).
- **Good Name:** "SAR-CH" (Search & Rescue).

Visual Command Rule: Use names that describe the *function* or the *location*, not the frequency.

Column 3: Tone Mode (The Gatekeeper)

This is the single most confusing column in CHIRP. This is where people mess up.

This setting controls the **Privacy Codes** (CTCSS/DCS). Think of the Privacy Code as a **Key**.

- **Setting: "(None)"**
    - **Visual:** An Open Door.
    - **Tx (Talk):** You shout to everyone.
    - **Rx (Listen):** You hear everyone.
    - *Use Case:* General comms, channel monitoring, emergency hailing (Channel 20).
- **Setting: "Tone" (Transmit Only)**
    - **Visual:** You have a Key to open *their* door, but your door is wide open.
    - **Tx:** You send a hidden "hum" (the tone) with your voice. This opens the squelch of the other radio (or repeater).
    - **Rx:** You hear *everything* on that frequency, even people not using the tone.
    - *Use Case:* **Accessing Repeaters.** Most repeaters require you to send a tone to "wake them up," but they transmit back to you without a tone.
- **Setting: "TSQL" (Tone Squelch - Transmit & Receive)**
    - **Visual:** You have a Key, and you locked your own door.
    - **Tx:** You send the tone.
    - **Rx:** Your radio stays silent *unless* the incoming signal has the matching tone.
    - *Use Case:* **Family Convoy.** You want to ignore the construction crew nearby using the same channel. You only want to hear your family.

SAFETY ALERT: If you set your radio to **TSQL** (Tone Squelch), you are filtering out the world. If a stranger is screaming "HELP HELP HELP" on Channel 16 but they aren't using your specific privacy tone, **you will not hear them.**

- *My recommendation:* Run "Tone" (Tx only) or None for most channels. Only use TSQL if you are in a highly congested urban area and the chatter is driving you crazy.

Column 4 & 5: Tone & ToneSql

These columns are where you actually pick the number of the key (e.g., 67.0, 141.3, etc.).

- If **Tone Mode** is "Tone", CHIRP looks at the **Tone** column.
- If **Tone Mode** is "TSQL", CHIRP looks at the **ToneSql** column.

Common Mistake: Setting Tone Mode to "Tone" but changing the number in the "ToneSql" column. The radio will ignore it. You have to fill in the column that matches the Mode.

Column 6: Duplex & Offset (The Repeater Math)

If you are programming a standard channel (Simplex), **Duplex** should be "(None)" or "off".

If you are programming a **Repeater**, you need to tell the radio to "Split" its personality. It listens on one frequency but talks on another.

The GMRS Standard Split:

- **Duplex:** "+" (Positive).
- **Offset:** "5.000" (5 MHz).

What this actually does: Let's say you are on **Channel 15** (462.550 MHz). If you set Duplex to "+" and Offset to "5.0", when you press the PTT button, the radio instantly jumps UP 5 MHz to **467.550 MHz**. This is the **Input Frequency** of the repeater. When you let go of the button, it instantly jumps back DOWN to **462.550** to listen.

Visual Check:

- Look at your CHIRP row.
- Is it a repeater channel (15-22)?
- Does it have a Repeater Tone? -> Set Tone Mode to "Tone" and pick the freq.
- Is Duplex set to "+"?
- Is Offset set to "5.000000"?
- *If yes, you are ready to hit the tower.*

Example: The "Golden Row"

Here is what a properly programmed Repeater channel looks like in CHIRP:

- **Translation:** Channel 23. Named "Denver Repeater". It sends a 100.0 Hz tone to wake up the repeater. It shifts UP (+) by 5.0 MHz when transmitting.

4.3 Creating Your Fleet

You have spent an hour refining your CHIRP spreadsheet. You have named the channels. You have deleted the junk channels (like the Chinese demo frequencies often found in Baofengs). You have added your local NOAA Weather channels (Channels 162.400 etc.) at the bottom of the list with "Duplex: Off".

This file is now your **Master Template**. Save it as Golden_Config_v1.img.

Now we achieve **Fleet Uniformity**.

The "Clone" Process

You want every radio in your group - your handheld, your wife's handheld, your truck mobile unit, your base station - to have the exact same map.

When you scream "Go to Channel Denver!", you don't want your wife asking "Which number is that?" You want it to be Channel 23 on *every single device*.

The Challenge: You cannot always upload a Baofeng file to a Midland radio. The file formats are different. **The Fix:** You use **Copy/Paste**.

1. Open your Golden_Config_v1.img (The Source).
2. Plug in the second radio. Download its (blank/messy) config. This opens a second tab in CHIRP.
3. Go to your Source tab. Highlight all rows (Ctrl+A). Copy (Ctrl+C).
4. Go to the Destination tab. Highlight Row 1. Paste (Ctrl+V).
5. Upload to Radio.

Now, even if the radios are different brands, the logic (Frequencies, Tones, Names) is identical.

The "Scan List" Strategy

One setting you must check before uploading is the **Scan Add** column (sometimes called "Skip").

When you press the "Scan" button on your radio, it cycles through the channels looking for traffic.

- **The Problem:** By default, it scans *everything*. It will scan FRS 1, then FRS 2, then FRS 3...
- **The Annoyance:** Channels 8-14 are low power and usually full of static. Channels 1-7 are full of kids screaming. If you leave these in your scan list, your radio will never shut up.
- **The Fix:** Go to the "Skip" or "Scan Add" column in CHIRP.
    - **Set Channels 1-14 to "S" (Skip).**
    - **Set NOAA Weather Channels to "S" (Skip).** (Otherwise, the scanner will stop on the weather constantly because it is a continuous broadcast).
    - **Set your Tactical Channels and Repeater Channels to " " (Empty/Scan).**

Now, when you hit Scan, your radio becomes a "Hunter." It ignores the garbage and only listens for your convoy or the repeaters you care about.

"Kid-Proofing" (The Keypad Lock)

If you are handing a radio to a child, or even a non-technical adult, they *will* accidentally change the settings. They will grab the radio by the body, and their thumb will mash the "Menu" button, or the "VFO/MR" button which switches the radio from "Channel Mode" (safe) to "Frequency Mode" (dangerous/confusing).

The Software Lock: In CHIRP, go to the **Settings** tab (on the left side). Look for **"Work Mode Settings"** or **"Keypad Lock"**.

- **Auto-Lock:** Set this to "ON". This means if no buttons are pressed for 10 seconds, the keypad locks. You have to hold the # key (usually) to unlock it.
- **Disable VFO:** Some radios allow you to completely disable the "VFO" button via software. Do this. Force the radio to stay in "Channel Mode." This prevents the user from accidentally typing in 440.000 and getting lost in static.

The Physical Lock: If you can't do it in software, teach the "Lock" command. On almost all Baofeng/Wouxun radios, **Press and Hold the # Key** for 2 seconds. A little padlock icon appears on the screen. Do this *before* you hand the radio to anyone.

Workbook Element: The "Visual Command CHIRP Template"

We have done the heavy lifting for you. We have created a standard .csv file that you can import into CHIRP.

This template includes:

- **Channels 1-22:** Standard GMRS (High Power channels set to Wideband).
- **Channels 23-30:** Common "Travel Tone" channels used by overlanders (462.550 with Tone 67.0, etc.).
- **Channels 100+:** NOAA Weather Frequencies (Rx Only).

Scan to Download Printable Copies.

How to Import the Template:

1. Open CHIRP.
2. File -> **Import**.
3. Select the .csv file you downloaded.
4. CHIRP will ask you to map the columns. The defaults are usually correct.
5. Click OK.

Action Step: Stop reading. Go to your computer. Download CHIRP. Order the FTDI cable if you haven't. Do not go into the field with a radio programmed by the factory. The factory doesn't know your mission. You are the Commander. The radio does what *you* tell it to do.

Once you have the fleet programmed, it's time to learn how to actually talk on it. The next chapter covers the "Language of the Field" - how to sound like a pro and clearer comms.

Supplemental: Deep Dive on "Why the Factory Config Sucks"

You might ask, "Why can't I just use the channels that came on the radio?"

If you buy a high-end Midland, you probably can. But if you bought a Baofeng UV-5G or similar:

1. **Narrowband vs. Wideband:** Factory settings often default to Narrowband (NFM) on all channels to be "safe" for FRS rules. GMRS is allowed Wideband (WFM) on the main channels. Wideband sounds louder and richer. If you don't change this in CHIRP, you will sound quiet compared to everyone else.

2. **Random Tones:** Some factory radios come with random privacy tones pre-set on certain channels. You might think you are on Channel 1, but you are actually on Channel 1 with a 67.0Hz tone. You won't hear anyone else who isn't using that tone. You are effectively deaf to the public. **Wipe the slate clean.**

Supplemental: The "Repeaterbook" Integration

A crucial CHIRP feature for the advanced user.

CHIRP has a built-in superpower: **It can talk to "Repeaterbook.com".**

1. In CHIRP, click 'Radio' -> 'Query Source' -> 'Repeaterbook'.
2. Select 'GMRS'.
3. Select your 'State' (e.g., Colorado).
4. It will pull a list of all public repeaters in your state.
5. You can select the ones near you and import them directly into your spreadsheet.

* **Note:** It automatically fills in the Frequency, Offset, and Tone. It saves you from typing errors.

* **Visual Command Rule**: Always verify the data. Sometimes Repeaterbook is outdated.

Listen before you transmit.

Chapter Summary Checklist

Before moving to Chapter 5, you should have:

1. A programming cable with a genuine 'FTDI chip'.
2. 'CHIRP Next' installed and running on your computer.
3. Successfully "Downloaded" the stock config from your radio (and saved it as a backup!).
4. Built your 'Golden Config':
    - Standard Channels 1-22 named.
    - Local Repeaters added with correct Offsets/Tones.
    - NOAA Weather channels added.
    - Scan list cleaned up (Skip low power/weather).
5. 'Cloned' this config to at least one other radio.
6. Verified communication between the two radios.

Chapter 5

Privacy Codes & Filters

The "Doorbell" Analogy

Objective: Shatter the dangerous illusion of "Privacy Codes" while mastering their actual utility as noise filters for convoy sanity.

Key Deliverable: A "Sanity Filter" protocol that blocks interference without compromising operational security awareness (avoiding the "Glass House" trap).

You are at a crowded campground. It's a holiday weekend. Every campsite is occupied by an RV, a tent, and three kids running around with bright yellow walkie-talkies. The airwaves are saturated.

You turn on your radio to Channel 1. *screech* "MOM! MOM! WHERE ARE THE MARSHMALLOWS?" *static* "Back it up... back it up... stop." *beeping* "Marco! Polo!"

It is a cacophony. It is unusable. You can't hear your own family because the noise floor is raised by fifty other users.

So, you look at your manual. It says: **"121 Privacy Codes for Secure Communication."** The box promises that this feature will create a private line just for you. You think, *"Excellent. I'll turn on Privacy Code 12. Now my channel is secure, and nobody can hear us."*

STOP.

You have just fallen for the second biggest lie in the radio industry (right behind the "35-Mile Range").

"Privacy Codes" provide absolutely zero privacy.

Let me repeat that because it is a matter of operational security: **A Privacy Code is not encryption.** It is not a scrambler. It does not hide your voice. It does not prevent others from hearing you.

If you are planning a surprise birthday party, "Privacy Codes" are fine. If you are discussing the location of your hidden supply cache, giving your credit card number to your wife over the air, or discussing a medical emergency involving personal data, "Privacy Codes" will get you compromised.

In this chapter, we are going to strip away the marketing fluff and look at the mechanical reality of **CTCSS** and **DCS**. We are going to replace the word "Privacy" with the word "Filter." By the end of this chapter, you will understand exactly who can hear you, who you can hear, and how to set up a "Family Tone" that actually works without giving you a false sense of security.

5.1 It Is NOT Encryption

To understand why the term "Privacy Code" is a dangerous misnomer, we have to visualize how a radio receiver works. The marketing department wants you to believe you are creating a private tunnel. Physics disagrees.

Mental Model: The Open House vs. The Doorbell

Imagine your radio is a **House**. The **Frequency** (Channel 1) is the **Street Address**.

When you turn your radio to Channel 1 with **NO codes** enabled (Open Squelch):

- **The State:** You have removed the front door, the windows, and the walls. You are living in a gazebo.
- **The Result:** Anyone standing on the street can shout into your living room. You hear everyone. If the wind blows (static), you hear it. If a kid screams three houses down, you hear it.
- **The Reverse:** When you shout, everyone on the street hears you.

Now, let's enable a **"Privacy Code"** (CTCSS/DCS).

What the Marketing says happens: You enter a soundproof bunker deep underground. Nobody can hear you, and you can't hear them.

What ACTUALLY happens: You put the walls and windows back on the house. You close the front door. You install a **Doorbell**.

- **The State:** You are sitting in your living room with the door shut.
- **The Result:** The kid screaming on the street? You don't hear him because your walls muffle the sound. The wind? You don't hear it. The construction crew? Silence. Your speaker remains quiet.
- **The Trigger:** Your wife walks up to the door. She has the specific button (The Tone) to ring your doorbell. *Ding-Dong.* The door opens (Squelch opens). You hear her voice clearly.
- **The Trap:** *While your door is closed, you cannot hear the street. BUT... the windows are made of single-pane glass. Anyone standing on the street can still hear YOU shouting inside.*

The "Glass House" Reality

When you transmit with a Privacy Code, you are shouting out of a house made of glass.

Any radio operator who has their "Door" open (No Codes set) can hear everything you say. This is the default setting for most scanners and many pro-users monitoring the bands.

- **Scanner Guy:** The guy with the police scanner? He hears you.
- **The Rookie:** The kid who just bought a radio and hasn't set any codes? He hears you.
- **The Pro:** The GMRS operator monitoring Channel 16? He hears you.

The **ONLY** thing a Privacy Code does is silence *your* speaker until it hears a matching code. It filters *incoming* audio. It does nothing to *outgoing* audio. It essentially puts earplugs in your ears, but it does not put a muzzle on your mouth.

SAFETY ALERT: The "Stalker" Scenario I have seen this happen in real life. A family is camping. They set their radios to "Channel 1, Privacy Code 5." They think they are on a private line because their radio is quiet. Dad transmits: *"Hey honey, I'm leaving the campsite now, the keys are under the mat, and the kids are alone in the tent."* Dad thinks only Mom heard that because only Mom can talk back to him. In reality, a creep in a van three sites over, scanning Channel 1 with no codes (Carrier Squelch), heard every single word. He knows Dad is gone. He knows where the keys are. He knows the tactical situation.

Visual Command Rule: Never say anything on a radio - encrypted or not - that you wouldn't yell across a crowded parking lot. Even military encryption can be broken. Your $40 Baofeng certainly isn't secure. Treat every transmission as a public broadcast.

So, Why Use Them?

If they don't provide privacy, are they useless? No. They are **Filters**.

They are tools for **Sanity Management**. If you are in a convoy of 4 trucks, and you are driving through Atlanta, the radio waves are full of chatter. Construction crews, kids, interference, and random static bursts. If you don't use a code, your radio will be constantly bursting with static and strangers' voices. You will eventually get annoyed and turn the volume down to save your headache. Then, when your lead truck warns of a tire on the road, you won't hear him because you muted the radio.

By using a code, you silence the chaos. Your radio sits quietly on the dash. It only makes noise when *your* team speaks. That is the utility. It is an **Audio Gatekeeper**, not a security guard. It ensures that when the radio talks, it is someone you actually want to listen to.

5.2 CTCSS vs. DCS

Now that we know *what* they are (Filters), let's look at the two different flavors of filters available on almost every GMRS radio. While they accomplish the same goal, they use different physics to get there.

1. CTCSS (Continuous Tone-Coded Squelch System)

This is the "Old School" analog system. It is often called "PL Tones" (Private Line - a Motorola trademark) or "Sub-Channels."

How it Works: When you hold down the PTT button, your radio transmits your voice. Underneath your voice, deeper than the human ear can hear (sub-audible), it transmits a constant low-frequency hum (a sine wave).

- *Example:* Tone 67.0 Hz.
- Your radio hums a 67.0 Hz note the entire time you are talking.
- The receiving radio listens. If it detects that specific 67.0 Hz hum, it unmutes the speaker. If the hum stops or changes, it mutes.

The Pros:

- **Universal:** Every radio, from a 1980s CB to a modern GMRS unit, supports CTCSS. It is the industry standard.
- **Range:** Because it is an analog hum, it tends to survive static better. If you are on the fringe of reception, the tone usually still gets through even if the voice is scratchy.

The Cons:

- **Slow:** It takes a split second for the receiving radio to detect the hum and open the speaker. This can cut off the first syllable of your sentence. (e.g., instead of "Stop!", they hear "...op!").
- **Limited Keys:** There are only about 38 to 50 standard tones. In a crowded city, chances are high that someone else is using 67.0 Hz, which leads to "False Opens" - where your radio wakes up for someone else's conversation.

2. DCS (Digital-Coded Squelch)

This is the "New School" digital system.

How it Works: Instead of a hum, your radio transmits a stream of binary data (1s and 0s) constantly while you talk. It's like a digital barcode embedded in the audio.

The Pros:

- **More Keys:** There are over 100 standard DCS codes. Less chance of overlap with strangers.

- **Instant:** Digital decoding is often faster/cleaner than analog detection (on good radios).
- **No "Tail":** With CTCSS, when you let go of the button, sometimes there is a brief blast of static (the "Squelch Tail") before the radio closes. DCS usually eliminates this, making the end of the transmission sound like a clean *click*.

The Cons:

- **Fragile (The Digital Cliff):** Digital signals are 'all or nothing.' If you are in a high-static environment (weak signal), the digital data might get corrupted.
    - *Analog Fade:* With CTCSS, as you get further away, the audio gets static-filled but remains readable.
    - *Digital Drop:* With DCS, as you get further away, the radio simply refuses to open. Or, the audio cuts in and out violently (like a bad cell phone call). This is the "Digital Cliff." You fall off it abruptly.

Visual Command Recommendation: Start with **CTCSS** (Analog Tones). It is more robust for field use where signals vary. Only switch to DCS if you are in a city and experiencing interference on every single CTCSS tone.

The "Rosetta Stone" of Bubble Pack Radios

Here is where the "Tactical Dad" runs into a wall. You have a Baofeng UV-9G (The "Pro" radio). Your brother-in-law shows up with a blister-pack Motorola Talkabout T800 (The "Consumer" radio).

He says: *"I'm on Channel 1, Privacy Code 12."* You look at your Baofeng menu. It asks for a frequency, like 100.0 or D023N. It does not have a "Code 12."

The Problem: Consumer manufacturers (Midland, Motorola, Cobra) decided that "67.0 Hz" was too confusing for the average user. So they created their own arbitrary numbering systems.

- **Motorola Code 1** = 67.0 Hz
- **Midland Code 1** = 67.0 Hz
- **Motorola Code 38** = 250.3 Hz
- **Midland Code 38** = ... might be different.

And then it gets worse. Once they ran out of the 38 Analog tones, they started adding Digital codes to the list to inflate the numbers on the box ("121 Codes!").

- **Code 39** might be the first Digital Code (D023N).
- Or Code 39 might be a non-standard Analog tone.

There is no universal standard for "Code 12." Also, if you want to turn the code **OFF** on a consumer radio, set it to **Code 0**.

Field Tactic: The "Tone Scan" Trick If you cannot figure out what "Code 12" maps to on your pro radio, do not guess. Use the **Tone Scan** feature.

1. Set your Baofeng to the same Channel (e.g., Channel 1).
2. Go to Menu -> R-CTCS (Receive CTCSS).
3. Hit "Scan" (* / Scan button).
4. Have your brother-in-law hold down his PTT button and talk continuously. *"Testing testing 1 2 3..."*

5. Your radio will cycle through the tones until it opens the squelch.
6. It will flash "**100.0**" (or whatever the tone is).
7. **Boom.** You found the translation. Lock that in.

Workbook Element: The Translation Chart We have included a generic translation chart in the Appendix, but *always* verify.

- *General Rule:* Codes 1-38 are usually the standard 38 CTCSS tones in ascending order (67.0 to 250.3).
- *General Rule:* Codes 39+ are usually DCS.

5.3 Tone Squelch vs. Tone Transmit

In Chapter 4, we touched on the CHIRP columns **Tone Mode**, **Tone**, and **ToneSql**. Now we are going to solidify this logic, because getting it wrong means you will be deaf or unheard. This is the single most common configuration error in GMRS.

This is the logic of the **Two-Way Door**.

The Four States of Tone

State 1: Carrier Squelch (CSQ)

- **CHIRP Mode:** (None)
- **Visual:** Both doors are removed.
- **Tx:** You send voice only. (No Key).
- **Rx:** You listen for voice only. (No Lock).
- *Result:* Maximum compatibility. You hear everyone; everyone hears you.
- *Best For:* Hailing frequencies, emergency comms, scanning. If you are lost, be in this mode.

State 2: Tone Transmit (T-TX)

- **CHIRP Mode:** Tone
- **Visual:** You have a Key in your hand, but your door is unlocked.
- **Tx:** You send Voice + Tone. (Using the Key).
- **Rx:** You listen for voice only. (No Lock).
- *Result:* You can open a locked door (like a Repeater), but you can still hear strangers.
- *Best For:* **GMRS Repeaters.** Repeaters almost always require a tone to enter (Input), but do not send a tone back (Output). If you lock your door (State 3) on a repeater channel, you won't hear the repeater responding.

State 3: Tone Squelch (T-SQL)

- **WARNING:** Do not SCAN while in this mode. If you scan with TSQL on, your radio will ignore every conversation that doesn't use your specific key. You will scan past active channels and hear nothing.
- **CHIRP Mode:** TSQL

- **Visual:** You have a Key in your hand, and you have Locked your door with the same key.
- **Tx:** You send Voice + Tone.
- **Rx:** You listen ONLY for Voice + Tone.
- *Result:* The "Private" Line. You only hear radios that have the exact same setting.
- *Best For:* Family camping, convoys in cities, blocking interference.

State 4: Receive Only Tone (Rare)

- **Visual:** You have no key, but your door is locked.
- **Result:** You can't open anyone's door, and nobody can open yours unless they have the key.
- **Utility:** Useless for 99% of users. Avoid this.

Visual Workflow: "Which Mode Do I Choose?"

Ask yourself two questions:

Question 1: Am I talking to a Repeater?

- **YES:** Use **State 2 (Tone Mode: Tone)**.
    - *Why:* Repeaters are usually "Public." They listen for a key to wake up, but they broadcast openly so that any scanner can hear them. If you lock your door (TSQL), you will filter out the repeater and think it is broken.
- **NO:** Go to Question 2.

Question 2: Am I in a "Noise Rich" Environment?

- **YES (City, Event, Crowded Camp):** Use **State 3 (Tone Mode: TSQL)**.
    - *Why:* You need to filter out the garbage.
- **NO (Deep Woods, Desert):** Use **State 1 (Tone Mode: None)**.
    - *Why:* In the middle of nowhere, hearing a stranger is actually useful. It means someone else is nearby (safety/intel). Don't filter out the only other human for 50 miles.

SAFETY ALERT: The "Double Lock" Fail A common mistake is having a mismatch in the group.

- **Dad's Radio:** Set to TSQL 100.0 (Locked).
- **Kid's Radio:** Set to (None) (Unlocked).

The Result:

- Kid talks -> Dad hears nothing. (Kid didn't use the key).
- Dad talks -> Kid hears Dad. (Kid has no lock).
- *The Green Light of Death:* Dad will look at his radio. He will see the Receive LED light up Green (indicating a signal is hitting the antenna). But the speaker will be dead silent. This is the universal sign of a **Tone Mismatch**. The radio is saying: "I hear a signal, but it doesn't have the secret handshake, so I am not letting it through."

- *Panic ensues.* Dad thinks Kid is lost or the radio speaker is broken. Kid thinks Dad is ignoring him.
- **Visual Command Rule:** Sync the fleet. If one radio is TSQL, ALL radios must be TSQL.

Action Step: The "Family Tone" Setup

You need to pick a standard tone for your family or group. This will be your default setting for your "Tactical Channels" (channels 23-30 in our Golden Config).

1. Avoid the "Default" Tones Humans are lazy. When they buy a Midland radio, they turn it to Channel 1, Code 1.

- **Code 1 (67.0 Hz)** is the most crowded tone in the world. Using this is like trying to have a private conversation in the middle of Times Square.
- **Code 12 (100.0 Hz)** is also very common because it's a nice round number.

2. Pick an "Obscure" Analog Tone Go for the weird numbers in the middle of the list. They are less likely to be used by the family next door.

- *Good Candidates:* 141.3 Hz, 151.4 Hz, 162.2 Hz.
- *Avoid:* 67.0, 88.5, 100.0.

3. Standardize Your "Travel Channel" In your CHIRP config, designate one channel (let's say Channel 25) as **"FAM-TAC"** (Family Tactical).

- **Frequency:** 462.650 (GMRS 19)
- **Tone Mode:** TSQL
- **Tone:** 141.3

Program this into every radio. Label it on your "Comms Card" (from Chapter 2).

The Protocol:

1. **Start on Channel 20 (National Call - No Codes).** Establish contact. *"Unit 1 to Unit 2."*
2. **Move to Secure.** *"Unit 2, switch to FAM-TAC."*
3. **Switch.** Both radios move to Channel 25 (TSQL 141.3).
4. **Verify.** *"Unit 2 on FAM-TAC, how copy?"*

Now you have moved from the "Open Street" to your "Private Room." You have filtered out the noise. You can communicate clearly. But you never, ever forget: **The glass walls are still there.**

Appendix: The "Rosetta Stone" (Abbreviated)

Print this and tape it to your Baofeng battery if you travel with "Bubble Pack" users.

Brand Code	Freq (Hz)	Brand Code	Freq (Hz)	Brand Code	Freq (Hz)
1	67.0	**14**	107.2	**27**	167.9
2	71.9	**15**	110.9	**28**	173.8
3	74.4	**16**	114.8	**29**	179.9
4	77.0	**17**	118.8	**30**	186.2
5	79.7	**18**	123.0	**31**	192.8
6	82.5	**19**	127.3	**32**	203.5
7	85.4	**20**	131.8	**33**	210.7
8	88.5	**21**	136.5	**34**	218.1
9	91.5	**22**	141.3	**35**	225.7
10	94.8	**23**	146.2	**36**	233.6
11	97.4	**24**	151.4	**37**	241.8
12	100.0	**25**	156.7	**38**	250.3
13	103.5	**26**	162.2	**39+**	*Usually Digital*

Note: This covers the standard 38 Analog tones used by Motorola/Midland/Cobra/Uniden. Always test to verify.

Troubleshooting: The Green Light of Death If you see your radio's Green LED light up (indicating a signal) but hear silence, you likely have a **Receive Tone (R-CTCS)** set that doesn't match the repeater.

- **The Fix:** Go into your menu and set **R-CTCS to OFF** (or 0.0). This opens your ears to everything the repeater sends.

Chapter 6

Repeater Operations

Extending Your Reach

Objective: Unlock regional communication range by mastering the "Split Highway" mechanics of repeaters and the +5MHz offset.

Key Deliverable: The ability to field-program a repeater interface in 60 seconds without a computer, turning a short-range handheld into a 50-mile asset.

You are standing in a deep, pine-choked valley. The air is cold, and the sun dipped below the ridgeline an hour ago. Your friend is in the next valley over, five miles away, waiting for your status update. Between you sits a massive granite ridge, 2,000 feet high and solid rock.

You press the PTT button on your 5-watt handheld radio. *Nothing.* Just the hiss of static. You try standing on a rock. You try holding the radio sideways. *Still nothing.*

The granite eats your signal. UHF waves, as we learned in Chapter 1, are Line-of-Sight. They hit the mountain and stop dead. In this moment, physics has defeated you. You are isolated.

But then, you remember the map. You switch your radio to Channel 23. You press the button and say, *"WRXP-555, radio check."*

Beep-Boop.

A clear, robotic "courtesy tone" chirps back at you. Then, a second later, your friend's voice comes through crystal clear, loud and authoritative, as if he is standing right next to you.

You have just unlocked the superpower of GMRS: The Repeater.

A repeater is a "Force Multiplier." It is a specialized machine that sits on top of that granite ridge, usually powered by solar panels and a bank of heavy batteries. It listens to your weak, 5-watt signal struggling up out of the valley, catches it, cleans it up, amplifies it to a massive 50 watts, and screams it down into the next valley - and often into the next three counties.

Using a repeater turns your $40 handheld radio into a 50-mile communication system. It transforms you from a lone hiker into a node on a regional network.

However, accessing a repeater requires more than just tuning to a channel. It requires understanding the **Split**. Most new users fail here because they treat a repeater like a normal radio. It isn't. It is a machine that demands a specific handshake. If you don't know the secret knock, the machine will ignore you.

In this chapter, we are going to visualize exactly how this machine works, how to find one near you, and how to program it into your radio while standing in the rain without a computer.

6.1 The "Split Highway" Concept

To understand a repeater, you have to stop thinking about "A Channel" and start thinking about "A Highway."

In standard **Simplex** communication (Unit-to-Unit):

- You talk on 462.550 MHz.
- Your friend listens on 462.550 MHz.
- It is a **One-Lane Road**. Only one car can go at a time. If you both try to talk at once, you crash (this is called "doubling"), and nobody hears anything but a garbled mess.

A Repeater is fundamentally different. A Repeater cannot listen and talk on the same frequency at the exact same time. If it did, its own powerful 50-watt transmitter would instantly blow out its own sensitive receiver. It would be like trying to listen to a whisper from across the room while you are screaming through a megaphone. You would deafen yourself.

So, the Repeater uses a **Split Highway**. It separates the traffic into two distinct lanes.

Mental Model: The High Road and The Low Road

Imagine the Repeater is a Giant stationed on a mountain peak. This Giant has excellent hearing and a booming voice.

1. The Listening Ear (Input Frequency) The Giant listens on a specific frequency called the **Input**.

- This is always in the **467 MHz** range (The "High" Road).
- When you talk *to* the repeater, your radio must transmit on this frequency. This is the only frequency the Giant's ears are tuned to.

2. The Shouting Mouth (Output Frequency) The Giant shouts on a different frequency called the **Output**.

- This is always in the **462 MHz** range (The "Low" Road).
- When you listen *to* the repeater, your radio must listen on this frequency. This is where the Giant broadcasts your message to the world.

The Magic Trick: Your GMRS radio handles this switch automatically - *if* and only *if* you program it correctly.

- **Standby State:** When you are just holding the radio, it sits on 462 MHz (The Output), listening for the Giant.
- **Active State:** The *millisecond* you press the PTT button, the radio instantly jumps UP to 467 MHz (The Input) to talk to the Giant.
- **Release State:** When you release the button, it jumps back DOWN to 462 MHz to hear the Giant's reply (or your friend's reply coming through the Giant).

The +5 MHz Offset Rule

The distance between the Input and the Output is exactly **5 MegaHertz**. This separation is critical. It provides enough "spectral distance" so that the Giant's mouth doesn't overpower his ears.

- **Output (Listen):** 462.550 MHz
- **Input (Talk):** 467.550 MHz
- **Difference:** +5.000 MHz

This is called the **Offset**. In the GMRS world, the Offset is *always* **Positive (+)** and it is *always* **5.000 MHz**.

(Note: Ham radio operators have to deal with negative offsets, variable splits, and complex band plans. We have it easy. In GMRS, if you are on a repeater channel, the answer is always +5).

Why This Breaks 90% of New Users Here is the classic scenario of the frustrated Rookie: You look up a repeater. It says "Frequency: 462.550." You type 462.550 into your radio. You press the button. You shout "Radio Check!" **Silence.**

Why? Because you forgot to turn on the Offset.

1. You pressed PTT.

2. Your radio transmitted on 462.550 (The Output).

3. The Repeater is listening on 467.550 (The Input).

4. **Result:** The Repeater didn't hear you. You were shouting at its mouth, not its ear. It's like throwing a ball at the back of the Giant's head. He doesn't know you are there.

5. You hear nothing back. You assume the repeater is broken. It isn't. You just knocked on the wrong door.

6.2 Finding Repeaters

So, where are these magical machines? They are everywhere. There are thousands of GMRS repeaters across the United States, usually located on high ground - mountaintops, water towers, skyscrapers, and even grain silos in the Midwest.

But unlike cell towers, your radio doesn't just "roam" onto them automatically. You have to hunt them down, identify them, and often ask for permission.

The Tool: MyGMRS.com

The absolute gold standard for finding repeaters is **MyGMRS.com**. It is a community-driven database where owners list their machines.

1. Go to the website (or download the app).

2. Create an account (You will need your valid GMRS Call Sign - this keeps the database clean of spammers).

3. Click "Map."

4. Zoom into your area of operations.

You will see pins on the map. Don't just look at the location of the pin; look at the **Coverage Map** circle around it. A repeater on a 10,000-foot peak might cover 80 miles. A repeater on a 100-foot water tower in a valley might only cover 5 miles.

Click on a pin near you. You will see a data sheet like this:

Field	Value	Meaning
Name	Denver 550	The local name (often Refers to the freq).
Output Freq	462.550	The channel you listen to (GMRS 15).
Input Tone	141.3	The "Key" required to open the door.
Output Tone	141.3	The tone the repeater sends back (for your TSQL).
Bandwidth	Wide	Set your radio to WFM.

Understanding "Permission"

This is the social contract of GMRS. Repeaters are privately owned. A regular guy named Dave spent $2,000 of his own money to buy the Motorola Quantar repeater, the hardline coax cable, and the 20-foot fiberglass antenna. He climbed the tower. He pays the monthly electric bill. He pays the site insurance. He is letting you use it out of the kindness of his heart (and for the benefit of community safety). **Respect his rules.**

When you look at a listing on MyGMRS, look for the **Access** status:

1. OPEN System

- **Status:** "Open" or "Public Placed."
- **Rules:** You can use it immediately. The tone is listed right there on the page. Just program it and talk.
- *Etiquette:* Be a good guest. Give your call sign. Don't hog the airtime with a 45-minute discussion about your sciatica. Leave gaps between transmissions so others can break in.

2. PERMISSION REQUIRED

- **Status:** "Permission Required" or "Request Access."
- **Rules:** You cannot just use it. You must click the "Request Access" button on the website.
- The owner (Dave) will get an email. He will check if you have a valid license and if you seem like a normal human being.
- If he approves, he will email you the **Input Tone**.
- *Note:* The listing usually hides the tone until you are approved. **DO NOT** try to "hack" or tone-scan a private repeater. Finding the tone and using it without asking is the fastest way to get banned by the local community. It is the digital equivalent of jumping a fence.

3. CLOSED / PRIVATE

- **Status:** "Private" or "Closed."
- **Rules:** Family and close friends only.
- *Why?* Maybe it's a solar-powered repeater with limited battery capacity. Maybe it's for a specific Neighborhood Watch. If it says Closed, move along.

Field Tactic: The "Travel Tone" Strategy What if you are driving cross-country and don't have time to look up every single repeater in every county? Many "Open" repeaters across the country adhere to a loosely agreed-upon standard called the **Travel Tone** (141.3 Hz). If you are driving through a new city and see a repeater on 462.675 but don't have internet to check the tone, try **141.3 Hz**. It works about 40% of the time on open systems. It's the "Skeleton Key" of the GMRS traveler.

6.3 Programming a Repeater Channel (Field Programming)

You are on a trail. The weather is turning bad. You meet a local wheeler in a Jeep. He rolls down his window and says, *"Hey, heads up, the main road is washed out. We are using the 'Bear Peak' repeater on 675 to coordinate recovery. Tone is 123.0."*

You nod. He drives off. You look at your radio. You don't have your laptop. You don't have CHIRP. You have to program this repeater into your radio **Right Now** using the keypad, while rain drips onto the screen.

This is the "Final Exam" of GMRS proficiency. The exact buttons vary by radio (Baofeng vs. Wouxun vs. Midland), but the **Logic** is universal. You need to map the inputs to the machine.

The "FPP" (Front Panel Programming) Workflow

Step 1: Switch to Frequency Mode (VFO) Most radios have two modes: "Channel Mode" (Memory) and "Frequency Mode" (VFO).

- **Channel Mode:** Shows names like "GMRS 15". It is safe. You can't mess it up.
- **Frequency Mode:** Shows numbers like 462.550. It is the "Wild West." This is where you build new channels.
- *Action:* Press the **VFO/MR** button until you see frequencies.

Step 2: Enter the Output Frequency

- Type 462.675. (This corresponds to GMRS Channel 20).
- *Why:* This is the frequency you will *listen* to.

Step 3: Set the Input Tone (The Key)

- Press **Menu**.
- Look for **T-CTCS** (Transmit Continuous Tone Coded Squelch) or **TX-CODE**.
- *Note:* Do NOT just set R-CTCS (Receive). That locks your ears but leaves your voice unlocked. You need **T-CTCS** to open the repeater's door.
- Select 123.0.
- Press Menu/Confirm.
- *Check:* You should see a little "CT" icon appear on the screen.

Step 4: Set the Offset Direction (SFT-D)

- Look for a menu item called **SFT-D** (Shift Direction).
- Select + (Positive).
- *Note:* Some GMRS-specific radios (like Wouxun KG-935G) hard-code this for you. If you can't find it, the radio might already know to shift up. But on a generic unlocked Baofeng UV-5G, you must set this manually.

Step 5: Set the Offset Amount (OFFSET)

- Look for **OFFSET**.
- Enter 005.000 (5 MHz).
- *Critical Warning:* Many radios display this as 00.000. You want 05.000. If you enter 00.500, you are setting a 0.5 MHz offset (Ham standard) and you will miss the repeater.

Step 6: The "Kerchunk" Test

- Before you save it, test it.
- Press the PTT button for one second. Release.
- **Listen.**
- **The Sound of Success:** You are listening for the **Repeater Tail**. This sounds like a short, sharp blast of static (*pshhht*) or a mechanical *click* that happens about 0.5 seconds *after* you release your PTT. It is the sound of the Giant's transmitter turning off.
- **The Result:**
  - **Tail / Courtesy Tone (Beep):** Success. You opened the machine.
  - **Silence:** You failed. Wrong tone, wrong offset, or out of range.
- **Etiquette Rule:** Do not just click the mic ("Kerchunking"). It is rude. Key up and say, *"WRXP-555 testing."* This identifies you as a respectful operator.

Step 7: Save to Memory

- Now that it works, save it so you don't lose it when you turn the radio off.
- Menu -> **MEM-CH** (Memory Channel).
- Select an empty slot (e.g., Channel 25).
- Press Confirm.
- Switch back to "Channel Mode" (MR) and go to Channel 25.

Visual Command Tip: Most "Bubble Pack" radios (Midland/Motorola) *cannot* be programmed with custom repeater channels in the field via a keypad. They are too simple. They usually have "Repeater Channels 15-22" pre-loaded in a separate bank.

- **Midland MXT Procedure:** You don't program offsets. You just switch to the "RP" channels (15RP through 22RP). The radio already knows the offset. You just go to Menu -> CTCSS and select the tone. It's easier, but less flexible.

Field Tactic: The "Reverse" Check (REV) Most radios have a button labeled **REV** or ***SCAN**. When you hold this button, your radio temporarily swaps the Input and Output frequencies. It makes you listen to the Input (467 MHz).

- **Why do this?** If you press REV and you can hear your friend's voice clearly *without* the repeater, you are close enough to talk Simplex.
- **The Pro Move:** Stop using the repeater. Switch to a Simplex channel (like Ch 16). Save the repeater for when you

actually need the range.

Checklist: "Repeater Pre-Check"

Before you try to talk on a repeater, run this 5-second mental audit. If you skip this, you will likely be the guy shouting into the void.

1. Frequency Match: Am I on the right Output Channel? (e.g., 462.xxx).
2. Tone Match: Do I have the correct **Transmit (Input)** tone set? (Double check T-CTCS vs R-CTCS).
3. Offset Active: Do I see a little "+" icon on my screen? (If not, I am talking Simplex and the repeater won't hear me. I am talking to its mouth).
4. Bandwidth: Is my radio set to **Wide (WFM)**?
    - *Why:* Most repeaters are Wideband. If your radio is set to Narrow (NFM), your audio will sound incredibly quiet, like you are whispering from inside a box. The repeater's "gate" might even cut you off because your audio isn't loud enough to keep it open.
5. Listen First: Is anyone using the repeater right now? Don't interrupt a conversation just to say "Radio Check." Wait for a break.

The Golden Rule of Repeaters: A Repeater is a shared resource. It is a "Community Campfire." When you step up to the campfire, don't start screaming. Introduce yourself. *"WRXP-555, monitoring."* This simple phrase tells the locals you are there, you are licensed, and you are listening. It invites conversation without demanding it. It marks you as a "Citizen Operator," not a "Lid" (a bad operator).

Troubleshooting: The Green Light of Death If you see your radio's Green LED light up (indicating a signal) but hear silence, you likely have a **Receive Tone (R-CTCS)** set that doesn't match the repeater.

- **The Fix:** Go into your menu and set **R-CTCS to OFF** (or 0.0). This opens your ears to everything the repeater sends.

Chapter 7

Protocol & Etiquette

Sounding Like an Operator

Objective: Transform from a mic-shy purchaser into a confident communicator using professional, concise logic.

Key Deliverable: Three "Glovebox Scripts" for instant reference.

7.1 The Anti-Sad Ham Manifesto

Let's get one thing straight before you key that microphone: You are not a trucker from 1978, and you are not playing a character in a bad Hollywood war movie. You are a parent securing your family, or a convoy leader guiding vehicles over an obstacle. The biggest hurdle to effective radio use isn't "Line of Sight" physics or CTCSS privacy tones - it is **Mic Fright**. This is the paralyzing fear that you're going to say the wrong thing and get yelled at by the "Radio Police."

In the amateur radio world, there is a subset of hobbyists often derisively called "Sad Hams." These are the gatekeepers. They are the guys who sit in their basements with $5,000 worth of equipment, listening to the silence, waiting for someone to make a minor procedural error so they can lecture them. They treat the airwaves like a country club where you have to wear a tie to get in.

We are not them.

In the *Visual Command System*, a radio is a tool, not a lifestyle. It sits in the hierarchy of survival gear right next to your fire extinguisher and your winch. You don't have a secret language for using a fire extinguisher, and you don't need one for GMRS.

The Problem with "10-Codes"

You have likely heard "10-4" (Message Received) or "10-20" (Location). These are called 10-Codes. They were popularized by law enforcement and CB radio users decades ago to shorten transmission times on crowded channels.

Do not use them.

Why? Because they are dangerous. 10-Codes vary by jurisdiction. In one county, a "10-50" might mean a car crash; in the next county over, it might mean an officer needs a bathroom break. When you are dragging a trailer down a scree slope or trying to locate a lost child in a state park, ambiguity is the enemy. If you say, "I've got a 10-33 here," and your wife thinks you mean "Emergency Traffic" but you actually meant "Spotter Required" based on a list you saw on the internet, you have introduced friction into a safety system. Friction gets people hurt.

The "Breaker Breaker" Trap

If you start a transmission with "Breaker Breaker, one-nine," you have immediately signaled to everyone within five miles that you have no idea what you are doing. This is Citizens Band (CB) slang from a bygone era. It has no place on GMRS. It's the verbal equivalent of wearing your shoes on the wrong feet.

The Solution: Plain Language

The United States government, specifically the Department of Homeland Security and FEMA, officially moved away from 10-Codes and jargon in 2006 (under the NIMS/ICS protocols). They mandated **Plain Language**.

If you want to sound like a Tier-1 operator, don't use code. Speak English.

- **Instead of:** "Can I get a 10-20 on the Lead Vehicle?"
- **Say:** "Lead Vehicle, what is your location?"
- **Instead of:** "That's a big 10-4, good buddy."
- **Say:** "Copy that." or "Received."
- **Instead of:** "Breaker One-Nine for a Radio Check."
- **Say:** "Radio Check."

Plain Language is the ultimate "anti-elitist" protocol. It works for your 6-year-old child, your spouse, and the Search and Rescue team that might be monitoring the channel. It strips away the ego and leaves only the information.

The Visual Command Rule of Brevity

While we use plain language, we do not use *excessive* language. A radio channel is a "single-file hallway." Only one person can walk through (talk) at a time. If you stand in the hallway having a casual conversation, you are blocking the flow for everyone else.

Think before you press the button (PTT). Visualize your sentence like a text message. If it would cost you $5.00 to send the message, how would you edit it down?

The "Sad Ham" Way (WRONG)	The Visual Command Way (RIGHT)
"Uh, yeah, this is WXYZ123, just calling out to the unit in the red Toyota, wondering if you guys saw that turn off back there, over."	"Red Toyota, you missed the turn. Turn around."
"Breaker, breaker, can I get a radio check? I just bought this new Baofeng and want to see if I'm hitting the repeater."	"Radio Check."
"10-4, I will proceed to your 10-20 effectively immediately."	"Moving to you now."

The Takeaway: Sounding like a pro doesn't mean sounding complicated. It means sounding *clear*.

7.2 Scripts for Anxiety-Free Comms

When adrenaline spikes - like when a vehicle is sliding sideways or a kid wanders off - your cognitive function drops. You don't rise to the occasion; you fall to the level of your training.

If you don't have a script, you will stutter, mumble, or freeze. We use the **"Hey You, It's Me"** protocol. This is the global standard for professional communications (aviation, military, marine).

1. **Hey You:** Address the person you want.

2. **It's Me:** Identify yourself.

3. **The Message:** The instruction or question.

Below are the three "Glovebox Scripts." Print these out. Tape them to the dash. Drill them.

Script A: Requesting a Radio Check

You've just turned your radios on. You need to verify they are working before you lose cell signal. Do not tap the microphone (this damages the element) and do not blow into it.

The Logic: You are throwing a ball into the dark. You are waiting for someone to throw it back.

The Script:

YOU: "Radio Check." (Wait 3 seconds). **YOU:** "Any station, Radio Check, this is [Your Call Sign]."

If someone replies:

THEM: "Loud and clear, [Call Sign]." **YOU:** "Thanks, [Your Call Sign] clear."

Note on Etiquette: If you hear someone ask for a radio check, answer them! A simple "I hear you, this is [Your Call Sign]" is all that is required. It is a courtesy that confirms their safety equipment is functional.

Script B: Calling a Specific Person

You are the "Chase Vehicle" and you need to talk to "Lead Vehicle" (Dad).

The Logic: Open the channel, establish the link, *then* send traffic. Do not dump the information in the first transmission, because if they weren't listening, they missed it.

The Script:

YOU: "Lead Vehicle, from Chase. Over."

(Wait for them to wake up and grab the mic)

THEM: "Go ahead, Chase."

YOU: "There is a large rock in the left track, suggest swinging wide right. Over."

THEM: "Copy. Wide right. Out."

Why this works:

- **"Lead Vehicle, from Chase":** This alerts *everyone* on the channel that the conversation is for Lead, and it is coming from Chase. Everyone else knows to stay quiet.
- **"Over":** This is a "handshake." It hands the frequency to the other person.
- **Read-back:** Notice "Lead" repeated the instruction ("Wide right"). This confirms they understood the specific safety instruction. This is called "Closed Loop Communication."

Script C: Emergency Break-In

This is the "Nuclear Option." Use this only when life, limb, or property is at immediate risk, and the channel is currently busy with other people talking. Perhaps a group of hikers is chatting on the channel you monitor, but your truck just caught fire.

You need to interrupt them.

The Logic: You must shatter the current conversation rhythm to gain attention.

The Script:

YOU: "BREAK, BREAK, BREAK. Emergency Traffic."

(The channel should go silent instantly).

YOU: "This is [Your Call Sign], I have a medical emergency at [Location]. Is anyone copying? Over."

Rules of the "Break":

1. **Never** use "Break" for non-emergencies (like asking for a mic check). It is like pulling a fire alarm to see if it works.
2. If you hear "Break, Break," stop talking immediately. Listen. If you are the strongest station, answer them and offer assistance.

The "Over" vs. "Out" Distinction

In movies, you hear "Over and Out." In reality, those are contradictory terms.

- **Over:** "I am done talking, but I expect a reply." (Handing the ball to you).
- **Out:** "I am done talking, and the conversation is finished." (Putting the ball in the box).
- **Clear:** "I am leaving this channel."

Correct: "Copy that, see you there. Out." **Incorrect:** "Copy that, Over and Out." (This sounds like "I'm expecting a reply, but I'm hanging up.")

7.3 The NATO Phonetic Alphabet

GMRS operates on UHF frequencies. While FM is clear, static happens. Obstacles happen. Wind noise happens.

The English language is terribly designed for radio.

"B" (Bravo) "C" (Charlie) "D" (Delta) "E" (Echo) "G" (Golf) "P" (Papa) "T" (Tango) "V" (Victor) "Z" (Zulu)

Nine letters in the alphabet end in the "EE" sound. If you are spotting your wife over an obstacle and shout "Go feet!" she might hear "Go Deep," "Go P," or "Go V."

If you say "My callsign is WRKC290," the "C" might sound like "Z" or "V" or "E". To solve this, we use the **NATO Phonetic Alphabet**. This is not militaristic cosplay; it is audio engineering designed to ensure that no two words sound alike, even in heavy static.

The Visual Command Cheat Sheet

Do not try to memorize this overnight. Print this page. Cut out this table. Tape it to the back of your handheld radio or your dashboard.

Letter	Word	Pronunciation	Why? (Audio Distinction)
A	Alpha	**AL**-fah	Hard 'L' sound distinguishes from others.
B	Bravo	**BRAH**-voh	Strong 'BR' start, distinct 'O' finish.
C	Charlie	**CHAR**-lee	'CH' is a hard consonant sound.
D	Delta	**DEL**-tah	Two hard syllables. Impossible to confuse.
E	Echo	**ECK**-oh	Short, sharp start.
F	Foxtrot	**FOKS**-trot	Two distinct words combined.
G	Golf	**GOLF**	Hard 'G', ends in 'F'.
H	Hotel	hoh-**TEL**	Distinct accent on second syllable.
I	India	**IN**-dee-ah	Three syllables, unmistakable.
J	Juliet	**JEW**-lee-ett	Soft 'J', sharp 'T' end.
K	Kilo	**KEY**-loh	Hard 'K'.
L	Lima	**LEE**-mah	Liquid 'L' sound.
M	Mike	**MIKE**	One syllable, sharp 'K' end.
N	November	no-**VEM**-ber	Three syllables, unlike 'Mike'.
O	Oscar	**OSS**-cah	Distinct 'O' start.
P	Papa	**PAH**-pah	Plosive 'P' sounds cut through noise.
Q	Quebec	keh-**BECK**	Hard 'Q' and 'K' sounds.
R	Romeo	**ROW**-me-oh	Rolling flow.
S	Sierra	see-**AIR**-rah	Sibilant 'S' sound.
T	Tango	**TANG**-go	Hard 'T' and 'G'.
U	Uniform	**YOU**-nee-form	Distinct 'U' start.
V	Victor	**VIK**-tah	Sharp 'K' in the middle.
W	Whiskey	**WISS**-key	Distinct 'W' and 'K'.
X	X-ray	**ECKS**-ray	Only one starting with 'E' sound besides Echo.
Y	Yankee	**YANG**-key	Twangy sound cuts static.
Z	Zulu	**ZOO**-loo	Deep vowels.

How to Practice Without Looking Weird

You don't need to speak in full phonetic sentences ("I AM GOING TO THE GOLF-RRR-OSC-EEE-RRR-Y STORE").

That is strictly "Sad Ham" territory.

You use phonetics **only for critical data**:

1. **Call Signs:** "This is W-R-K-C-2-9-0" becomes "Whiskey Romeo Kilo Charlie Two Nine Zero."
2. **Map Coordinates:** "Grid Square Alpha Four."
3. **Spelling clarifications:** "Meet at the trail head. That is trail B as in Bravo."

The "Say Again" Rule: If you don't understand a transmission, do not say "What?" or "Huh?". Say: **"Say Again."** If you need them to use phonetics because the signal is bad, say: **"Say Again, verify with phonetics."**

Action Step: The "Driveway Drill"

Reading this chapter does not make you a competent operator. Pushing the button makes you a competent operator.

Anxiety comes from the unknown. If the first time you try to use "Script B" is when you are stuck in a mud hole with your wife yelling at you and the rain pounding on the roof, you will fail. You will revert to shouting "CAN YOU HEAR ME?" and getting frustrated.

We are going to kill that anxiety today, in the comfort of your driveway.

The Setup

1. **Participants:** You (The Tactical Dad) and your Partner (The skeptical spouse or eager child).
2. **Gear:** Two handheld radios (HTs).
3. **Location:** Your driveway and inside your house (or down the street).

The Drill Execution

Step 1: The Separation Give one radio to your partner. Send them inside the house or to the end of the block. You stay at the vehicle. *Ensure both radios are on the same channel.*

Step 2: The Radio Check (Script A)

- **You:** "Radio Check."
- **Partner:** "Loud and clear. This is [Partner Name]."
- **You:** "Received. Out."

Why: This confirms battery, frequency, and volume.

Step 3: The Specific Call (Script B)

- **You:** "Base, this is Mobile 1. Over."
- **Partner:** "Go ahead, Mobile 1."
- **You:** "I am initiating the convoy test. Requesting current time. Over."
- **Partner:** "Time is [Current Time]. Over."
- **You:** "Copy [Current Time]. Out."

Why: This practices the "Hey You, It's Me" handshake. It feels awkward at first. Do it anyway. It builds the muscle memory of *waiting* for the reply before dumping data.

Step 4: The Phonetic Spelling

- **You:** "Base, Mobile 1. I need you to record a license plate number. Ready to copy?"
- **Partner:** "Go ahead."
- **You:** "License plate is: Alpha, Bravo, Seven, Niner, Zulu, Tango. Over."
- **Partner:** "I copy Alpha Bravo Seven Nine Zulu Tango?"
- **You:** "Correct. Out."

Why: This forces you to glance at the cheat sheet and use the words "Alpha" and "Zulu" instead of "Apple" and "Zebra."

The "Hot Mic" Awareness Check

Before you finish the drill, teach your family about the "Hot Mic." Have your partner hold the PTT button down and try to listen to you. They won't hear anything. **Rule:** *When you are talking, you are deaf.* GMRS is "Simplex" (one way at a time). If you hold the button down to think ("Uhhhhh...."), you are jamming the frequency and nobody can tell you to stop. **Teach:** Push. Pause (1 second). Talk. Release immediately.

Debrief

Once the drill is done, put the radios back in the chargers. You have now established a baseline of competence. The next time you hand that radio to your spouse on a trail, they won't look at it like it's an alien artifact. They will remember the "Driveway Drill." They will know the flow.

You have moved from "Person with a Radio" to "Operator."

Chapter 8

The Family PACE Plan

SOPs for Safety

Objective: Replace "panic and guessing" with a pre-loaded decision algorithm.

Key Deliverable: A fill-in-the-blank PACE Card for your visor.

8.1 PACE Defined: The Algorithm of Survival

You are five miles down a washboard trail. The sun is setting, casting long, confusing shadows across the desert floor. You pick up the radio to tell your spouse (driving the chase vehicle) that you are taking the left fork toward the campsite. You press the button. Silence. You press it again. Static.

What do you do?

Do you stop in the middle of the trail? Do you turn around and risk a multi-point turn on a narrow ledge? Do you wait? Does she know to wait? Or does she take the right fork because "it looked smoother," separating your convoy by two ridges and ten miles of darkness?

In that moment of silence, your heart rate spikes. Your IQ drops. This is not the time to invent a plan.

Without a plan, you are gambling. In the *Visual Command System*, we do not gamble with family safety. We use **PACE**.

PACE is an acronym borrowed from Special Operations, used by Green Berets and Forward Air Controllers. But you don't need a beard and a suppressed rifle to use it. It is simply a logic tree for communication failure. It answers the single most important question in the backcountry: *"What exactly do we do when this thing stops working?"*

The Hierarchy of Comms

Most people have a Plan A. Smart people have a Plan B. A PACE plan gives you four layers of redundancy, sorted by friction and reliability. It is a contract between everyone in your group.

P: Primary (The "Easy Button")

This is your default method of communication. It is the path of least resistance.

- **In the Suburbs:** Your Primary is likely your **Cell Phone**. It's easy, high-bandwidth, and you already have it. However, cell towers are fair-weather friends. In a disaster or a valley, they are the first to fail.
- **In the Convoy:** Your Primary is your **GMRS Radio on a specific simplex channel** (e.g., Channel 16). It provides instant, one-to-many communication without dialing. It creates a "party line" where everyone hears everything, providing total situational awareness.
- *Visual Command Analogy:* The Primary is the paved highway. It's fast, smooth, and comfortable. But if there's a roadblock or a washout, you cannot just sit there. You need an exit ramp.

A: Alternate (The "Detour")

The Alternate is a backup system that provides similar capability but on a different path. It requires a deliberate, pre-briefed action to switch.

- **Scenario:** Channel 16 is suddenly flooded with a construction crew using high-power commercial radios, or a group of kids playing hide-and-seek. Or perhaps you accidentally enabled a "Privacy Tone" that your partner doesn't have, locking them out.
- **The Fix:** You switch to **GMRS Channel 20 (Simplex)**.
- **The Key:** Everyone must know *before* you leave the driveway that Channel 20 is the Alternate. If you switch and your spouse doesn't, you are talking to yourself. This isn't just about changing numbers; it's about having the discipline to make the switch in unison.
- *Visual Command Analogy:* The Alternate is the dirt service road running parallel to the highway. It's a bit bumpier (you have to stop and change channels), but it gets you to the same destination efficiently.

C: Contingency (The "Force Multiplier")

The Contingency is used when the environment has defeated your Primary and Alternate. Usually, this means terrain has blocked your Line of Sight. Simplex radios cannot punch through a granite mountain.

- **Scenario:** You drove over a ridge. Your spouse is still in the valley. Your simplex signal is hitting the dirt and dying. You cannot hear them, and they cannot hear you.
- **The Fix:** You switch to the **Local Repeater** (e.g., The "Saddleback 700" Repeater).
- **Why it works:** The repeater is high up on a tower or peak. You talk *up* to it; it blasts your voice *down* to the valley, bridging the gap that simplex couldn't cross. This requires foresight - you must have the repeater frequencies and

input tones programmed before you lose service.

- *Visual Command Analogy:* This is the helicopter. When the road is washed out and the service road is flooded, you fly over the obstacle.

E: Emergency (The "Nuclear Option")

The Emergency method is for when the world has truly gone wrong. Radios are dead, broken, or batteries are flat. Or perhaps someone is injured and you need outside help immediately. This is the layer of last resort.

- **Scenario:** Total electronic failure, vehicle rollover, or medical critical.
- **The Fix: Satellite Messenger** (Garmin inReach/Zoleo) or **Physical Rendezvous** (Walking back to the last known intersection).
- **The Mental Anchor:** Establishing a physical "Meet Point" (e.g., "If we lose all comms, we meet back at the trailhead") provides a massive psychological advantage. It stops people from wandering aimlessly.
- *Visual Command Analogy:* This is the flare gun or the hiking boots. It is slow, expensive, or physically demanding, but it works when nothing else does.

The "If-This-Then-That" Logic Tree

The beauty of PACE is that it removes emotion. You don't get frustrated when Channel 16 fails; you simply execute the logic. You don't scream at the radio; you switch the channel.

The Convoy Protocol:

"Lead vehicle to convoy. Primary is compromised. Executing PACE. Moving to Alternate (Channel 20). Acknowledge."

If you get no response on Primary, you blindly switch to Alternate and try again. If the plan was briefed correctly, your team will be there waiting for you.

8.2 Kid Protocols: Managing the "Junior Operators"

Handing a GMRS radio to a child is a rite of passage. It gives them autonomy, a sense of responsibility, and a tether to safety. However, an untrained child with a radio is a liability - and potentially a beacon for trouble.

We need to establish the "Junior Operator" Rules of Engagement to ensure the radio remains a tool, not a toy.

Rule 1: The "Whistle Rule" (Analog Backup)

Batteries die. Knobs get bumped. Kids drop electronics in creeks. If your child's safety relies 100% on a $30 piece of plastic and a lithium-ion battery, you have failed the PACE plan.

The SOP: Every child issued a radio is also issued a high-decibel safety whistle (pealess, like a Fox 40). Lanyarded to their neck or clipped to their pack (never loose in a pocket).

- **The Rule:** "If you push the button and Daddy doesn't answer three times, stop pushing. Blow the whistle three times. Stay put."
- **Why:** A whistle cuts through wind, river noise, and forest canopy differently than a radio voice. It requires no electricity and no Line of Sight. It is the 'Emergency' layer of their personal PACE plan.
- **The Universal Signal:** Teach them that three blasts means "SOS" or "I need help." Two blasts means "Come here." One blast means "Where are you?"

Rule 2: "Creepy Chatter" Protocol

GMRS is a public band. You are sharing the airwaves with everyone within 5 to 50 miles. This includes helpful grandpas, bored truckers, and, unfortunately, creeps.

If your child has a "cute" voice, they may attract attention. A stranger might jump on the channel and say, *"Hey there, little ranger, where are you camping?"* or *"What's your name? I have a dog here."*

This is not paranoia; it is operational security (OPSEC). A predator uses information to close distance. Don't give it to them.

The Training: Teach your children to identify "Stranger Traffic."

1. **The Trigger:** Anyone asking for their name, location, or who they are with (other than Mom/Dad).
2. **The Action: Do Not Reply.** Silence is the best armor. A stranger cannot find you if they cannot hear you.
3. **The Notification:** The child immediately brings the radio to you, or switches to the designated "Safe Channel" (Alternate) if they are old enough to manage it.

The "Code Word" Concept: Give your child a code word that means "Switch channels immediately."

- **Dad says:** "Jurassic Park."
- **Kid does:** Clicks radio knob one click to the right (to the pre-set Alternate channel).
- **Why:** This allows you to move the conversation away from the creeper without engaging the stranger or alerting them that you are moving. It maintains your control over the situation.

Rule 3: Radio Discipline for Kids (The "No DJ" Rule)

Kids love to hear their own voice. They love to push the "Call" button that makes the ringing sound. They love to key the mic and breathe, or narrate their hike.

- **The Consequence:** If they jam the frequency with nonsense, you cannot hear the warning about the bear on the trail, the approaching Jeep, or the weather alert.
- **The Fix:** "The radio is a seatbelt. We don't play with seatbelts. We wear them for safety. If you play DJ, the radio goes in Daddy's pack, and you have to hold hands the rest of the hike."
- **The Mechanical Fix:** Enable the 'Keypad Lock' before handing the radio to a child. This prevents them from accidentally changing the channel while hiking or playing.
- **The "Listen First" Habit:** Teach them to listen for 3 seconds before they talk. If someone else is talking, they must wait. This prevents "doubling," where two signals cancel each other out, resulting in a horrible screeching noise.

8.3 The "Lost Comms" Procedure

The most terrifying sound in the backcountry is absolute silence when there shouldn't be. You are waiting for your scout to check in. You call. Nothing. You call again. Nothing.

The human brain immediately jumps to catastrophe. *"They fell off a cliff." "They are hurt."* Usually, they just drove behind a rock, their battery died, or they bumped the volume knob.

To prevent panic-induced errors (like driving off a trail to search or splitting up the group), we use a **Visual Timeline** for Lost Comms.

The Timeline of Silence

T-Minus 0:00 to 0:01 (The "Idiot Check")

Before you panic, check your own gear. 90% of radio failures are user error.

- **Is *your* radio on?**
- **Is *your* volume up?** (Did you bump it getting out of the truck? Turn it all the way up until you hear the beep/static).
- **Are you on the correct channel?** (Did you accidentally sit on the button and switch to Channel 4?)
- **The Keypad Lock Betrayal:** Did you accidentally lock the keypad? Or worse, is the keypad *unlocked* and you bumped a setting? Many radios have a 'Lock' icon. Check it.
- **Action:** Turn radio off and on. Check channel. Check volume. Unlock/Lock keypad.

T-Minus 0:01 to 0:05 (The "Static Check")

If your gear is good, assume the environment is the problem. Your vehicle is a cage of metal and electrical noise.

- **Action:** Stop the vehicle or stop walking.
- **Action: Kill the engine and accessories.** Alternators, spark plugs, and **cheap LED light bars** generate RF interference that can drown out weak signals. You might not hear the noise, but your radio does.
- **Action:** Turn your squelch *down* to zero (or press the 'Monitor/Moni' button) to listen to raw static. A weak signal might be breaking the squelch threshold but not opening it. You are looking for a faint voice in the white noise.
- **Call:** "Convoy, this is Lead. Radio check. Over."

T-Minus 0:05 to 0:15 (The "Seek Higher Ground")

Radio waves are light waves. They cannot go through mountains, dense wet foliage, or rock walls. If you have lost comms, you likely lost Line of Sight. You are in a "Radio Shadow."

- **Action:** Do not drive deeper into a valley or canyon. Drive *up*. Get to a ridgeline or a clearing. Even 10 feet of elevation can restore a signal. Standing on the roof of your truck adds 6 feet to your antenna height - that can double your horizon.
- **Protocol:** "Convoy, Lead is moving to high ground for comms. Hold position." (Say this even if you think they can't hear you - they might be receiving but unable to transmit).

T-Minus 0:15+ (The "Blind Transmit" & Rendezvous)

If 15 minutes pass with no contact, you are no longer in a "glitch." You are in a "situation."

- **Action:** Execute the PACE plan logic.
    - Try Alternate Channel.
    - Try Contingency Repeater.
- **The Blind Call:** Transmit your intentions *in the blind*. You assume they can hear you, even if you can't hear them.

"Chase, this is Lead. Lost comms. I am holding at the trail intersection marked 'Scenic Overlook' for 30 minutes. If no contact, I will backtrack to the trailhead. Out."

- **Why:** They might hear you perfectly but have a broken microphone (or a shy child holding the radio). Telling them exactly where you are and what you are doing prevents them from driving in circles looking for you. It gives *them* a plan, even if they can't talk back.

Workbook Element: The PACE Card

You will not remember the repeater frequency when your truck is overheating, the sun is down, and your kids are crying. Write it down.

Instructions:

1. Print this card (or copy it onto an index card).
2. Fill it out with a Sharpie.
3. **Weatherproof it:** Cover the card in clear packing tape to laminate it. A soggy card is useless.
4. Tape it to your dashboard or visor.
5. Make a copy for every radio user in your group.

```
+------------------------------------------------------------+
|                 VISUAL COMMAND :: PACE CARD                |
+------------------------------------------------------------+
| MISSION/TRIP: ____________________________________________ |
| GROUP CALL SIGN: _________________________________________ |
+------------------------------------------------------------+
|  P  | PRIMARY (The Main Channel)                           |
|     | Channel: ______  (Name: ___________)                 |
|     | Tone: _________                                      |
+------------------------------------------------------------+
|  A  | ALTERNATE (The Backup / Simplex)                     |
|     | Channel: ______  (Name: ___________)                 |
|     | Tone: _________                                      |
+------------------------------------------------------------+
|  C  | CONTINGENCY (The Repeater / High Ground)             |
|     | Channel: ______  (Name: ___________)                 |
|     | Rptr Input Tone: _________                           |
+------------------------------------------------------------+
|  E  | EMERGENCY (Total Failure)                            |
|     | [ ] Sat Messenger   [ ] Cell Phone (911)             |
|     | [ ] Physical Meet Point: ___________________________ |
+------------------------------------------------------------+
| LOST COMMS PROTOCOL:                                       |
| 1. Check Vol/Ch/Lock.                                      |
| 2. Kill Engine / Check Static.                             |
| 3. Move High.                                              |
| 4. Wait 15 Mins.                                           |
| 5. Return to: ____________________________________________ |
+------------------------------------------------------------+
```

The "Briefing" Habit

Before you put the car in 'Drive', hold up this card. Point to it. Ask your passenger: *"What is our Alpha channel today?"* If they don't know, tell them.

This takes 30 seconds. It saves 3 hours of panic. You are now operating with a system, not just a gadget.

Chapter 9

The Mobile Build

Overland Install Guide

Objective: Upgrade from a handheld "toy" to a 50-watt "Command Center" without burning down your truck.

Key Deliverable: A vehicle that communicates as well as it crawls.

Introduction: The Difference Between a Radio and a System

If you have followed the previous chapters, you likely have a handheld radio (HT) clipped to your sun visor or rattling around in a cup holder. That HT is a great tool. It's your "scout" weapon - perfect for spotting a driver over an obstacle or hiking up a ridge. But if you are serious about the *Visual Command System* - if you are the Convoy Lead, the Chase Vehicle, or the primary safety node for your family - the handheld is not enough.

You need a Mobile Unit.

A Mobile Unit (or "Base Station") is a high-power (20 to 50 watts) radio hard-mounted into your vehicle, connected to the external antenna we mounted in Chapter 6, and powered directly by your vehicle's electrical system. This isn't just an accessory; it's infrastructure.

This upgrade changes the game physically and tactically.

- **Wattage:** You go from 5 watts (throwing a baseball) to 50 watts (firing a cannon). This punches through foliage and bounces signals off canyon walls with authority.
- **Reliability:** No more dead batteries in the middle of a trail. Your power source is now a 60-pound lead-acid battery constantly recharged by the engine.
- **Audio Authority:** A loud, external speaker that cuts through the noise of mud-terrain tires, diesel engines, and screaming kids. You will hear every call, even with the windows down.
- **The Faraday Cage Effect:** Your vehicle is a metal box. Using a handheld *inside* the car is like trying to shout through a steel bucket. A mobile unit moves the "mouth" (antenna) outside the bucket, instantly doubling your effective range before you even add power.

However, installing a mobile radio is not like plugging in a USB charger or wiring a dashcam. You are integrating a high-power Radio Frequency (RF) transmitter into a complex web of automotive electronics. Do it wrong, and you'll have a radio that whines like a dying cat every time you accelerate. Do it really wrong, and you'll burn out the transmitter, melt your wiring harness, or drain your starter battery in the backcountry.

This chapter is the mechanic's manual for the Tactical Dad. Put on your gloves. We are going under the hood.

9.1 Power & Noise: The Lifeblood of Your Comms

The single most common mistake new operators make is treating a 50-watt radio like a low-draw accessory. They see a fuse box, they see an "Accessory" slot, and they lazy-wire the radio into the nearest 12V source.

Do not do this.

To understand why, we have to look at the physics of your truck's electrical system.

The Theory: Dirty Power vs. Clean Power

Your vehicle's electrical system is a chaotic storm of electromagnetic noise.

- **The Alternator:** This is a generator that spins at thousands of RPM, creating electricity to charge your battery. It generates Alternating Current (AC) which is rectified into Direct Current (DC). This process is messy. It creates a "ripple" in the voltage - a high-pitched electronic noise that rises and falls with your engine RPM.
- **The Fuel Pump & Injectors:** High-speed electric motors and solenoids that click open and closed hundreds of times a second, sending spikes back down the power lines.
- **The LED Light Bars:** This is the silent killer of radio performance. Cheap off-road lights and USB chargers use "switching power supplies" that chop electricity up rapidly. They act like miniature radio jammers, spewing Radio Frequency Interference (RFI) everywhere.

If you wire your radio to the fuse box inside the cabin, you are tapping into a "dirty" circuit shared by the blower motor, the turn signals, and the fuel pump. You are asking your radio to listen for a weak signal (a whisper) while plugging its ears into a jackhammer.

The Symptom: "Alternator Whine." You press the gas pedal, and your radio emits a high-pitched *wheeeeeeee* sound that gets higher as you rev the engine. It makes you sound like an amateur, drives everyone on the channel insane, and masks incoming weak signals.

The Golden Rule: Direct to Battery

The cleanest power source in your vehicle is the **Battery Lead Acid Plate**. The battery acts as a massive chemical capacitor. It absorbs the ripples from the alternator and smooths out the spikes from the injectors. It is a buffer - a calm pool of water next to a raging river.

The Protocol:

1. **Wire Gauge Matters:** Do not use thin speaker wire. For a 50-watt radio, use **10 AWG or 12 AWG** automotive-grade wire. Thin wire causes "Voltage Drop." When you transmit, the radio demands 10-12 amps instantly. If the wire is too thin, the voltage drops below 11V, and the radio reboots or shuts down mid-sentence.
2. **Positive (+):** Run the red power wire from the radio directly to the Positive terminal of the vehicle's battery.
3. **Negative (-):** Run the black ground wire from the radio directly to the Negative terminal of the battery. If you must use a chassis ground point, **you must scrape the paint down to bare metal**. A bolt through paint is not a ground; it is a capacitor.

Note regarding modern vehicles with Battery Management Systems (BMS): Some newer trucks (F-150s, Toyotas, Broncos) have a sensor on the negative terminal clamp. This sensor tells the computer how much energy is leaving the battery. If you wire directly to the clamp, bypassing the sensor, the computer might get confused and fail to charge the battery correctly. In these cases, follow the negative cable from the battery to where it bolts to the frame (usually 6 inches away). Ground your radio *there*.

The Fuse: Your Fire Insurance

Every wire that leaves the positive terminal of your battery is a potential welding rod. If that wire chafes against the firewall and shorts out, it will glow red hot, melt the insulation, and set your carpet on fire.

Rule: The fuse protects the *wire*, not the radio. You must place a waterproof fuse holder on the positive wire **within 12 inches of the battery**, rated for your specific radio (usually 15A or 20A). If the wire shorts out anywhere between the battery and the radio (like where it passes through the firewall), the fuse pops, and your truck doesn't burn down. Do not rely on the fuse built into the back of the radio chassis or the one inline near the radio body. Those protect the radio from internal faults; they do nothing for a short in the engine bay.

Ferrite Chokes: The "Magic Beads"

Sometimes, even with direct wiring, you still pick up static or interference - often from your own accessories. This is where we use **Ferrite Chokes** (Mix 31 or Mix 43 snap-on cores).

The Physics: A ferrite choke is a chunk of magnetic ceramic. When you clamp it around a wire, it acts as a resistor for high-frequency energy (RF noise) but passes low-frequency energy (DC power) without issue.

The "Wattage is Water" Analogy: Imagine your power wire is a hose carrying water (electricity) to your radio. RFI noise is like air bubbles trapped in the water, causing sputtering. A ferrite choke is a filter that traps the bubbles but lets the water flow freely.

Deployment:

1. **Snap-on:** Buy 5mm or 7mm snap-on chokes.
2. **Loop it:** Don't just clip it on. **Loop the wire through the choke twice** if possible. The physics of inductance means that two turns through the core quadruples the noise-suppressing effectiveness.
3. **Location:** Place one choke on the power cable as close to the radio body as possible. Place another on the power cable as close to the battery as possible.

Result: A "black background" on your audio. Silence when no one is talking, allowing you to hear distant stations that would otherwise be buried in the noise floor.

9.2 Routing the Coax: The Arteries of the System

In Chapter 6, we mounted the antenna. Now we have to get the coaxial cable (the thick black wire) from the roof/bumper into the cab without destroying it.

Coax cable (usually RG-58, RG-8X, or LMR-240) is not a rope. It is a precision-engineered pipe consisting of a center copper conductor, plastic insulation, a braided metal shield, and an outer jacket. The spacing between the center wire and the shield must remain perfect.

The Crush Rule: If you pinch the cable hard enough to deform it (turn it from round to oval), you have ruined it. You have changed the "impedance" at that spot from 50 ohms to something else. This acts like a kink in a garden hose - the signal hits the kink and bounces back (SWR), heating up your radio instead of transmitting.

The "Pinch Point" Danger Zones

1. The Door Jam Guillotine

The lazy installer runs the coax through the door opening and slams the door shut on it. The rubber weather stripping might protect it for a week, but eventually, the metal door frame will work-harden the copper shield and snap it.

- **Verdict:** Never run coax through a primary door jam. It *will* fail, usually when you need it most.

2. The Tailgate Hinge

If mounting on a tailgate, you must route the cable alongside the factory wiring harness. Watch the "scissor action" of the hinge.

- **The Fix:** Use zip ties to secure the coax to the existing wire loom. Cycle the tailgate open and closed slowly multiple times. Watch the cable. Does it pull tight? Does it kink? It needs a "service loop" - a bit of slack that allows movement without tension.

3. The Hood Hinge

For "ditch light" mounts or lip mounts, you often route cable into the engine bay.

- **The Trap:** The hood hinge mechanism is a metal crushing machine.
- **The Fix:** Route the cable *behind* the hinge mechanism, securing it to the firewall cowl, ensuring it never touches the moving arms. Use split-loom tubing to protect the coax from engine heat.

Penetrating the Firewall: The Right Way

You need to get the power wire and the antenna coax from the engine bay into the cabin. The barrier is the Firewall - the steel plate separating you from the engine.

Method A: The "Daylight" Trick (Best) Look under the dashboard with a flashlight (up near the brake pedal). Look for the factory rubber boot where the main wiring harness passes through. Have a helper shine a bright flashlight from the engine side at that same rubber boot.

- **The Goal:** If you see light, or a nipple on the rubber boot, that is your path.
- **The Tool:** Use a wire coat hanger (straightened out) or a "fish tape." Tape your wire to the end of the coat hanger (use electrical tape, make a smooth, tapered cone so it doesn't snag).
- **The Move:** Poke the coat hanger carefully through the side of the rubber boot (do not cut the main wires!). Push it through until your helper grabs it in the engine bay. Pull the wire through.
- **The Seal:** If the hole is loose, dab a little RTV silicone sealant on it. Engine fumes (carbon monoxide) can enter the cabin through unsealed firewall holes.

Method B: The Drill (Last Resort) If you must drill a new hole:

1. **Check the other side:** Ensure you aren't drilling into the brake booster or a wiring loom.
2. **Drill Pilot:** Start small.
3. **Step Bit:** Use a "Unibit" step drill to make a clean hole.
4. **The Grommet: MANDATORY.** You cannot run a wire through a bare metal hole. The vibration will act like a saw, cutting through the insulation in 100 miles and shorting out the car. You *must* install a rubber grommet (available at any auto parts store) in the hole before passing the wire.

The Drip Loop

Water travels down wires. If your coax runs down from the roof and straight into a firewall grommet, rain will follow the wire, go through the grommet, and drip onto your feet.

- **The Fix:** Create a "Drip Loop." Before the wire enters the hole, bend it down into a 'U' shape and then back up into the hole. Water will run down to the bottom of the 'U' and drip off onto the ground, rather than climbing up into the cabin.

9.3 SWR (Standing Wave Ratio): The Engine Tune-Up

You have power. You have an antenna. You have coax. You are ready to transmit, right?

STOP.

If you key the mic now, you might destroy your new $200 radio. You must check the **SWR**.

The Physics of the "Parking Brake"

Radio Frequency (RF) energy flows out of your radio, down the coax, and into the antenna. The antenna's job is to radiate that energy into the air.

- **In a perfect world (SWR 1.0:1):** 100% of the energy leaves the antenna.
- **In the real world:** The antenna is never perfectly tuned to the exact frequency you are using. Because of this mismatch, some of the energy hits the end of the antenna and *bounces back* down the coax, returning to the radio.

The Analogy: Imagine you are spraying a fire hose (the radio) through a nozzle (the antenna).

- If the nozzle is open, water flies out.
- If you put your thumb over the nozzle (bad tuning), the water sprays back in your face.
- If you plug the nozzle completely, the water pressure backs up and bursts the hose or blows up the pump.

High SWR (Standing Wave Ratio) is that back-pressure.

- **Heat:** The energy that comes back has to go somewhere. It turns into heat inside your radio's "final transistors."
- **Damage:** If the SWR is too high (above 3.0), the radio will overheat and burn out its transmitter components in seconds.
- **Performance:** Even if it doesn't break, High SWR means your signal isn't getting out. An SWR of 2.5 means you might only be transmitting 30 watts instead of 50. You are driving with the parking brake on.

The Tool: The SWR Meter

You need an SWR meter covering the UHF (GMRS) frequencies (462-467 MHz). A standard CB radio meter *will not work* (wrong frequency range).

- **Budget:** Surecom SW-102 (Digital, easy). Highly recommended for the glovebox.
- **Pro:** NanoVNA (Complex, but powerful). Shows you the antenna curve across the whole band.
- **Analog:** Daiwa CN-901 (Old school, reliable). No batteries required, but bulky.

The Tuning Procedure (The "Measure Twice, Cut Once" Protocol)

Setup:

1. Drive your vehicle to an open area. **Do not** tune inside a garage. Metal walls, concrete ceilings, and nearby cars will reflect the signal and give you false readings. You need at least 20 feet of clear space.

2. Close all doors and windows. Open doors change the ground plane shape.

3. Connect the Meter:

    ◦ Radio -> Patch Cable -> Meter Input (TX).

    ◦ Meter Output (ANT) -> Coax -> Antenna.

The Test:

1. Set your radio to a middle channel (GMRS Channel 19).

2. Set radio power to **Low** (to protect it if the SWR is terrible).

3. Key the microphone (Press PTT).

4. Read the meter.

Interpreting the Number:

- **1.0 to 1.5: Excellent.** Do not touch anything. You are a wizard.
- **1.5 to 2.0: Good.** Acceptable for daily use.
- **2.0 to 2.5: Marginal.** You are losing power, and the radio will run warm. Try to improve it.
- **2.5 to 3.0+: DANGER.** Do not transmit. Something is broken or badly mistuned. Check for a short in the coax or a bad ground.

How to Tune (Adjusting the Length)

An antenna is a bell. We want it to ring at a specific pitch (462 MHz). We adjust the pitch by changing the length of the metal whip.

- **If SWR is lower on Channel 1 than Channel 22:** The antenna is **too long**. (It resonates at a lower frequency).
    - *Action:* Loosen the set screw (grub screw) on the antenna whip. Pull the metal whip out. **Mark the current depth with a Sharpie** so you know where you started. Snip off **1/8th of an inch** (3mm) using linesman pliers or a dremel. Re-insert. Tighten. **Step away from the vehicle.** Your body reflects RF signals. If you lean over the antenna to watch the meter, you will get a false reading. Test again.
- **If SWR is lower on Channel 22 than Channel 1:** The antenna is **too short**. (It resonates at a higher frequency).
    - *Action:* Raise the whip up in the mount if possible. If you already cut it too short... congratulations, you just bought a new replacement whip ($15). This is why we cut in tiny increments. You can't put metal back on.

The "Ground Plane" Factor: If you cannot get the SWR below 2.5 no matter what you do, you likely have a **Ground Plane** issue. The antenna whip is only half the antenna. The other half is the "mirror image" reflected by the metal surface under it.

- **Diagnosis:** Your antenna doesn't have enough metal under it (e.g., you mounted it on a plastic roof rack, a composite camper shell, or a tube bumper).
- **The Fix:** You must run a ground strap (braided wire) from the antenna mount to the vehicle's chassis, or add a metal plate/disc (at least 6-8 inches diameter) under the antenna base.

Action Step: The "Install Inspection" (The 10-Point Check)

You've wired, routed, and tuned. Before you declare the rig "Mission Ready," execute this inspection. Loose wires start fires, and vibration finds every weakness.

The Checklist:

1. [ ] **Battery Connections:** Are the ring terminals tight on the battery? Can you rotate them by hand? (They should be rock solid).
2. [ ] **Fuse Placement:** Is the fuse within 12 inches of the battery? Is the cap on the fuse holder sealed against water/mud?
3. [ ] **Wire Chafe (Engine Bay):** Follow the red wire. Does it touch the hot engine block, the exhaust manifold, or sharp metal edges? Zip tie it away from heat and abrasion. Use split-loom tubing for extra armor.
4. [ ] **Grommet Integrity:** Check where the wires enter the firewall. Is the grommet seated? Is the silicone dry?
5. [ ] **Interior Routing:** Look under the dash. Are wires dangling near the brake or gas pedals? (Zip tie them up high! A wire catching your shoe can cause an accident).
6. [ ] **Coax Kinks:** Check the coax run. Are there any sharp 90-degree bends or pinched spots? Ensure the "Minimum Bend Radius" is respected (usually the size of a soda can).
7. [ ] **Connector Tightness:** Check the PL-259 connector on the back of the radio. Hand-tighten, then give it a tiny nip (1/8 turn) with pliers. Vibrations loosen these over time.
8. [ ] **Radio Mount:** Grab the radio body. Shake it violently. If it moves, the mounting screws are loose. Off-road driving creates earthquake-level vibration.
9. [ ] **SWR Verification:** Final test. Is SWR under 2.0 on Channel 1 and 22?
10. [ ] **Audio Check:** Call for a radio check. Ask specifically: "How is my signal? Is there any whine or static when I rev the engine?"

The "Shake-Down" Run: The final test is the road. Go find a washboard dirt road. Drive it for 20 minutes. Stop. Check the mount. Check the SWR again. If your antenna mount is loose, washboards will find it. If your power connection is intermittent, washboards will find it. If nothing fell off and nothing is smoking, you are cleared for operations.

Chapter 10

Troubleshooting

The Visual Command Flowcharts

Objective: Diagnose and fix communication failures in under 60 seconds using binary logic.

Key Deliverable: Three field-ready diagnostic trees for your glovebox.

Introduction: Logic, Not Magic

You are on the trail. It's late afternoon, the shadows are getting long, and the "check engine" light of the lead vehicle just flickered. You pick up the mic to coordinate a stop. You press the button. Nothing happens. You press it harder (because clearly, hydraulic pressure fixes electronics). Still nothing.

At this moment, the "Tactical Dad" usually splits into two distinct, unhelpful personalities:

1. **The Magician:** This operator believes that if he randomly mashes buttons, twists knobs back and forth, and shakes the radio like a Polaroid picture, he will accidentally stumble upon the solution. He is hoping for magic. He usually ends up changing a setting that was actually correct, compounding the problem.

2. **The Defeatist:** This operator tries once, fails, and immediately spirals. He throws the radio on the passenger seat, declares the entire GMRS system "junk," and drives the rest of the day in angry silence, stewing over the money he spent.

Both approaches are failures of command. They are emotional reactions to a mechanical problem.

Radio problems are rarely mysterious. They are mechanical, electrical, or procedural. They follow the laws of physics, not Murphy's Law. In the *Visual Command System*, we do not guess. We trace the signal path. We ask "Yes/No" questions until we find the failure point.

Troubleshooting is simply the art of eliminating variables.

- If the radio turns on, the battery is good.
- If it hisses when you drop the squelch, the speaker is good.
- If it receives but can't transmit, the antenna is likely physically intact, but the settings are wrong.

This chapter replaces the frantic "Magician" with the calm, methodical **Diagnostician**.

10.1 The "I Can't Hear You" Flowchart (Receiver Failure)

The Scenario: You can see the lead vehicle through your windshield. You can see the driver talking into his hand mic, gesturing wildly. Your radio is dead silent. You feel isolated, and that isolation breeds anxiety.

The Logic: The signal is leaving *their* truck (we assume) but being blocked before it hits *your* eardrum. We must check the "Gates" in order, from the easiest (human error) to the hardest (hardware failure).

Step 1: The "Idiot Check" (Volume & Power)

Do not skip this. Do not be offended. 30% of all radio failures on the trail are simply the volume knob getting bumped against a hip (HT) or a center console (Mobile).

- **The Tactile Check:** Turn the knob all the way down until it clicks off. Then turn it on and halfway up.
- **The Auditory Check:** Listen for the startup beep or voice prompt ("Channel Mode!"). If you don't hear the beep, the radio isn't making sound.
- **Visual Command:** Look at the screen. Is it lit?
    - *HT Fix:* If the screen is dead, remove the battery and wipe the contacts with your thumb. Re-seat it firmly.
    - *Mobile Fix:* Check the inline fuse near the battery. A loose fuse holder can disconnect power on bumpy roads.

Step 2: The Channel Match

GMRS channels are standardized by the FCC, but channel *names* and memory slots are not. This is a massive source of confusion when mixing radio brands (e.g., Midland vs. Wouxun vs. Baofeng).

- **The Problem:** Your buddy says, "Go to Channel 5." He means the 5th slot in *his* radio's memory, which he custom-programmed as the "Rubicon Repeater" (462.675 MHz). You go to the standardized GMRS Channel 5 (462.6625 MHz). You are now existing in parallel universes.
- **The Fix:** Confirm the **Frequency**, not just the Channel Number.
    - *Command:* "Lead, this is Chase. Confirm you are on frequency 4-6-2 point 6-6-2-5?"
    - If they don't know how to check their frequency, you must switch to a "Universal Standard" channel like Channel 16 (462.575) to re-establish comms.

Step 3: The "Privacy Tone" Mismatch (The Silent Killer)

This is the #1 cause of "I can see you talking but I can't hear you." It happens when someone accidentally enables a "Receive Tone" (RX CTCSS/DCS) on their radio.

- **The Physics:** Remember, a Privacy Tone is an electronic key. If your radio is set to expect "Tone 100.0 Hz," it creates a locked door. It will *only* open that door if the incoming signal carries the 100.0 Hz key. If the sender is not transmitting that tone (or is sending a different one), your radio hears the signal but refuses to open the speaker.
- **The Diagnostic:** Press the **MONITOR** (Moni) button on the side of your radio.
    - *What this does:* It forces the Squelch gate open and temporarily ignores all privacy tones. You will hear loud white noise (static).
    - *The Test:* Hold 'Monitor' while the other person talks.
    - *Result A:* **You hear them clearly through the static.**
        - *Verdict:* Your settings are wrong. You have a Receive Tone set that shouldn't be there. Go to Menu -> Rx Tone -> Set to OFF (or 0).

- *Result B:* **You still hear only static while they talk.**
  - *Verdict:* The issue is not the Tone. It is likely Frequency (you aren't on the same channel) or Range (they are too far away).

Step 4: The Squelch Trap

Squelch is the gatekeeper that keeps the background cosmic radiation and electrical noise quiet. If the gate is set too high (too strong), it won't open for a weak signal.

- **The Scenario:** You are separated by a large hill. The signal is weak. Your squelch is set to Level 5 (Factory Default). The weak signal hits the gate but isn't strong enough to push it open.
- **The Check:** Look at your screen. Is the **"RX" (Receive) icon** lighting up green (or showing signal bars), but no sound is coming out? This indicates the radio *hears* something but the Squelch is suppressing it.
- **The Fix:** Lower your Squelch setting (SQL). Drop it to 1. This makes the "gate" lighter, allowing weaker signals to pass through.

Step 5: The "Headset" Ghost

Did you have an earpiece or hand-mic plugged in earlier? Did you drop your HT in a puddle?

- **The Problem:** The headphone jack is a mechanical switch. When you insert a plug, it physically disconnects the main speaker. Sometimes, dirt, lint, or water gets jammed in the jack, tricking the radio into thinking an accessory is still plugged in. It routes audio to a non-existent headset.
- **The Fix:** Plug and unplug the accessory jack 3-4 times rapidly to clear debris. Blow it out with compressed air (or a sharp breath).

Step 6: The "RFI Wall" (The Overland Special)

This is unique to modern modified vehicles. You might be jamming *yourself*.

- **The Scenario:** You hear constant static, or your range has dropped to 50 feet.
- **The Test: Turn off your engine.** Turn off your LED light bars. Unplug your cheap USB chargers, Dash Cams, and AC Inverters (which are notoriously noisy in the RF spectrum).
- **The Reveal:** If the radio suddenly comes to life and you can hear your group, your vehicle is generating Radio Frequency Interference (RFI). Cheap LED drivers and USB buck converters scream across the UHF spectrum, drowning out incoming signals.
- **The Fix:** Buy high-quality electronics (e.g., ferrite-shielded lights) or accept that you cannot use the radio and the lights simultaneously.

The "I Can't Hear You" Decision Tree

```
START
  |
  +-> [1] Is Radio ON and Volume UP? (Did you hear the beep?)
  |       NO  -> Check Battery / Fuse.
  |       YES -> Continue.
  |
  +-> [2] Do Frequencies Match? (Ignore Channel Names)
  |       NO  -> Match Frequencies (e.g., 462.5625).
  |       YES -> Continue.
  |
  +-> [3] Press & Hold MONITOR Button. Do you hear them now?
  |       YES -> TONE MISMATCH. Turn off your Rx Tone (CTCSS/DCS).
  |       NO  -> Continue.
  |
  +-> [4] Is the Green Rx Icon lighting up?
  |       YES -> Check Squelch (Lower it to 1) or check for "Headset Ghost."
  |       NO  -> Continue.
  |
  +-> [5] Turn off Engine/Lights. Did reception improve?
  |       YES -> RFI Interference. Your truck is the problem.
  |       NO  -> They are likely out of range or not transmitting.
```

10.2 The "You Can't Hear Me" Flowchart (Transmitter Failure)

The Scenario: You are talking, giving directions, telling jokes. But nobody is answering. You feel like a ghost. You suspect your radio isn't "getting out," or perhaps they are just ignoring you.

The Logic: The signal is dying somewhere between your voice box and the antenna tip.

Step 1: The PTT (Push-To-Talk) Discipline

Are you pushing the button *before* you speak, or *as* you speak?

- **The Problem:** Digital radios, and especially Repeaters, take a few milliseconds to "wake up" and open the audio path. If you say "Rock in the road," but you push the button on the "o" in Rock, they hear "...ck in the road." This leads to confusion and constant "Say again?" requests.

- **The Fix:** The "One-Mississippi" Rule. Push. Pause (inhale). Talk. This ensures the carrier wave is established before your voice modulation begins.

Step 2: The Microphone Connection

On mobile units, the hand mic connection is the physical weak link. It gets pulled, twisted, stepped on, and yanked.

- **The Diagnostic:** While holding the PTT button, wiggle the connector where it enters the radio body. Watch the **Red TX Light**.

  - Does it flicker?
  - Does it go out?

- **The Fix:** Unplug it. Check for bent pins. Blow it out. Re-seat it firmly until it clicks. On HTs (Handhelds) with the 2-pin Kenwood connector, you often have to press *very* hard to get the plug fully seated past the weather seal.
- **Check Mic Gain:** Ensure your "Mic Gain" setting in the menu hasn't been inadvertently lowered. If set too low, your modulation will be so quiet that others will think you aren't transmitting.

Step 3: TX Power (High vs. Low)

Are you whispering when you should be shouting?

- **The Problem:** You are trying to hit a repeater 20 miles away, or a vehicle on the other side of a ridge. Your radio is set to "L" (Low Power / 5 Watts) instead of "H" (High Power / 50 Watts).
- **The Fix:** Look at the screen for a small "L" or "H". Go to Menu -> TX Power -> Switch to High.
  - *Regulatory Note:* Remember that Channels 8-14 are legally restricted to Low Power (0.5W) on all radios. If you are on Channel 12, you *cannot* switch to High. If you need range, move the convoy to Channel 16.

Step 4: Repeater Offsets (The "Input" Error)

This is the classic Rookie Repeater Mistake.

- **The Scenario:** You are trying to talk on a Repeater (e.g., The "700"). You can *hear* the repeater traffic perfectly. But when you talk, nobody hears you.
- **The Physics:** Repeaters are "Duplex" machines. They listen on one frequency (Input) and talk on another (Output). Your radio must be programmed to *shift* its transmission to the Input frequency automatically when you press PTT. This is called the "Offset" (+5.000 MHz).
- **The Diagnostic:** Look closely at your screen's frequency display. Press the PTT.
  - *Rx Frequency (Listening):* 462.700
  - *Tx Frequency (Pressing PTT):* **Does it change to 467.700?**
  - If the number *stays* 462.700 when you push the button, you are talking on the Output. You are shouting at the repeater's mouth instead of its ear.
- **The Fix:** Enable the "+5M", "OFFSET", or "RPT" function in the menu (depending on your radio brand).

Step 5: Repeater Input Tones

The Repeater "Password."

- **The Scenario:** You have the offset correct (the frequency shifts). You key up. You hear nothing back (no "ker-chunk" or "tail").
- **The Problem:** Most repeaters are kept private (or semi-private) to prevent interference. They require a specific PL Tone (e.g., 141.3 Hz) to "wake up." If you aren't sending that tone, the repeater ignores your signal entirely.
- **The Fix:** Verify the **Tx Tone** (Transmit CTCSS) in your menu matches the repeater's requirement. *Note: Many regional repeater networks use a standard "Travel Tone" of 141.3 Hz. If you are guessing, try that first.*

The "You Can't Hear Me" Decision Tree

```
START
  |
  +-> [1] Does the Red 'TX' Light turn on when PTT is pressed?
  |       NO  -> Check Mic Connection (Push hard) / Check Battery.
  |       YES -> Continue.
  |
  +-> [2] Are you on the correct Channel/Frequency?
  |       NO  -> Match Group.
  |       YES -> Continue.
  |
  +-> [3] Are you trying to use a Repeater?
  |       NO (Simplex) -> Check Range. Switch to High Power. Try moving vehicle 5 feet.
  |       YES (Repeater) -> Continue below.
  |
      +-> [3a] Does frequency shift +5MHz when PTT pressed?
      |       NO  -> Wrong Setting. Enable Offset (+5.000).
      |       YES -> Continue.
      |
      +-> [3b] Is the correct Tx Tone programmed?
              NO  -> Enter Tone (Try 141.3 if unknown).
              YES -> Repeater might be down or out of range.
```

10.3 The "Radio is Hot/Smells" Flowchart (Hardware Failure)

The Scenario: You pick up the hand mic and the radio body burns your hand. Or, you smell the distinct, sharp scent of "Magic Smoke" (burning ozone/plastic). Or, your transmission starts strong but cuts out or gets "robotic" after 10 seconds.

The Logic: Energy is not leaving the antenna. It is bottling up inside the radio and turning into heat. This is an immediate hardware emergency.

Step 1: The "Touch Test"

Mobile radios get warm. That is normal. They act as their own heatsinks, dissipating 50 watts of power.

- **Normal:** Warm coffee mug. Uncomfortable but holdable.
- **Danger:** Hot stove. You reflexively pull your hand away.
- **Action:** If it is "Hot Stove" hot, **STOP TRANSMITTING IMMEDIATELY.** Turn the radio off. Let it cool for at least 20 minutes.

Step 2: The SWR Check (Reflected Power)

Refer back to Chapter 9. Heat is almost always caused by High SWR (Standing Wave Ratio).

- **The Physics:** The antenna is a nozzle. If the nozzle is clogged (bad tuning, broken wire), the energy hits the clog and bounces back down the cable. It returns to the radio and cooks the final transistors.
- **The Field Fix:**

  a. **Visual Inspection:** Is the antenna still there? (Did a low tree branch snap it off?) Is the coil bent?

b. **Cable Trace:** Is the coax cable pinched in the door jamb? Is the connector loose on the back of the radio?

c. **Meter Check:** If you have an SWR meter, check it. If SWR is > 3.0, **DO NOT TRANSMIT.** You are dead in the water until you fix the antenna system.

Step 3: The "Duty Cycle" Abuse

GMRS radios are not chat rooms. They are not designed for "podcasting."

- **The Rating:** Most mobile radios have a "Duty Cycle" of 10-20%. This means for every 1 minute of talking, you need 4 minutes of listening/cooling.
- **The Abuse:** If you (or your kids) hold the button down for 5 minutes straight telling a story or reading a book, the radio will overheat and engage thermal shutdown.
- **The Fix:** Brief brevity. Keep transmissions under 30 seconds. If you have a "talker" in the group, gently remind them: "Break for cooling."

Step 4: The Voltage Drop (The "Brownout")

Is the radio cutting out, rebooting, or dimming when you press PTT?

- **The Cause:** Bad wiring. When you transmit, the radio tries to pull 10-12 Amps of current. If you used thin wire, or if your ground connection is loose/corroded, the pipe is too small. The voltage drops from 13.8V to 9V. The radio detects the drop and shuts off to protect itself.
- **The Fix:** Check your fuse holder (is the plastic melted?). Check your battery terminals. Ensure you wired directly to the battery, not a cigarette lighter plug.

The "Hardware Crisis" Decision Tree

```
START
  |
  +-> [1] Is there smoke or a burning smell?
  |       YES -> KILL POWER IMMEDIATELY. Pull the fuse. Do not turn on.
  |       NO  -> Continue.
  |
  +-> [2] Is the radio too hot to touch?
  |       YES -> Stop Transmitting. Let cool for 20 mins. Check Duty Cycle.
  |       NO  -> Continue.
  |
  +-> [3] Does the radio reboot/dim when you press PTT?
  |       YES -> Check Battery Voltage / Ground Wire / Fuse Holder.
  |       NO  -> Continue.
  |
  +-> [4] Check SWR / Antenna
          Is the antenna broken? Is the coax pinched?
          If SWR is High -> Fix Antenna before transmitting.
```

Action Step: The "Broken Arrow" Drill

Troubleshooting charts are useless if you can't read them because you are panicking or frustrated. You need muscle memory.

The Exercise: Next time you are out with your group (or just in the driveway with the kids), initiate a "Broken Arrow" drill.

1. **Sabotage:** Have your spouse take your radio and change *one* setting without you looking.
   - Turn on a random Receive Tone.
   - Turn Squelch to Max.
   - Change the Offset.
   - Unscrew the antenna connector halfway so it's loose.
2. **The Test:** Get in the truck. Close the door. Try to call them.
3. **The Solve:** Do not guess. Use the Flowcharts mentally.
   - *"I have power."*
   - *"I am on the right channel."*
   - *"They are talking but I can't hear them."*
   - *"Press Monitor... ah, I hear them now. It must be a Tone mismatch."*
   - *"Checking Tone settings... yes, Tone is on. Turning it off."*

Time Standard: You should be able to diagnose and fix a "Soft Fail" (settings issue) in under 60 seconds.

By mastering these trees, you stop being the guy who bangs on the dashboard when things don't work. You become the guy who fixes the comms, calms the group, and gets the convoy moving again. You become the **Visual Command**.

Chapter 11

Advanced Operations

Scanning & Monitoring

Objective: Transform your radio from a passive telephone into an active intelligence-gathering tool.

Key Deliverable: A "Sentinel" configuration that watches the weather, the convoy, and the threat environment simultaneously.

Introduction: The Intelligence Officer

Up to this point in the book, we have treated the radio primarily as a **Tactical Transmitter** - a device for you to send commands to your family or convoy. You push the button, you speak, they listen. It is a tool of output.

But in the *Visual Command System*, transmission is only half the battle. The other half - and often the more critical half - is **Reception**, or more accurately, **Intelligence Gathering**.

Imagine you are driving into a box canyon in southern Utah. The sky is turning a bruised purple, signaling a pressure drop. Your convoy is chatting happily on Channel 16 about lunch. But what is happening five miles ahead? Is there a flash flood warning issuing for this specific slot canyon? Is there a logging crew blocking the exit route on Channel 20? Is there a rescue operation happening on the local repeater, implying the trail is closed?

If you sit silently on Channel 16, waiting for someone to call you, you are driving blind. You are reactive, not proactive.

You need to become the **Intelligence Officer**. You need to configure your radio to act as an electronic sentry, sweeping the airwaves for threats, weather changes, and logistical information while you keep your eyes on the trail. This is where we move beyond "Push-to-Talk" and master the art of Scanning, Dual Watch, and Information Dominance.

11.1 Scanning Strategies: The Electronic Sentry

Scanning is simply the radio's ability to cycle through a list of channels rapidly, stopping only when it hears a voice. It sounds simple, but most users get it wrong. They press "Scan" and get blasted by white noise, or they get stuck listening to a drive-thru order at a Taco Bell three miles away, missing their own team's check-in.

To scan effectively, you must master three variables: the **Scan List**, the **Scan Mode**, and the **Tone Scan**.

The Scan List: Curating Your Feed

Your radio likely has 200+ memory channel slots. You do not want to scan all of them. If you scan every possible frequency

(GMRS, FRS, MURS, Marine, Local Police), the cycle time - the time it takes to get back to Channel 1 - might be 30 seconds. In those 30 seconds, your lead driver could call out "Rockfall!" and you would miss it because your radio was listening to a garbage truck on Channel 2.

Rule: A Scan List should be mission-specific. You don't take a snow shovel to the desert; don't put snow plow frequencies in your desert scan list.

- **The Convoy List (Group A):** This is tight and fast. It includes only the Convoy Channel (e.g., Ch 16), the Alternate (Ch 20), and the primary Emergency Repeater. The cycle time is under 1 second. You will never miss a word.
- **The Area List (Group B):** Includes all GMRS channels, plus local Sheriff/Fire frequencies (listen-only) relevant to the specific county you are traversing. This is for situational awareness when the convoy is parked or quiet.

How to "Skip" (The Noise Filter): When programming your radio (especially in CHIRP software), every channel has a "Skip" column. This is your primary filter.

- **"S" or "Skip":** The scanner ignores this channel completely.
- **"Off" or Blank:** The scanner includes this channel in the cycle.

Strategy: Permanently "Skip" the FRS/GMRS hybrid channels (8-14). These channels are legally restricted to low power (0.5 watts) and have fixed antennas. They are the playground of blister-pack radios sold at big-box stores. They are usually clogged with children playing tag or families coordinating parking at the mall. They are noise, not signal. Remove them from your scan list immediately to increase your scan speed.

"Nuisance Delete": The Field Fix

Here is the scenario: You are scanning your Area List. You are looking for emergency traffic or trail reports. Suddenly, the scan stops on Channel 22. It's a construction crew fixing a bridge five miles away. *"Bring the loader... back it up... beep beep beep..."* The radio stays locked on them. You press "Scan" to resume. It cycles for 2 seconds, hits Channel 22 again, and stops. *"Yeah, dump that load here."*

You are now in **Scan Hell**. You can't hear your team because the construction crew is dominating the airwaves, and you can't walk away from the radio.

Do not go into the menu to delete the channel. That is too slow and dangerous while driving. **Use "Nuisance Delete."**

On almost all modern GMRS radios (Wouxun, Midland, Baofeng, BTech), there is a shortcut feature specifically for this. When the scanner stops on a channel you don't want to hear:

- **The Move:** Press and hold the **"*" (Star)** key or the designated **Side Key** (check your manual).
- **The Result:** The radio beeps and *temporarily* removes that channel from the active scan list. It will not stop there again until you cycle power.
- **The Reset:** When you turn the radio off and on again, the channel is restored to the list.

Visual Command Logic: This allows you to aggressively filter your feed in real-time. If it's not relevant to the mission, Nuisance Delete it. Clear the deck so you can hear what matters.

Scanning Modes: CO vs. TO vs. SE

This setting confuses everyone because the acronyms are non-intuitive. It dictates *how* the radio behaves when it finds a signal. Choosing the wrong one breaks your flow.

1. **TO (Time Operation):** The radio stops on a signal for a set time (e.g., 5 seconds), then resumes scanning *even if the person is still talking.*

  - *Scenario:* You hear: "Be advised, there is a massive washout at mile marker..."
  - ***Scan Resumes***
  - You missed the location.
  - *Verdict:* **Useless.** Avoid this setting.

2. **SE (Search/Terminate):** The radio stops on a signal and stays there forever. It stops scanning completely.
   - *Scenario:* You find a signal. You listen. The conversation ends. You drive for 20 minutes, realizing too late that your radio stopped scanning and you are now listening to silence on a dead channel.
   - *Verdict:* **Bad.** This defeats the purpose of a sentry.
3. **CO (Carrier Operation):** The radio stops when it hears a voice. It stays there as long as the voice is talking. Once the voice stops (and a 2-second buffer passes), it resumes scanning.
   - *Scenario:* You hear the full washout report. The transmission ends. The radio waits 2 seconds for a reply. Hearing none, it goes back to patrolling the frequencies.
   - *Verdict:* **The Standard.** Set your radio to CO (or "Busy" on some models). This mimics natural listening.

Priority Scanning: The "Check-In"

Some advanced radios (Wouxun KG-1000G, Midland MXT500) offer **Priority Scanning**. This is a game-changer for convoy leaders.

- **The Setup:** You set Channel 16 as your "Priority Channel."
- **The Action:** You start scanning Channels 1-22.
- **The Logic:** Even if the scanner stops on Channel 5 to listen to hikers, the radio will microscopically "check" Channel 16 every 0.5 seconds. If someone talks on Channel 16, the radio brutally cuts off the hikers and switches to Channel 16.
- **The Benefit:** You can monitor the world without fear of missing your primary team.

Tone Scanning: The Key to the City

You are in a new town. You scan and find a repeater on Channel 20 (462.675). You hear locals chatting. You try to talk to them, but they can't hear you. Why? You are missing the 'Privacy Tone' (CTCSS/DCS).

Instead of guessing random numbers, use **Tone Scan**.

1. **Tune** to the busy channel (Ch 20).
2. **Menu:** Find 'SC-CTC' or 'Tone Scan'.
3. **Activate:** The radio will listen to the incoming voice and cycle through all possible tones.
4. **Lock:** When it matches the tone (e.g., 141.3), the screen will flash 'CTCSS 141.3' and stop.
5. **Save:** Save that tone to your Transmit setting. You now have the key to the city.

NOAA Weather Alerts: The Voice of God

Every "Overland Adventurer" needs to monitor the weather. Mountains generate their own weather systems. A sunny morning can turn into a hail-spiking thunderstorm in 20 minutes, turning a dry wash into a deadly river.

Your GMRS radio likely has 7 dedicated **NOAA Weather Channels** (frequencies 162.400 through 162.550 MHz). These are broadcast by the National Weather Service (NWS). They are automated, robotic voices reading the forecast on a loop.

The "Monitor" vs. "Alert" Dilemma:

- **Monitor Mode:** You tune to the weather channel manually and listen to the robot. Useful for morning briefings while making coffee.
- **Alert Mode:** You put the radio in "Weather Scan" or "Alert Standby." The radio is silent. It consumes very little power. However, if the NWS issues a Severe Thunderstorm Warning or Tornado Warning, they transmit a specific **1050 Hz Tone**. Your radio detects this tone and sounds a siren alarm (like a smoke detector), then opens the audio so you can hear the warning.

The Protocol: When you get to camp, do not turn your radio off.

1. Switch to the local NOAA channel that comes in clearest.
2. Activate "WX Alert" mode.
3. Leave the radio on overnight (connected to the Aux battery or a solar generator).

If a flash flood is coming down the canyon at 3:00 AM, that radio is your only warning system. It is the sentry that never sleeps.

11.2 Dual Watch: The Two-Eared Operator

Scanning is great for finding random traffic. But what if you have two specific, high-priority channels?

1. You need to listen to **Channel 16** (The Convoy) so you don't miss a turn.
2. You need to listen to **Channel 20** (The Scout) who is two miles ahead checking the bridge integrity.

If you are on Ch 16, you miss the Scout. If you are on Ch 20, you miss the Convoy. If you Scan, you might miss the start of a sentence.

Enter **Dual Watch** (often labeled TDR, DW, or Dual Wait).

The Concept: Time-Division Multiplexing

Most GMRS radios have one receiver. They cannot *truly* listen to two frequencies at the exact same nanosecond. However, Dual Watch switches the receiver back and forth between Frequency A and Frequency B roughly 5 to 10 times per second.

- Check A... Check B... Check A... Check B...

To the human ear, it sounds instant. If someone talks on A, the radio locks on A. If someone talks on B, it locks on B.

The Setup: A/B Configuration

On your radio screen, you usually have a Top Line (A) and a Bottom Line (B).

1. **Set Line A:** Channel 16 (The Convoy Main).
2. **Set Line B:** Saddleback Repeater (Emergency Comms / Scout).

3. **Activate:** Turn on "TDR" or "Dual Watch" in the menu. An icon (usually "DW" or "S") appears.

The Synchronization Trap

There is a flaw in Dual Watch: **Simultaneous Transmission.** If the Scout talks on Channel 20 *at the exact same moment* the Convoy talks on Channel 16, the radio will lock onto whichever signal is split-second faster or stronger. You will miss the other one completely.

- *Mitigation:* This is why "Radio Discipline" and brevity are important. If everyone chatters endlessly, Dual Watch fails. Keep messages short to leave airtime for the other channel to break through.

The Danger Zone: PTT Priority Here is where people get in trouble socially and tactically. You are listening to both A and B. You hear the Scout talk on Line B. You grab the mic and press the button to reply.

- *Which channel does your radio transmit on?*

This depends on a menu setting called **"TX Priority"** (Transmit Priority).

- **Option 1: Selected (A/B):** The radio transmits on whichever line has the "Arrow" or is highlighted on the screen.
  - *Risk:* You might be listening to B, but the arrow is parked on A. You press the button and blast your reply to the Convoy (A) instead of the Scout (B). You say "Bridge looks sketchy," and the Convoy on Ch 16 gets confused because they aren't at the bridge yet.
- **Option 2: Busy (Last Active):** The radio automatically transmits on the channel that *last received a signal.*
  - *Benefit:* If the Scout talks on B, and you press PTT within 5 seconds, you reply on B automatically.
  - *Risk:* If the Convoy (A) chips in right after the Scout stops talking, your reply might switch to A mid-thought.

The Visual Command Recommendation: Set TX Priority to **"Selected" (Manual)**. Train yourself to look at the screen before you press the button. It creates a deliberate OODA loop:

1. "I hear traffic on B."
2. "Look at screen."
3. "Press 'A/B' button to move arrow to B."
4. "Press PTT."

It adds one second to your reaction time, but it prevents "Cross-Channel Pollution," where you accidentally scream at the wrong group of people.

11.3 Digital/Data (The Future)

Historically, GMRS was strictly analog. Just voice waves modulating a carrier signal. But in 2017 and 2021, the FCC updated the rules. We are now allowed to send short data bursts (limited to seconds) on GMRS channels.

This has opened the door for **Text Messaging** and **GPS Location Sharing**.

The Hybrid Radios (BTech GMRS Pro / Motorola T800)

Radios like the **BTech GMRS Pro** or the **Motorola T800** are "App-Centric" radios. They connect to your smartphone via Bluetooth.

How it Works:

1. **The Input:** You open the companion app on your phone. You see a map. You type a text message: *"Camp site found. 34.556, -112.443."*

2. **The Transfer:** The phone sends the data to the radio via Bluetooth.

3. **The Burst:** The radio converts that data into a screeching audio burst (like an old dial-up modem or a fax machine) and blasts it over the GMRS frequency.

4. **The Decode:** The receiving radio hears the screech, decodes it, and pops the text message up on *their* phone screen or radio display.

The Analog Annoyance (Etiquette Warning): To you, it's a silent text. To everyone else on Channel 16 using a normal radio, it sounds like a screeching modem from 1999. **Rule:** Warn the group before sending data. 'Sending GPS coords, stand by for data burst.' Do not spam data on a busy voice channel.

The Tactical Advantage: Silent Command

Why would you text when you can talk? Why complicate a simple tool?

1. **Coordinate Accuracy:** Reading GPS coordinates over voice is a nightmare. *"Three four... did you say four or five?... point five five..."* It is prone to error. Sending a data burst is 100% accurate. The recipient just clicks the map pin and their navigation starts.

2. **Silence:** You are hunting. You are observing wildlife. Or you are in a security situation where you do not want to make noise. Texting allows for silent coordination without breaking noise discipline.

3. **Persistence:** If you call someone on voice and they are in the bathroom or changing a tire, they miss the message. A text message waits on their screen until they return. It creates a log.

The Limitation: The Walled Garden

The catch is compatibility. Analog voice is universal; a Midland can talk to a Wouxun. **Data is proprietary.**

- A BTech GMRS Pro cannot send a text to a Motorola T800. They speak different digital languages (modulation schemes).

- A Garmin Rino cannot send location data to a Baofeng.

The Verdict: For a tight-knit family or a dedicated squad, Digital GMRS is a superpower. It provides a "Mini-Internet" completely off-grid. But for a general convoy with mixed gear, stick to Analog Voice (Channel 16). Use Data as your "Inner Circle" comms layer.

Battery Warning: Using Bluetooth and GPS features drains your radio battery significantly faster. If you are running digital, bring spare batteries.

Action Step: The "Storm Watch" Drill

You need to trust your sentry before the storm hits. You need to know what the alarm sounds like so you don't sleep through it.

The Scenario: You are at home. A storm front is moving in, or it is simply a Wednesday.

The Drill

1. **Find the Frequency:** Use your radio's scan function to find the strongest NOAA weather channel in your area.

2. **Set Alert:** Go into your menu and enable "WX Alert" (or hold the WX button, depending on the model). The radio should go silent. It is now "Armed."

3. **The Wait:** Leave the radio on the kitchen counter while you cook dinner.

4. **The Test:** Every Wednesday between 11:00 AM and 12:00 PM (local time), the National Weather Service runs a **Weekly Test**. They send the 1050 Hz tone.

5. **The Result:** Your radio should scream like a banshee (a siren tone distinct from normal traffic) and then open the audio channel to the test message.

If it works, you have verified your "Electronic Sentry." Pack that radio for your next trip with the confidence that it will watch the sky while you watch the trail.

Conclusion

Status - Clear

Objective: Cement the transformation from "Consumer" to "Commander."

Key Deliverable: A sustainable maintenance ritual to ensure your system works when the world breaks.

The Transformation: From Amazon Box to Airwave Authority

Take a moment. Look at the radio sitting on your desk or mounted in your truck. Pick it up. Feel the weight of it.

Three months ago, or whenever you started this book, that device was just a piece of plastic and circuitry. It was a gadget you bought because you saw a YouTuber use one, or because you had a vague, gnawing anxiety about cell phone coverage in the mountains. It was an object of mystery, shrouded in intimidating acronyms like CTCSS, SWR, and UHF. You likely looked at it with a mix of hope and confusion, wondering if this black box was actually going to save you or just complicate your life.

You were a **Buyer**. You were operating on the consumer premise that simply owning the object would buy you safety. You thought the transaction was the solution.

Today, looking at that same radio, you should see something different. You don't see a gadget. You see a node in a network. You see a tool that extends your influence beyond the horizon. You see the ability to call for help when the sky turns black, to coordinate a rescue when a truck slides off the trail, or to simply tell your kids to come back to camp for dinner without screaming your lungs out. The mystery is gone, replaced by utility.

You are no longer a Buyer. You are an **Operator**.

This is not a semantic distinction. It is a fundamental shift in psychology and responsibility.

- **The Buyer** hopes the radio works. They rely on luck and factory settings.
- **The Operator** *knows* the radio works because they tuned the SWR, programmed the repeater offset, and drilled the PACE plan with their family. They have verified the physics.
- **The Buyer** fears the silence of a dead channel, interpreting it as a failure.
- **The Operator** knows how to scan, how to break squelch, and how to move to high ground to re-establish the link. They view silence as a variable to be solved, not a dead end.

You have adopted the *Visual Command System*. You have rejected the gatekeeping of the "Sad Hams" who want you to memorize circuit diagrams from 1950 before you are allowed to push a button. You have embraced the pragmatic, "if-this-then-that" logic of the field guide. You have learned that wattage is just water pressure, that antennas are just nozzles, and that a radio without a plan is just a paperweight.

The Core Principles: A Final Review

Before we sign off, let's lock in the core doctrines one last time. These are the pillars that will hold your system up when stress and adrenaline try to knock it down.

1. The Hardware is Secondary; The Mindset is Primary

We spent chapters on Baofengs, Midlands, and Wouxuns. We discussed superheterodyne receivers and quarter-wave antennas. But the specific brand doesn't matter as much as the brain behind it. A trained operator with a $30 blister-pack radio who understands line-of-sight is more effective than a confused novice with a $5,000 base station who doesn't know what a repeater offset is. You now know that "the best radio" is not the most expensive one; it is the one you know how to program in the dark, in the rain, with gloves on.

2. Plain Language is Power

We killed the "10-Codes." We buried "Breaker Breaker." We established that sounding like a professional means speaking English. Under stress, the human brain sheds complex codes. If you scream "10-33" at a frightened spouse, they will freeze. If you say "Emergency, I need help," they will act. "Dad to Mom, location?" is infinitely superior to "Unit Alpha requesting 10-20 from Base." Clarity is speed. Speed is safety.

3. PACE is Your Safety Net

Technology fails. Batteries die. Physics wins. You cannot argue with a mountain blocking your signal. You now have a **PACE Plan** (Primary, Alternate, Contingency, Emergency). You know that when Channel 16 fails, you don't panic - you simply execute the next step in the flowchart. You switch to Channel 20. When Channel 20 fails, you hit the Repeater. When the Repeater fails, you blow the whistle or check the GPS. You have replaced panic with a pre-loaded decision tree, allowing you to act while others are paralyzed.

4. Intelligence Over Noise

You learned that the radio is not just for talking; it is for listening. You are the Sentry. You monitor the NOAA weather channels to detect storms before you see clouds. You scan for local traffic to detect logging trucks before you meet them on a blind corner. You use the radio to build a mental map of the battlefield - whether that battlefield is a chaotic trail run or a stormy campsite. Information dominance prevents emergencies.

The Challenge: The Law of Entropy

Here is the hard truth: **Systems decay.**

Nature hates order. The second you finish installing your mobile unit, the universe begins working to dismantle it. If you leave your truck parked in a field for a year, the tires go flat, the battery dies, and mice build nests in the air filter.

Your radio system is no different. It is under constant attack from the elements and time.

- **Lithium-Ion batteries** self-discharge. Their internal chemistry is unstable. If you leave your handhelds in a drawer for six months, they will be dead when the power goes out. Worse, if left at 0%, they may chemically lock and never charge again.
- **Connectors corrode.** That PL-259 connector on your truck's antenna is exposed to rain, mud, road salt, and UV radiation. It will eventually oxidize, creating high resistance that chokes your signal.
- **Repeater details change.** The owner of the "Saddleback 700" might change the privacy tone, move the antenna, or take the system offline entirely. A frequency list from 2023 is a history book, not a field guide.
- **Skills rust.** If you don't use the NATO phonetic alphabet for a year, you will forget whether "B" is Bravo or Baker.

You will forget how to unlock the keypad. You will forget which channel is the Alternate.

The "Buyer" puts the radios in a plastic bin, shoves them in the garage, and forgets them until an emergency strikes. Then, in the moment of crisis, they pull out a dead radio with corroded terminals and scream, "Why doesn't this work?!"

The **Operator** fights entropy with a ritual.

The Solution: The Monthly Radio Check

You need a Standard Operating Procedure (SOP) for maintenance. You need a "Radio Shabbat" - a designated time when you inspect and refresh your comms.

I recommend the **First Saturday of the Month**. It takes 15 minutes. It ensures 100% readiness.

The "First Saturday" Checklist

1. The Power Cycle (5 Minutes)

- Gather every handheld radio you own. Dump them on the kitchen table.
- Turn them on. Check the battery voltage indicator. **Remove the battery and inspect for swelling.** If the battery spins like a top on a flat table, it is swollen (spicy pillow). Discard it safely and order a replacement.
- If they are below 80%, put them on the charger.
- *Crucial Step:* If they are fully charged, leave them *on* for an hour to cycle the chemicals, then top them off. Lithium batteries hate sitting at 0%, but they also degrade if left at 100% for years without use. Using them is good for them; it wakes up the ions.

2. The Mobile Inspection (3 Minutes)

- Walk out to your vehicle. Don't just look at it; touch it.
- Grab the antenna. Wiggle it. Is the mount tight? Did the last car wash loosen the set screws? If the antenna wobbles, your ground plane is compromised.
- Check the coax cable. Is it pinched in the door jamb? Is the insulation cracked by the sun?
- Turn the mobile unit on. Check the voltage readout (usually on the screen). Is your truck battery healthy (12.6V engine off, 13.8V engine on)? A radio that dims when you key the mic is warning you of a failing truck battery.

3. The Comms Check (The Family Drill) (5 Minutes) This is the most important part. It keeps your family involved and comfortable with the gear.

- Give a handheld to your spouse or kid. Send them to the backyard, the basement, or the end of the driveway.
- Get in your truck (or use another handheld).
- **Execute the Protocol:**
    - **You:** "Base, this is Mobile. Radio Check."
    - **Them:** "Mobile, Base. Loud and clear."
    - **You:** "Switch to Alternate Channel 20."

- **Them:** "Switching 20."
- **You:** (On Ch 20) "Radio Check on Alternate."
- **Them:** "Loud and clear on 20."

- **Why do this?**
  - It verifies the hardware works on multiple frequencies.
  - It verifies the batteries work under load.
  - It verifies your family remembers how to unlock the keypad and change channels without you standing over their shoulder.
  - It normalizes the radio. It stops being a "scary emergency device" and becomes a normal household tool like a toaster or a flashlight.

4. The Weather Sentry (2 Minutes)

- Tune to your local NOAA weather frequency.
- Verify you can hear the robot clearly.
- This confirms your receiver sensitivity is good. If you can't hear the weather station you usually hear, your antenna system has a problem.

5. The "Software" Update (As Needed)

- Once a year (maybe every January), plug your radios into the computer and open CHIRP.
- Check RepeaterBook.com. Have any new repeaters popped up in your area? Have old ones changed tones?
- Update your radio memories. A radio with 5-year-old repeater data is a paperweight.

Final Encouragement: You Are the Hub

We live in a fragile world. We have built a society that relies entirely on a delicate web of cell towers, fiber optic cables, and server farms. When that web works, it is miraculous. But when it breaks - due to a hurricane, a wildfire, a cyberattack, or just a lack of funding in a rural county - it breaks completely.

When the bars on your cell phone disappear, the illusion of safety vanishes with them. For the average person, this is a moment of terror. They feel cut off, alone, and vulnerable.

But you are not the average person anymore.

When the grid goes down, you do not lose your voice. You simply reach for the microphone. You have an independent, physics-based, decentralized communication network that you own, maintain, and command. You are not dependent on a subscription service or a corporation to talk to your family.

You are the **Communications Hub** for your tribe.

- You are the one who will know the weather is turning before the sky goes dark, because you monitored the NOAA alert.
- You are the one who will coordinate the convoy when the lead truck breaks an axle in the dead zone, because you

built the PACE plan.

- You are the one who will hear the call for help from a stranger and have the power to answer.

Own that responsibility. It is a heavy weight, but it is a good weight. It is the weight of competence.

Don't let the skills rust. Don't let the batteries die. Teach your kids. Teach your neighbors. Be the "Elmer" (mentor) that the "Sad Hams" refused to be. Share your knowledge freely and build a community that is resilient, capable, and connected.

The airwaves are free. The equipment is ready. The rest is up to you.

Radio Check. Signal: 5 by 5. Status: Clear.

Whiskey Sierra Lima Charlie Two Three Zero

End of Transmission

Appendix: The Visual Command Toolkit

Quick Reference Cards

The Visual Command Toolkit: Quick Reference Cards

Print, Laminate, and Keep in Your Kit

GMRS Frequency Chart

Ch	Freq (MHz)	Type	Max Pwr
1	462.5625	Simplex	5W
2	462.5875	Simplex	5W
3	462.6125	Simplex	5W
4	462.6375	Simplex	5W
5	462.6625	Simplex	5W
6	462.6875	Simplex	5W
7	462.7125	Simplex	5W
8-14 *(Avoid)*	467.5625+	FRS ONLY	0.5W
15	462.5500	Simplex/RPT	50W
16	462.5750	Simplex/RPT	50W
17	462.6000	Simplex/RPT	50W
18	462.6250	Simplex/RPT	50W
19	462.6500	Simplex/RPT	50W
20	462.6750	Simplex/RPT	50W
21	462.7000	Simplex/RPT	50W
22	462.7250	Simplex/RPT	50W

Note: Ch 15-22 are the main GMRS/Repeater Output channels.

NATO Phonetic Alphabet

A - Alpha
B - Bravo
C - Charlie
D - Delta
E - Echo
F - Foxtrot
G - Golf
H - Hotel
I - India
J - Juliet
K - Kilo
L - Lima
M - Mike
N - November
O - Oscar
P - Papa
Q - Quebec
R - Romeo
S - Sierra
T - Tango
U - Uniform
V - Victor
W - Whiskey
X - X-ray
Y - Yankee
Z - Zulu

Numbers:
0 - Zero 1 - Wun 2 - Too
3 - Tree 4 - Fower 5 - Fife
6 - Six 7 - Seven 8 - Ait
9 - Niner

Pro-Words:
OVER: I am done talking, reply expected.
OUT: I am done talking, no reply expected.
COPY: Message received and understood.
SAY AGAIN: Please repeat your last message.

Relevant 10-Codes (GMRS)

Note: Plain language is preferred in GMRS, but these are common shorthand.

- **10-1:** Receiving poorly / Signal weak.
- **10-2:** Receiving well / Signal strong.
- **10-4:** Message received (OK).
- **10-6:** Busy, stand by.
- **10-9:** Repeat message.
- **10-20:** What is your location? ("What's your 20?")
- **10-21:** Call by telephone.
- **10-33: EMERGENCY TRAFFIC ONLY.**
- **10-77:** Negative contact.

Visual Command Rule:
If you are stressed, forget the codes. Speak normally. *"Unit 1, I can't hear you, move to higher ground."* is better than a misused 10-code.

Repeater Quick Guide

The Golden Rule: Offset is ALWAYS **+5.000 MHz**.

Input (Talk): 467.xxx MHz
↓
Output (Listen): 462.xxx MHz

Troubleshooting Checklist:

1. **Right Channel?** (15-22 only).
2. **Right Tone?** (Check T-CTCS / Transmit Tone).
3. **Offset On?** (Do you see a '+' symbol?).
4. **Wide Band?** (Set to WFM, not NFM).

Field Programming Steps:

1. Go to VFO (Freq Mode).
2. Enter 462.xxx frequency.
3. Menu -> T-CTCS -> Set Tone.
4. Menu -> SFT-D -> Set (+).
5. Menu -> OFFSET -> Set 005.000.

Etiquette:
"[Your Call Sign] monitoring." to announce presence.
Don't "Kerchunk" (click mic) without ID.

The Visual Command Toolkit

Repeater Map Log

INSTRUCTIONS: Use this log to record repeaters for your local area or travel route.

- **Source:** Find data on *MyGMRS.com* or local club listings.
- **Input Tone:** The "Key" your radio transmits to open the repeater (T-CTCS).
- **Output Tone:** The tone the repeater transmits back (R-CTCS). Optional, but filters noise.
- **Offset:** Remember, GMRS Offset is ALWAYS **+5.000 MHz**.

Location / City	Name	Output Freq	Input Tone	Output Tone	Status	Notes / Range

The Visual Command Toolkit

GMRS Frequency & Bandwidth Chart

LEGEND:

- Shared / Low Power: FRS & GMRS mixed. 5 Watts Max. Narrow or Wide.
- **Restricted (FRS Only):** Low power (0.5W). Narrowband only. **Avoid for tactical use.**
- Main GMRS / High Power: The 'Command Lanes.' 50 Watts. Wideband. Repeater Capable.

CH	Frequency (MHz)	Max Power	Bandwidth	Usage / Notes
1	462.5625	5 Watts	NFM / WFM	Interstitial / Shared
2	462.5875	5 Watts	NFM / WFM	Interstitial / Shared
3	462.6125	5 Watts	NFM / WFM	Interstitial / Shared
4	462.6375	5 Watts	NFM / WFM	Interstitial / Shared
5	462.6625	5 Watts	NFM / WFM	Interstitial / Shared
6	462.6875	5 Watts	NFM / WFM	Interstitial / Shared
7	462.7125	5 Watts	NFM / WFM	Interstitial / Shared
8	467.5625	0.5 Watts	**NFM ONLY**	**FRS ONLY** (Avoid)
9	467.5875	0.5 Watts	**NFM ONLY**	**FRS ONLY** (Avoid)
10	467.6125	0.5 Watts	**NFM ONLY**	**FRS ONLY** (Avoid)
11	467.6375	0.5 Watts	**NFM ONLY**	**FRS ONLY** (Avoid)
12	467.6625	0.5 Watts	**NFM ONLY**	**FRS ONLY** (Avoid)
13	467.6875	0.5 Watts	**NFM ONLY**	**FRS ONLY** (Avoid)
14	467.7125	0.5 Watts	**NFM ONLY**	**FRS ONLY** (Avoid)
15	462.5500	**50 Watts**	WFM (Pref)	GMRS Main / RPT Output
16	462.5750	**50 Watts**	WFM (Pref)	GMRS Main / RPT Output
17	462.6000	**50 Watts**	WFM (Pref)	GMRS Main / RPT Output
18	462.6250	**50 Watts**	WFM (Pref)	GMRS Main / RPT Output
19	462.6500	**50 Watts**	WFM (Pref)	GMRS Main / RPT Output
20	462.6750	**50 Watts**	WFM (Pref)	**National Emergency / Travel**
21	462.7000	**50 Watts**	WFM (Pref)	GMRS Main / RPT Output
22	462.7250	**50 Watts**	WFM (Pref)	GMRS Main / RPT Output

Note: Repeaters listen on 467.xxx (Input) and talk on 462.xxx (Output). The Offset is always **+5.000 MHz**.

Scan to Download Printable Copies.

The Baofeng Radio Family Lifeline

The Real-World Protocol to Design a Mission-Ready Communication Plan, Master CHIRP Programming, Configure Repeater Offsets, and Train Your Household

MorseCode Publishing

THE PROMISE

In 14 days, you will stop guessing and start communicating.

If you follow the steps in this book, you will move from confusion to competence. You will:

1. **Build a working channel plan** that actually makes sense for your area, filtering out the digital noise and dead frequencies that clog up your memory banks.
2. **Program repeaters correctly** (so people can actually hear you). No more "chunking" the repeater without getting a response because you missed a tone setting.
3. **Run 3 real-world drills** to prove your gear works when cell service dies. You will know exactly how far you can talk in your specific neighborhood, through your specific trees and buildings.
4. **Troubleshoot failures calmly** instead of panicking when you hear static. When the radio acts up (and it will), you'll know if it's a battery issue, an antenna issue, or user error.

You bought the radio because you wanted to be prepared. Right now, you just own a piece of plastic with too many buttons. Two weeks from now, you will own a **capability**. You will know exactly which button to press, even in the dark.

WHO THIS BOOK IS FOR

- **The "Just in Case" Guy:** You bought a UV-5R off Amazon three years ago, threw it in a go-bag next to a first aid kit, and haven't touched it since. It's probably sitting next to a battery that hasn't been charged since 2021. You know you *should* learn how to use it, but the manual reads like stereo instructions.
- **The "Tactical Dad":** You aren't trying to play soldier. You just want a backup way to talk to your family during storms, power outages, or crowded events like the county fair when cell towers get jammed. You want a "Gray Man" setup, effective, discreet, and reliable.
- **The Frustrated Beginner:** You got your license (or are studying for it), but the classes only taught you electrical theory, not how to actually use the radio in your hand. You can hear the local repeater, but nobody ever answers when you call.

WHO THIS BOOK IS NOT FOR

- **The Elitist:** If you want to argue about why cheap radios are ruining the RF spectrum or why everyone should buy a $400 Yaesu, put this back on the shelf. We know it's a cheap radio. That's the point. It's the radio people actually have.
- **The Cosplayer:** If you think buying the radio makes you an operator, you're in for a rude awakening. Hanging a radio on a plate carrier doesn't mean you can communicate. Skill matters more than gear.

- **The Pirate:** If you want to jam police frequencies, troll on GMRS channels, or act like an idiot on the air, we can't help you (and we won't). This book teaches responsible, effective, legal communication.

HOW TO USE THIS BOOK (CHOOSE YOUR PATH)

You don't have to read this cover-to-cover right now. This is a field guide, not a novel. Jump to the problem that is making you want to throw the radio across the room.

- **"I can hear people, but nobody hears me."**
  - *Go to Chapter 3 (Repeater Mastery)* or *Chapter 1 (The 5 Settings That Kill You).* We'll fix your offsets and tones in ten minutes.
- **"My computer won't talk to the radio (Driver/Cable Hell)."**
  - *Go to Chapter 4 (CHIRP 2.0).* We'll bypass the bad drivers and get your spreadsheet loaded.
- **"I can't reach my house from the grocery store."**
  - *Go to Chapter 5 (Range Reality Lab).* Physics is stubborn, but we can optimize your antenna and placement to squeeze every mile out of 5 watts.
- **"I want a plan for my family that isn't complicated."**
  - *Go to Chapter 9 (The Family Comms Plan).* We'll turn "Here, take this" into a structured, stress-tested communication protocol.

A Note on "The Gray Man" Approach Throughout this book, we prioritize being **effective** over being loud. The goal isn't to look cool with a giant whip antenna and a tactical vest. The goal is to have a discrete, reliable lifeline when the phones go down. We follow the rules, we respect the airwaves, and we keep a low profile. The best operator is the one nobody notices until they need help.

INTRODUCTION

THE UPGRADE MINDSET

The Driveway Moment

It's Saturday afternoon. You've finally convinced your wife (or buddy) to help you test those radios you bought six months ago. The sun is shining, and you feel ready. You hand her one unit and say, "Just stay in the kitchen. I'm going to drive to the end of the block. If things get weird, we'll use these." You feel capable. Prepared. You drive three streets away, park, and key the mic.

"Check, check. Can you hear me?"

Silence. You try again, speaking louder this time. *"Honey, radio check. Over."*

Static. Then, a weird garbled noise that sounds like a fax machine dying. Then, silence again. You drive back to the house, frustrated, feeling the heat rise in your cheeks. You spent $60 on radios and $20 on a cable, and right now, shouting out the window would be more effective. She hands you the radio and says, "I heard a noise, but I couldn't understand you. And then it started scanning by itself and picking up the weather station."

Congratulations. You have just experienced the **Baofeng Reality Check.**

Most people stop here. They assume the radio is junk (it's not), or that range is a lie (it's complicated), or that they just aren't smart enough to figure it out (false). The problem isn't the radio. The problem is that you treated a professional tool like a toy walkie-talkie. You expected it to work like an iPhone, you push a button and it just works. But RF (Radio Frequency) doesn't care about your expectations.

The 4 Truths of The Baofeng Radio Family Lifeline

To move from "guy with a radio" to "capable operator," you have to accept four truths. These are the pillars of the 2.0 mindset:

1. Expectations: It is a $25 Radio. It has a slow scanner. The receiver filters noise about as well as a screen door holds back water. It is not a Motorola public safety radio. But if you understand its quirks and work around them, it is 90% effective for 10% of the price. That is a trade we will take every time.

2. Practice is a Perishable Skill. You cannot learn CHIRP software during a blackout. You cannot figure out "Simplex vs. Duplex" while a hurricane is rattling your windows. Adrenaline dumps your IQ by about 20 points. If you have to look up the manual to change a frequency, you are dead in the water. You verify *now* so muscle memory takes over *later*.

3. Planning > Frequencies. Having 100 random channels programmed in your radio is useless. A list of repeater frequencies for a state you'll never visit isn't preparedness; it's digital hoarding. Having 5 channels that you and your family know by heart, and act on instinctively, is a superpower.

4. Diagnostics is the Skill. Things will break. Batteries die. Settings get bumped in your bag. Connectors corrode. The "2.0

Operator" doesn't panic; they troubleshoot. This book teaches you how to look at a silent radio, run a 60-second diagnostic, and fix it in the field.

The Operator Upgrade Ladder

This book is designed to move you up this ladder. Where are you right now?

- **Level 1: The Listener (Passive).** You can turn it on and hear the weather channel. You are essentially holding a police scanner that is hard to program. You have zero transmit capability.
- **Level 2: The Programmer (Technical).** You can use software to put channels in. You have a spreadsheet, but you don't really know if those channels work. You are technically set up, but operationally untested.
- **Level 3: The Communicator (Active).** You have made contact. You know your range limits. You can hit a repeater and get a signal report. You trust the gear because you've used it.
- **Level 4: The Troubleshooter (Resilient).** When it fails, you know why. You have a backup plan (and a backup battery). You are the "comms guy" for your family or group.

A Note on Hardware (The "Hero" Device)

In this book, we use the classic **Baofeng UV-5R** for our diagrams and instructions. It's the Honda Civic of radios, ubiquitous, cheap, customizable, and capable.

VARIANT CALLOUT: "Is my radio different?" If you own a **UV-82** (the one with the rocker switch), a **GMRS-V1**, or a **UV-5G**, your radio is 95% the same as the UV-5R inside. However, you might have locked menus (due to FCC rules) or different buttons. Look for these **Variant Callout** boxes throughout the book. We will tell you exactly when your specific model requires a different step so you never feel left behind.

ACTION STEP: The 10-Minute Assessment

Before you turn the page, put your radio on the table. Be honest with yourself. We need to establish a baseline before we can upgrade.

Your Mission Checklist:

- **Power:** Does the battery hold a charge? (If it's been in a drawer for a year, charge it *now*. A dead radio is just a paperweight).
- **Connection:** Do you have the programming cable? (Not just the charging cable, the USB cable that connects to your PC. You cannot act as a Level 2 Operator without this).
- **Legal:** Do you have your GMRS license or Ham Technician license? (If not, we aren't the police, but we strongly suggest getting legal. It protects the hobby and lets you practice without looking over your shoulder. GMRS is the "easy button", $35, no test, covers the whole family. Go do it.)
- **The "Baseline":** Turn the radio on. Press the VFO/MR button (orange button on many models). Does the radio speak "Channel Mode" or "Frequency Mode"?

If you don't know what that last bullet point means, don't worry. It's the single most confusing button on the device, and we're going to master it in Chapter 1.

Let's calibrate.

Jared Johnson (KF0RTU) Jared@MorseCodePublishing.com

Chapter 1

THE 2.0 CALIBRATION

WHAT MATTERS NOW

YOU HAVE THE RADIO in your hand. Maybe you've already mashed some buttons and heard static. Maybe it's fresh out of the box and still smells like factory plastic.

Before we program a single frequency, we need to calibrate the device, and more importantly, calibrate *you*.

Most people skip this step. They tear open the box, Google "police frequencies," and try to program their local dispatch immediately. Then they get frustrated when the radio beeps at them, refuses to transmit, or transmits on a frequency that interferes with actual emergency services.

We are going to strip this down to the "Known Good" state. We are building a foundation so solid that when things go sideways in the dark, when the power is out, the rain is hammering the roof, and your family is looking to you for answers, you know for a fact that it's not the radio's settings that are fighting you.

THE 5 SETTINGS THAT CAUSE 80% OF THE PAIN

If you understand these five concepts, you understand the Baofeng. If you ignore them, you will forever be "the guy with the radio that doesn't work," frantically pressing buttons while everyone else is communicating.

1. VFO vs. MR (The Two Brains)

Your radio has a split personality. It has two distinct operating modes, and you must know which brain is in charge at all times. If you don't, you will try to save a channel and fail, or try to tune a frequency and get locked out.

- **Frequency Mode (VFO - Variable Frequency Oscillator):** This is the **manual** brain. It is your scratchpad. You type in numbers directly (e.g., 1-4-6-5-2-0). The radio goes there immediately.
    - *When to use it:* When you are searching for a new signal, testing a frequency you just found online, or meeting a buddy on a random frequency that isn't in your saved list.
    - *Behavior:* Changes here are temporary unless you explicitly save them to memory.
- **Channel Mode (MR - Memory Recall):** This is the **saved** brain. It is your contact list. You cannot change the frequency numbers here; you can only scroll through what you have previously saved (Channel 1, Channel 2, etc.).
    - *When to use it:* In 99% of real-world scenarios. Under stress, you don't want to type 4-6-2...; you just want to click to "Channel 1: HOME."
    - *Behavior:* This mode protects you from accidentally bumping the keypad and changing your frequency.

The Trap: You try to type a specific frequency while in Channel Mode. The radio ignores you, or worse, it interprets your button presses as a command to jump to "Channel 146," which is empty. You think the keypad is broken. It's not; you're just in the wrong brain.

The Fix: The VFO/MR button (orange button on the UV-5R) toggles between them.

- **Visual Check:** Look at the right side of the display.
  - If you see small channel numbers (001, 002, 127), you are in **Channel Mode**.
  - If there are *no* small numbers on the right, you are in **VFO Mode**.
- **Audible Check:** The radio literally announces "Frequency Mode" or "Channel Mode" when you press it (unless you turned the voice off, which we'll discuss later).

2. Duplex/Offset (The Repeater Math)

Simplex means you talk and listen on the exact same frequency (like a normal walkie-talkie or FRS radio). Repeaters are different. They are "deaf" on the frequency they broadcast on. To talk to a repeater, you must listen on one frequency but transmit on another (usually 0.6 MHz away for Ham, or 5.0 MHz for GMRS).

The Logic: Imagine a translator standing on a hill. He shouts loudly to everyone in the valley (Output Frequency). But to get him to shout your message, you have to whisper in his ear (Input Frequency). If you shout at his mouth, he won't hear you. The "Offset" is simply the distance between his mouth and his ear.

The Trap: You program a repeater frequency, but you forget the offset. You press the PTT (Push-to-Talk), and you transmit on the listening frequency (shouting at the translator's mouth).

- **Symptom:** You can hear people talking clearly. When you transmit, nobody responds. You might hear a "kerchunk" (the repeater activating) but your audio doesn't go through.

The Fix:

- **Menu 25 (SFT-D):** Direction of the shift.
  - **OFF:** Simplex (Direct radio-to-radio).
  - **+ (Plus):** Transmit frequency is *higher* than receive.
  - **- (Minus):** Transmit frequency is *lower* than receive.
- **Menu 26 (OFFSET):** The amount of shift.
  - Standard Ham (2m): 00.600 (Type all five digits)
  - Standard GMRS: 05.000
  - Standard Ham (70cm): 05.000

3. Tone Logic (The Key to the Door)

Think of a standard frequency as a house. Anyone can stand on the sidewalk and yell (transmit). A "Tone" (CTCSS or DCS) is a lock on the door.

- **Encode Tone (T-CTCS / Menu 13):** This puts a "key" in your pocket. When you speak, your radio sends a sub-audible tone (invisible key) along with your voice. The repeater (the house) checks for this key. If you have it, it opens the door and broadcasts your message. If you don't, it ignores you. **Most repeaters require this.**

- **Decode Tone (R-CTCS / Menu 11):** This puts a lock on *your* speaker. Your radio will stay 100% silent, even if the frequency is busy, unless the incoming signal has the matching key.

The Trap: You set a "Receive Tone" (R-CTCS) when you didn't mean to. This is the "Cone of Silence" error.

- **The Scenario:** You and your wife are trying to test radios. She is transmitting, but your radio is silent. You see the green light on your radio light up (indicating a signal is there), but no sound comes out.

- **The Cause:** Your radio is waiting for a specific tone key that she isn't sending. You locked your own door and didn't give her the key.

The Golden Rule: Unless you have a specific reason (like living near a interference-heavy area or blocking out other repeater traffic), **leave R-CTCS (Menu 11) and R-DCS (Menu 10) set to OFF.** Only set the **T-CTCS (Menu 13)** when a repeater requires it.

4. Scan Behavior (CO / TO / SE)

You will want to scan channels to monitor traffic, maybe to listen for storm spotters or find activity during an emergency. But *how* the radio stops scanning determines if you catch the info or miss it entirely.

The Trap: You set it to scan. The radio lands on a conversation, maybe a storm report. "Tornado spotted near..." and then suddenly, the radio starts scanning again, cutting off the location. Why? Because you told it to move on after 5 seconds, regardless of whether they were done talking.

The Fix (Menu 18 - SC-REV):

- **TO (Time Operation):** The "Sample" Mode. It stops for 5 seconds to let you hear a snippet, then moves on no matter what. **Terrible for following conversations.**

- **CO (Carrier Operation):** The "Listening" Mode. It stops while someone is talking. It stays there as long as the signal is present. It resumes scanning only after the signal drops for a few seconds. **This is what you want for general monitoring.**

- **SE (Search Operation):** The "Find It" Mode. It stops when it finds a signal and *stays there*. It stops scanning completely. Use this when you are hunting for active channels and want to lock onto the first one you find.

- **Action:** Go to Menu 18 and set this to **CO** right now.

5. Squelch Realities (The Gatekeeper)

Squelch is the noise gate. It decides how strong a signal must be before the speaker turns on.

- **Too Low (0):** The gate is open. You hear constant static (white noise).

- **Too High (9):** Only a massive signal (someone standing next to you) will open it. You will miss weak signals from distant repeaters.

The Trap: The factory settings on Baofengs are notoriously bad. The difference between Level 1 and Level 9 is often negligible on stock firmware. They don't scale linearly.

The Fix (Menu 0 - SQL):

- Set this to **1** for now. (Note: In a dense city with lots of interference, you might need **3** or **5** to stop the static pops.) It's the most sensitive setting that isn't constant static.
- *Pro Tip:* If you are driving and hear constant "PFFT... PFFT..." static pops, bump it to 2 or 3. If you leave it at 5 or 6, you might miss a weak emergency transmission.

YOUR BASELINE: CREATE A KNOWN-GOOD STARTING POINT

We are going to wipe the slate clean. Out of the box, these radios often come with random Chinese test frequencies, marine channels, or interference-prone garbage programmed into the memory. We don't want those ghost channels.

The Reset Strategy

*Warning: This deletes all saved channels. If you bought a customized "Pre-Programmed" radio from a vendor like BetterSafeRadio or Rugged Radios, **SKIP THIS STEP**. You paid extra for that programming; don't wipe it out. Just do the "Settings Check" below.*

VARIANT CALLOUT: GMRS & Locked Radios If you have a **UV-5G, GMRS-V1, or UV-82C**, your radio firmware may block the "Reset All" function to comply with FCC Part 95 rules (you cannot delete the primary GMRS channels). That's fine. Just manually verify the settings in the Menu using the worksheet below.

The "Tactical Dad" Factory Reset (Standard UV-5R):

1. Turn radio ON.
2. Press Menu, type 40 (RESET).
3. Press Menu, select ALL.
4. Press Menu again to confirm. The radio will go black, reboot, and likely speak a Chinese greeting.
5. **Don't panic.** We anticipated this.
6. Press Menu, type 14 (VOICE), press Menu. Use the arrow keys to select ENG (English). Press Menu to save. Now it speaks your language.

The "Baseline Channel" (Channel 000)

We are going to program one channel manually right now using the keypad. **Pro Tip:** Type fast. If you pause for more than 5 seconds, the radio gets bored and exits the menu. This is your "sanity check" channel. If you ever get confused, lost in menus, or think the radio is broken, you go to Channel 000.

Your Mission: Find the National Simplex Frequency for your service. This is the "Town Square" of radio frequencies.

- **Ham:** 146.520 MHz
- **GMRS:** 462.5625 MHz (This is universally Channel 1 on GMRS radios)

Why? This is the common meeting place. If you are lost, you go here. If you are testing, you go here. If you are trying to make contact with a stranger in an emergency, this is your best bet.

REALITY CHECK: WHAT THE BAOFENG IS (AND ISN'T)

Before we start building your plan, let's be honest about the hardware. Managing expectations is the key to happiness.

What it is GREAT at:

- **Disposable Utility:** If you drop it in a creek, step on it, or hand it to a neighbor who loses it, you are out $25. This makes you willing to actually *use* it. You won't hesitate to toss it in a glovebox or a backpack.
- **Battery Life:** With a decent battery (and not scanning constantly), these will sit on standby for days. The idle power consumption is very low.
- **Versatility:** It listens to FM radio, Marine bands, NOAA weather, and more. It is a Swiss Army Knife.

What it SUCKS at:

- **The "Front End" (Receiver Quality):** This is the biggest weakness. The Baofeng receiver chip is easily overwhelmed by strong signals.
    - *The Scenario:* You are driving near a cell tower or a high-power paging transmitter. Suddenly, your radio goes silent. Or maybe it blasts static.
    - *Desense:* The radio is "desensitized." It's like trying to hear a whisper while someone is screaming in your other ear. The radio protects itself by going deaf.
    - *The tell:* If you are in the city and can't hear anything, unscrew the antenna. If you suddenly hear a split second of clear audio before the signal drops, your radio was overloaded.
- **Scan Speed:** It is slow. Painfully slow. Do not try to scan 100 channels. By the time it cycles through, the conversation will be over. Keep your scan lists short (15-20 channels max).
- **Stock Antenna:** It's often called a "rubber ducky." It is durable, but electrically, it is a compromise. It works for 1-2 miles, but don't expect miracles. (We fix this in Chapter 5).

ACTION STEPS

1. The "Before You Troubleshoot" Checklist

Copy this onto an index card, laminate it with packing tape, and tape it to the back of your battery.

If the radio isn't working, check these in order:

1. **Battery:** Is the voltage over 7.4V? (Hold the 0 key for 2 seconds to check voltage. On some older models, hold the SCAN key instead). A low battery causes weird glitches before it actually dies.
2. **Mode:** Are you in VFO or MR? (Press VFO/MR). If you are trying to change channels, be in MR. If you are typing frequencies, be in VFO.
3. **Tone:** Do you see "CT" or "DCS" on the left of the screen? If you are trying to talk to a new radio and you see these letters, turn them OFF. You are likely locking yourself out.
4. **Offset:** Do you see a little + or - at the top of the screen? If you are trying to talk Simplex (radio-to-radio), this MUST be off.

5. **Antenna:** Is it actually screwed on tight? (Connector SMA-Female vs Male confusion is real, make sure the center pin actually connects).

2. My Baseline Settings Worksheet

Go into your menu and set these NOW. This is your foundation. Do not pass Go until these match.

Menu #	Name	Setting	Why?
0	SQL	1	**Max sensitivity.** We want to hear everything. If it's too noisy, bump to 2.
1	STEP	**2.5K or 5.0K**	**Precision.** Allows you to tune to specific narrowband frequencies without rounding errors.
2	TXP	**HIGH**	**Power.** Default to max power. You can always switch to Low (tap the # key) to save battery later.
3	SAVE	**OFF**	**CRITICAL.** Battery save mode sleeps the receiver. It causes you to miss the first 1-2 seconds of a transmission (the "Baofeng delay"). Turn it OFF for reliability.
7	TDR	**OFF**	**Focus.** Dual Watch lets you listen to two channels. It is confusing for beginners. Keep it OFF so you know exactly who you are talking to.
14	VOICE	**ENG**	**Feedback.** Helps you know what mode you are in without looking.
18	SC-REV	**CO**	**Carrier Operation.** Ensures the scanner resumes only after the signal drops, so you don't miss the end of the sentence.
25	SFT-D	**OFF**	**Simplex Default.** We start with no offset. We add offsets only when programming specific repeaters.
39	ROGER	**OFF**	**Etiquette.** The "Roger Beep" screams, "I am a newbie with a cheap radio." It is annoying to other operators. Turn it off.

Once these are set, your radio is calibrated. You are now a Level 1 Operator. You have a known-good device. Turn the page.

Chapter 2

BUILD YOUR CHANNEL PLAN

BEFORE YOU TOUCH CHIRP

Most people program their radios like they pack for a vacation: they throw everything they might possibly need into the suitcase until the zipper bursts.

They go to a database website, download a CSV file of "All Repeaters in the State," upload 127 random frequencies into the radio, and call it good. They feel prepared because their memory banks are full.

This is a tactical disaster.

Data hoarding is not preparedness. When you are stressed, cold, or driving in the rain with a power outage looming, you cannot scroll through 80 channels labeled K5PLA, W3ABC, and N5XYZ trying to remember which one covers your neighborhood and which one is three counties away.

You will scroll past the one you need, or you will be scanning so much junk that you miss the urgent call from your wife. You will be stuck listening to a repeater 40 miles away while your local team is trying to reach you on Simplex.

Before we open the software or connect a cable, we are going to grab a piece of paper and design a **System**. We are going to build a mental map that works even when you are tired and confused.

THINK LIKE A SYSTEM DESIGNER

The Baofeng UV-5R is dumb hardware. It does not have "Memory Banks," "Groups," or "Zones" like a $5,000 Motorola public safety radio. It just has a big, disorganized bucket of 128 channels (000–127).

If you mix Simplex channels, Repeaters, Marine bands, and NOAA weather stations randomly in that bucket, you have created a mess. We need to impose order on chaos. We need to turn that bucket into a filing cabinet.

1. The "Virtual Zone" Strategy

Professional radios have a physical knob to switch between "Zones" (e.g., Police Zone, Fire Zone, Tactical Zone). Since the Baofeng doesn't have that knob, we have to create **Virtual Zones** using channel number ranges.

By reserving specific blocks of numbers for specific tasks, you reduce the mental load required to find a frequency.

- **Zone 1 (Channels 000–019): The "Hot" Zone.**
  - These are the channels you use 90% of the time. This includes your Simplex frequencies, family channels, and your one primary repeater.

    - *The Tactical Advantage:* You can access these by typing 0-0-1 or 0-1-0 instantly. Muscle memory works here. If you are in the dark, you know the low numbers are the safe numbers.

- **Zone 2 (Channels 020–059): The Repeater Bank.**
    - All the secondary repeaters in your county and the surrounding counties.
    - Organize these geographically (e.g., North repeaters first, then South) or by priority. These are the channels you switch to when the "Hot Zone" repeater is down or out of range.
- **Zone 3 (Channels 060–089): Interoperability (Interop).**
    - MURS, GMRS, FRS (if you are scanning them). The channels "other people" use.
    - If you hear a neighbor calling for help on a Walmart blister-pack radio, they will likely be here. You keep these separate so they don't clutter your main operation.
- **Zone 4 (Channels 090–127): The Boneyard.**
    - Weather stations, Police/Fire listen-only frequencies, and travel channels.
    - These are "Read Only" channels. You put them at the end so you don't accidentally scroll past them when looking for your family channel.

Why? Because if you need to talk to your family, you know instinctively to type 0-0-1. You don't have to hunt. You don't have to look at the screen. You just know.

2. Naming Conventions Under Stress

The UV-5R display gives you **7 characters**. That is it. If you name a channel NorthMountain, the radio will likely truncate it to NorthMo or just show the frequency numbers.

We need a naming language that tells you **Who** and **What** at a glance. Under stress, your IQ drops. You won't remember that 146.880 is the "Club Repeater." You need the screen to tell you.

The "Bad" Way:

- K5ABC (This is a callsign. Who is it? Where are they? Is this a repeater or a person?)
- RPTR 1 (Which repeater? The one near your house or the one near work?)
- 146.520 (Just the raw frequency. Unless you are a math savant, this means nothing in an emergency.)

The "Tactical Dad" Way:

- HOME-S (Home Simplex. It tells you exactly who it is for and that it is Simplex.)
- CONVOY (Car-to-Car. Universal channel for travel.)
- GMRS-15 (Standard GMRS Ch 15. Standardized naming helps when talking to others.)
- CITY-R (City Repeater. Tells you the location and the type.)
- SHRF-L (Sheriff Listen-Only. Reminds you *not* to transmit.)

The Spouse Test: If you hand the radio to your spouse or teenager and say, "Go to the Home channel," can they figure it out? If the screen says 146.520, probably not. If it says HOME-S, yes. The radio should speak human, not robot.

THE 3-LAYER PLAN

We are going to fill your "Virtual Zones" using a three-layer approach. This ensures you have layers of redundancy. If Layer 2 fails, you fall back to Layer 1. If Layer 1 is out of range, you try Layer 3.

Layer 1: Simplex Local (The Lifeboat)

Reliability: High | Range: Low (1–3 miles)

This is direct radio-to-radio communication. It relies on zero infrastructure. It works when repeaters are down. It works when power is out. It works when you are hiking in a dead zone. It is the only layer you can 100% control.

- **Channel 001:** CALL-S (National Calling Freq: 146.520 Ham or 462.5625 GMRS). This is the "Town Square."
- **Channel 002:** HOME-S (Your chosen family frequency). Pick a quiet frequency away from the National Calling channel. **Pro Tip:** This is the ONE place where adding a Tone (CTCSS) is smart. If you set both radios to use Tone 100.0Hz, you won't hear the random trucker driving by on the same frequency. It creates a "semi-private" line for your house.
- **Channel 003:** TAC-1 (A secondary frequency).
    - *Scenario:* You are coordinating a convoy, but HOME-S is crowded with chatter from another group. You simply say, "Switch to Tac 1," and everyone clicks one click up.

Variant Callout: *If you are using GMRS, you are legally restricted to specific channels. Use Channel 15-22 for this layer. If you are Ham, look up the "Band Plan" for your area to find a dedicated simplex frequency (e.g., 146.580).*

Layer 2: Primary Repeaters (The Grid)

Reliability: Medium | Range: High (10–50 miles)

These are the big towers on mountains. You use them to talk across the county. However, repeaters are owned by people. They can lose power. They can break. They can be shut down. Never rely on them as your *only* option.

- **Selection Criteria:** Do not program every repeater in the state. Pick the **Top 3**:
    a. The one that covers your **House**.
    b. The one that covers your **Work**.
    c. The one used by the local **ARES/RACES** (emergency response) group.
- **Naming:** CNTY-R (County Repeater), ARES-P (Emergency Net Primary).

Layer 3: Travel/Backup (The "Just in Case")

Reliability: Low | Range: Variable

This is the stuff you scan just to know what's going on. Information gathering is just as important as transmitting.

- **NOAA Weather:** Program your local NOAA frequency. (You will set this to *Skip Scan* later, but you need it in the memory).

- **Local EMS/Fire:** (If they aren't encrypted/digital). Hearing the fire dispatch can tell you where the storm damage is before the news does. Check RadioReference.com for your county. If the Mode column says "FM" or "NFM", you can listen. If it says "P25", "DMR", or "NXDN", your Baofeng is unfortunately deaf to it.
- **MURS/FRS:** The "General Public" frequencies. Good for listening to see if anyone nearby is calling for help.

SCAN STRATEGY THAT DOESN'T BETRAY YOU

Scanning on a Baofeng is slow. It scans about 3 channels per second. Compare that to a police scanner that does 50+ channels per second.

If you scan all 128 channels, it takes **40+ seconds** to loop around once. A typical radio transmission ("This is Base, checking in.") takes 3 seconds. You will mathematically miss 90% of the conversation.

The "Exclude" Tactic

In Chapter 4, we will use CHIRP to set "Skip" flags. But for your plan right now, mark which channels you actually want to scan.

The Golden Rule of Scanning: Only scan channels you are willing to answer.

- **Do Scan:** Your Home channel, the main Repeater, the National Calling frequency. Keep this list tight, under 15 channels if possible.
- **Do Not Scan:** The 50 other repeaters you rarely use.
- **The "NOAA Trap":** Never, ever add the NOAA weather channel to your scan list. It transmits a continuous signal 24/7. Your scanner will hit it, stop, and stay there forever. You will never hear anything else.

Priority Scan vs. Dual Watch (TDR)

You have two ways to "watch" your home channel while listening to others.

1. **Dual Watch (TDR):** The radio listens to Channel A (Top Line) and Channel B (Bottom Line) at the same time, switching back and forth rapidly.
    - *Pro:* Great if you only care about two channels (e.g., Home + Repeater). It is faster than scanning.
    - *Con:* It can be confusing to know which one you are transmitting on if you aren't paying attention to the little arrow on the screen.
2. **Scanning:** The radio cycles through a list of enabled channels.
    - *Pro:* Covers more ground (Home, Repeater, Interop, and Fire Dispatch).
    - *Con:* You might miss the start of a transmission because the radio was busy checking a quiet channel.

Recommendation: For daily carry, use **Dual Watch**. Set Top Line (A) to your primary Repeater (for situational awareness). Set Bottom Line (B) to your Home Simplex (for family). This guarantees you never miss a call from home.

WORKBOOK TOOLS

Action Step: Fill Out Your Paper Plan

Do not skip this. Write it down before you touch the computer. Having this on paper is your backup when the computer crashes.

The "Tactical Dad" Channel Worksheet

Ch #	Name (7 Char Max)	Freq	Tone?	Purpose	Scan?
000	CALL-S	146.520	None	National Call (Ham)	Yes
001	HOME-S	______	______	Family Primary	Yes
002	TAC-1	______	______	Backup / Hiking	No
003	MURS-3	151.940	None	Public Interop	Yes
010	CNTY-R	______	103.5	Main County Rptr	Yes
011	SKYWRN	______	______	Weather Spotters	Yes
020	NOAA	162.550	None	Weather Audio	NO
127	POLICE	155.xxx	None	Listen Only	No

Your Assignment:

1. Identify your **1 Simplex Frequency** for the family.
2. Identify your **2 Primary Repeaters** (Check RepeaterBook.com).
3. Write them in the chart.
4. Create 7-character names for them.

Once this is written down, programming in Chapter 4 will take 5 minutes. If you try to decide while programming, it will take 5 hours.

Chapter 3

REPEATER MASTERY

PROGRAM IT. VERIFY IT. TRUST IT.

You've stood in your driveway, keyed the mic on 146.520, and heard nothing but static. You walked to the end of the block, tried again, and got the same result. You realized the hard way that 5 watts of power doesn't go very far through suburban sprawl, hills, and pine trees.

Enter the **Repeater**.

A repeater is simply a massive "force multiplier" for your radio. It is a high-power radio installed on top of a mountain, a skyscraper, or a water tower. It listens for your weak, scratchy signal on one frequency and simultaneously screams it out on another frequency to the entire county.

It turns your 2-mile "line of sight" range into a 50-mile "county-wide" footprint. It is the difference between talking to your neighbor and coordinating with a family member three towns over.

But repeaters are picky. They are not open channels like FRS walkie-talkies. They are like a secure facility: if you don't knock on the door correctly (Tone) and enter through the right gate (Offset), they will ignore you completely. You can scream at them all day, and they won't even wake up.

This chapter turns "I think I hit the repeater" into "I have solid copy." We are going to move from guessing to knowing.

FIND REPEATERS THE RIGHT WAY

Don't guess. Don't scan randomly hoping to find one; you'll mostly hear digital noise or interference. You need actionable intelligence.

The Source: Download the **RepeaterBook** app (free) or go to RepeaterBook.com. It is the gold standard database for amateur radio repeaters.

The Data You Need (Write this down): To successfully hit a repeater, you need four specific pieces of intel. If you are missing one, it won't work. It's like a combination lock, miss one number, and the shackle stays closed.

1. **Output Frequency (The "Listen" Freq):** This is the big number listed in the app (e.g., 146.880). This is what you listen to. It's the voice of the repeater.
2. **Offset (The Math):** The repeater cannot listen and talk on the exact same frequency at the exact same time (it would deafen itself). It listens on one frequency and talks on another. The "Offset" is the distance between them.
    - 2 Meter Ham is usually 0.600 MHz.
    - 70cm Ham is usually 5.000 MHz.

  - GMRS is always 5.000 MHz.

3. **Direction (The Shift):** Do you add (+) or subtract (-) the offset to find the input frequency? The app will tell you.
   - *Example:* If Output is 146.880 and Offset is -0.600, your radio will transmit on 146.280.
4. **Tone (The Key):** Look for "Tone," "PL," or "CTCSS." If it says 103.5, that is your electronic key card. You must program your radio to send this sub-audible tone. If you don't send it, the repeater ignores you. If it says "Open" or "CSQ" (Carrier Squelch), there is no lock, and you don't need a tone.

PROGRAM REPEATERS TWO WAYS

We will start with the "Easy Way" (Software) to get your bulk list done, but you must master the "Rescue Method" (Keypad) for the field. If you are in a blackout, CHIRP won't save you.

Method 1: The CHIRP Method (Desk Duty)

We cover the full CHIRP workflow in Chapter 4, but here are the columns that specifically matter for repeaters. Getting these wrong is the source of 90% of forum posts asking for help.

1. **Frequency:** Enter the **Output** frequency (e.g., 146.880). Do not enter the input frequency here.
2. **Duplex:** Set to + or - (matches the Direction). If you leave this as "(None)", you are programming a Simplex channel and the repeater won't hear you.
3. **Offset:** Enter the amount (e.g., 0.600). Double-check your decimal points. 6.000 will put you totally out of band.
4. **Tone Mode:** This is the tricky one.
   - If the repeater needs a tone: Set to Tone. This turns on the **Encode** (Transmit) tone.
   - If no tone: Set to (None).
   - *Warning:* Do NOT select ToneSql unless you are 100% sure you need it. ToneSql sets both Transmit AND Receive tones. If the repeater doesn't transmit a tone back (many don't), your radio will stay silent, and you won't hear the reply.
5. **Tone:** Enter the value (e.g., 103.5).
   - *Note:* The "ToneSql" column is ignored by the radio if "Tone Mode" is set to just Tone.

Method 2: The Manual "Rescue Method" (Field Duty)

It's 2 AM, the power is out, trees are down, and you need to hit a repeater you didn't plan for. You don't have your laptop. You have your thumbs and a dim backlight.

The "Tactical Dad" Keypad Sequence (UV-5R): *Follow this exactly. If you pause for 5 seconds, the menu times out and drops you back to the main screen. Speed counts.*

1. **Go to VFO Mode:** Press VFO/MR until the radio speaks "Frequency Mode" and you see frequency numbers (no channel numbers on the right).
2. **Select A-Band:** Press EXIT/AB so the little black arrow is on the top line. Always program on the top line to avoid confusion.

3. **Enter Frequency:** Type the Output Freq directly (e.g., 1-4-6-8-8-0).

4. **Set the Tone (The Key):**

    - Press MENU, type 13 (T-CTCS), press MENU.
    - Use the Up/Down arrows to find your tone (e.g., 103.5).
    - Press MENU to save (Verify the arrow drops to the next line). Press EXIT.
    - *Why T-CTCS?* Because we only want to *transmit* the tone. We want to listen to everything.

5. **Set the Offset Amount (The Distance):**

    - Press MENU, type 26 (OFFSET), press MENU.
    - Type 00600 (for 2m) or 05000 (for 70cm/GMRS).
    - Press MENU to save. Press EXIT.

6. **Set the Offset Direction (The Shift):**

    - Press MENU, type 25 (SFT-D), press MENU.
    - Select + or - using the arrows.
    - Press MENU to save. Press EXIT.

7. **Save the Channel:**

    - Press MENU, type 27 (MEM-CH), press MENU.
    - Use arrows to find an empty channel (e.g., 020). Note: If the channel number has CH- in front of it (like CH-020), it is full. **The radio will not overwrite it.** You must first go to **Menu 28 (DEL-CH)** to delete it, then come back here to save.
    - **Crucial Step:** Press MENU to confirm. The voice says *"Receiving Memory."*
    - Press EXIT.

8. **Switch to Channel Mode:** Press VFO/MR. Use arrows to go to Channel 20.

The "Quick Check": While on the channel, tap the SCAN (*) button quickly (don't hold it). You should see an R appear on the screen, or the frequency number should shift briefly (e.g., from 146.880 to 146.280). If the numbers shift, your radio is doing the math correctly.

VERIFICATION PROTOCOL

Just because it's programmed doesn't mean it works. I have seen men carry "programmed" radios for years that were incapable of transmitting because of a simple typo in the offset.

1. Listen First (The 15-Minute Rule)

Before you talk, listen. Amateur radio isn't a telephone; it's a party line.

- Is there a "Net" (scheduled meeting) happening?

- Are two people deep in a conversation about their hip surgery?
- Don't barge in. Wait for a break. If you interrupt a directed emergency net, you become the problem.

2. The "Kerchunk" Test

This is the non-verbal handshake.

- **Action:** Tap your PTT button for **one second** and release.
- **Listen:** Do you hear a burst of static, a "click," or a "beep-boop" courtesy tone return immediately after you let go?
- **Result:** That return sound is the "Squlech Tail." It means the repeater heard your signal, verified your tone, opened its receiver, and stayed on for a second before shutting down.
    - *If you hear the tail:* You are making it into the machine. Your Tone and Offset are correct.
    - *If silence:* You did not hit it. Check your settings or move to higher ground.
- **Etiquette:** Don't do this constantly. Better yet, key up and say **'Testing'** instead of just clicking. It keeps the cranky old-timers happy and is technically required by the rules.

3. The Signal Report (Radio Check)

Once you verify the repeater wakes up, ask for a human check. A repeater might hear your carrier wave (the kerchunk) but your audio could be too quiet or full of static.

The Script: "This is 'Your Callsign/Name', testing a new radio setup. Can anyone give me a signal report?"

The Response (Decoding what they say):

- **"Full Quieting":** This is the gold standard. It means your signal is so strong it is fully "quieting" the background static. You sound like you are standing next to them.
- **"Scratchy but readable":** You are understandable, but there is background hiss (bacon frying sound). You are likely on the edge of the repeater's range. Don't move while talking.
- **"You're into the machine, but no audio":** You hit the repeater (the light went green), but your mic gain is too low, you are whispering, or your headset is broken. Speak closer to the mic.
- **"Picket Fencing":** You are moving (driving or walking), and your signal is chopping in and out rapidly. Stop moving.

COMMON FAILURE PATTERNS

1. The "Silent Treatment" (Wrong Tone Mode)

- *Symptom:* You transmit, the repeater wakes up (you hear the tail), but nobody answers you even though you hear them talking.
- *Cause:* You likely set **R-CTCS** (Receive Tone) on your radio. This puts a lock on *your* speaker. If the repeater output doesn't have a matching tone (and many don't), your radio keeps the speaker muted. You are deaf to the world.
- *Fix:* Turn **R-CTCS** (Menu 11) to OFF. Only use **T-CTCS** (Menu 13).

2. The "Wrong Door" (Wrong Offset Direction)

- *Symptom:* You transmit, and absolutely nothing happens. No "kerchunk," no tail, no angry feedback.
- *Cause:* You set + instead of -. You are transmitting on 147.480 instead of 146.280. You are knocking on the back wall of the building instead of the front door.
- *Fix:* Check RepeaterBook. Band plans usually dictate this: If the frequency is below 147.000, it's usually -. If above, it's +. (But always verify!).

3. The "Alligator" (All Mouth, No Ears)

- *Symptom:* You can hear the repeater perfectly, it sounds strong. But when you talk, they say you are "in the mud" or unreadable.
- *Cause:* The repeater is a 100-watt monster on a 200-foot tower. It can "shout" to you easily. You are a 5-watt handheld in a valley. You can hear the shout, but you can't shout back loud enough to be heard.
- *Fix:* This is a physics problem, not a settings problem. Get a better antenna (Chapter 5), get outside of your car, or get to higher ground.

ACTION STEPS

Repeater Bring-Up Checklist

- **Data:** Verified Output Freq, Offset, Direction, and Tone on RepeaterBook (Checked "Last Update" date to ensure it's not abandoned).
- **Program:** Entered into CHIRP or Keypad correctly.
- **Reverse Check:** Tapped * (Scan) button on the radio face to watch the frequency shift numbers.
- **Kerchunk:** Tapped PTT for 1 second. Heard the "tail" (static/beep) return.
- **Voice:** Made a radio check. Received a readable report from a human.

"First Successful Repeater Contact" Script

Use this today. Don't wait for an emergency to feel awkward on the mic. Get a win.

You: "Monitoring the 'Repeater Name' machine. This is "Your Callsign" *(Wait 10 seconds. This is polite. It lets people know you are there without demanding a response).*

You: "This is 'Your Callsign', looking for a quick signal report. I'm testing a UV-5R from "Your Location."

Them: "Your Callsign," this is "Their Callsign". You're loud and clear into the repeater."

You: "Thanks for the report, 'Their Callsign.' Just getting my settings dialed in. 'Your Callsign' clear."

Chapter 4

CHIRP 2.0 - THE BULLETPROOF PROGRAMMING WORKFLOW

If programming a Baofeng via the keypad is like writing a novel via text message on a flip phone, using CHIRP is like sitting down at a desktop workstation with a full keyboard and a mouse.

Keypad programming is a survival skill; CHIRP programming is a lifestyle skill. You need the survival skill for when the world falls apart, but you need the lifestyle skill to prepare for it.

CHIRP is free, open-source software that allows you to interface with your radio, manage hundreds of channels in a spreadsheet view, and unlock settings that are physically impossible to change from the radio's faceplate. It turns a tedious three-hour chore into a five-minute drag-and-drop operation.

However, CHIRP is also the place where most "Tactical Dads" quit. They buy the $20 cable, plug it in, get a Windows error message, fight with it for three hours, and then throw the cable in a drawer, never to be seen again. They assume they aren't "tech-savvy" enough.

This chapter is your technical support hotline. We are going to navigate the "Driver Hell" (and explain why it happens so you don't panic), fix the squelch settings that the factory broke, and build a channel management system that can adapt to hurricanes, road trips, and daily life without breaking a sweat.

SECTION 1: THE GAUNTLET (CABLE & DRIVER SANITY CHECKS)

Let's address the elephant in the room immediately. You plugged the cable in, and it didn't work. You aren't crazy, and your computer isn't broken. You are a casualty of a corporate war.

The "Counterfeit Chip" Conspiracy

Here is the reality: 99% of the cheap programming cables sold on Amazon (the $10 specials) use a cloned (counterfeit) "Prolific" chip. These are unauthorized copies of a chip designed by a company called Prolific.

Prolific got tired of losing money to cloners, so they did something aggressive: they updated their official Windows drivers to intentionally sabotage counterfeit chips. When Windows automatically updates your drivers (which it does constantly in the background), it grabs the latest official driver. That driver asks the chip, "Are you real?" The chip fails the test, and the driver shuts it down.

The result? Your computer sees the cable, but it displays a yellow warning triangle in Device Manager and refuses to talk to it. It might say "Code 10: Device cannot start."

The Fix is NOT buying a new cable. If you buy another $10 cable, you will have the same problem. The fix is telling

Windows to ignore the new, aggressive driver and use the older, dumber driver (from 2008) that doesn't care about counterfeit chips.

The "5-Minute Fix" (Step-by-Step for Windows)

Do not skip a step. Do exactly this. If you are on a Mac, you likely won't have this specific issue, but Windows users suffer this 90% of the time.

1. **Plug it in:** Connect your programming cable to the USB port. Do NOT connect the radio yet.
2. **Open Device Manager:** Right-click the Windows Start button and select **Device Manager**.
3. **Find the Error:** Look for the section labeled "Ports (COM & LPT)". Click the arrow to expand it. You will likely see "Prolific USB-to-Serial Comm Port" with a yellow warning triangle icon next to it. That triangle is the sabotage in action.
4. **The Rollback:**
   - Right-click the device. Select **Update driver**.
   - Select **Browse my computer for drivers** (Do not search automatically).
   - Select **Let me pick from a list of available drivers on my computer**.
5. **The Magic Step:** You will see a list of driver versions.
   - The "Bad" Driver: Usually version 3.8.x or newer (dated 2015, 2019, etc.). This is the one stopping you.
   - The "Good" Driver: Look for version **3.2.0.0** (dated **2008** or **2007**).
   - *Troubleshooting:* If you do *not* see the 2008 driver in this list, Windows has deleted it. You need to go to a reputable site like Miklor.com, download the "Prolific 3.2.0.0 Installer," run it, and then restart this step.
6. **Select & Install:** Click the 3.2.0.0 version and click Next. Windows will install it instantly.
7. **Verification:** The yellow triangle should disappear immediately. Note the **COM Port Number** (e.g., COM3, COM4). You will need this number to tell CHIRP where to look.

The Connection Ritual (Avoiding the "Radio Not Found" Error)

Once the driver is fixed, you must follow the ritual physically. The Baofeng connector (the two-prong K-plug) is notoriously tight on new radios. The plastic casing often prevents the pins from seating fully.

1. **Cable First:** Plug the USB end into the computer.
2. **Radio OFF:** Turn the radio volume knob OFF. (Connecting a live radio can sometimes trigger the Push-to-Talk).
3. **Plug In:** Insert the two-prong connector into the radio. **PUSH HARD.**
   - *The "Zero Gap" Rule:* You might feel one click. That is not enough. Push harder until there is **zero gap** between the cable head and the radio body. If you see metal pins, it's not in.
   - *Diagnosis:* If CHIRP says "Radio Not Found" or "Failed to communicate," 9 times out of 10, you simply need to push the plug in harder.
4. **Radio ON:** Turn the volume knob to 100%. A fully powered radio ensures the data signal is strong.

5. **Open CHIRP:** Go to Radio -> Download from Radio.

SECTION 2: THE GOLDEN RULE OF PROGRAMMING

You are now connected. The green light is flashing on your radio. Before you type a single frequency, you must memorize the Golden Rule.

READ > SAVE > EDIT > WRITE

1. READ (Download from Radio)

Always start by pulling the current data *from* the radio. Even if the radio is brand new and empty.

- *Why?* This "handshake" tells CHIRP exactly what firmware version your radio is running (e.g., N5R-20, BFB-297). CHIRP needs to know the map of the radio's brain before it can write to it.

2. SAVE (The "Factory Image")

The moment the download finishes, a tab will open showing your channels. **STOP.** Go to File -> Save As. Name it Baofeng_Factory_Backup_DATE.img.

- **Do not touch this file again.** Save it to your Dropbox, Google Drive, or a dedicated folder.
- *The "Soft Brick" Savior:* If you ever mess up your settings so badly the radio stops working, or you mess up the squelch settings and can't hear anything, this file is your time machine. Re-uploading this file returns the radio to the exact state it was in when you bought it.

3. EDIT

Now, make your changes. Delete the factory junk channels. Add your local repeaters. Change your settings. This is where the work happens.

4. WRITE (Upload to Radio)

Send the data back to the device.

The "Firmware Rot" Warning

Do not take an image file you saved for a radio you bought in 2020 and try to force-upload it to a radio you bought in 2025. Even if they look identical and are both "UV-5Rs," Baofeng changes the internal memory locations constantly. If you force an old memory map onto a new chip, you can "soft brick" the radio (it lights up but screen is blank, or keys don't work).

- *The Safe Workflow:* Always download from the *new* radio to open a fresh tab. Then, open your *old* file, copy the channels (Ctrl+C), and paste them (Ctrl+V) into the new radio's tab. This moves the *data* without overwriting the *firmware map.*

SECTION 3: TEMPLATES & MASS EDITING (YOUR LIFE RAFT)

Looking at a blank CHIRP spreadsheet is intimidating. Let's fill it using the **Channel Plan** we designed in Chapter 2.

The Columns That Matter

Ignore the columns that don't matter. CHIRP exposes everything, but you only need these:

- **Loc:** The channel number (0-127). This matches your "Virtual Zones."

- **Frequency:** The "Listen" frequency (Output).
- **Name:** 7 Characters max. Use ALL CAPS. (e.g., HOME-S, FIRE-L).
- **Tone Mode:** (See Chapter 3). Use Tone for Repeaters (Encode only). Use (None) for Simplex.
- **Tone:** The specific Hz value (e.g., 103.5).
- **Duplex:** + (Plus), - (Minus), or (None) for Simplex.
- **Offset:** The math (0.600, 5.000).
- **Mode:** Always NFM (Narrow FM) for Ham/GMRS/FRS. Using "FM" (Wide) is technically illegal on these frequencies now and will make you sound quiet to others.
- **Power:** High or Low.
- **Skip: Crucial.** If you check this box ("S"), the Scan function will skip over this channel.

Importing External Data (The Fast Way)

Don't type repeaters manually. You will make a typo. CHIRP links directly to databases.

1. **Weather:** Go to Radio -> Import from Stock Config -> US NOAA Weather Alerts. This loads all 7 NOAA channels.
2. **Repeaters:** Go to Radio -> Query Source -> RepeaterBook.
3. **Filter:** Select your State and County. Select "2 Meters" or "70 Centimeters".
4. **Execute:** Click OK.

The "Import" Window Strategy: CHIRP will show you a list of 50 found repeaters. Do **not** just click OK. That clutters your radio with junk.

- **Uncheck All.** Start clean.
- Manually check the 3-5 repeaters you actually identified in your Chapter 2 plan.
- **Adjust the "To" Column:** This tells CHIRP where to put them. If you want them in your "Repeater Zone" (starting at Channel 20), change the first channel's "To" value to 20. The rest will automatically renumber (21, 22, 23). This keeps your memory map clean.

Mass Editing (The Power Move)

This is the single best feature of CHIRP. Let's say you imported 20 channels, but they are all set to "Wide" FM and "Low" power, and you want them all "Narrow" and "High".

1. Click the first channel number row (e.g., Row 20).
2. Hold Shift and click the last channel number row (e.g., Row 40). The whole block is highlighted.
3. **Right-Click** anywhere in the highlighted block -> **Properties**.
4. A "Bulk Edit" window pops up.
5. Check the box for **Power** and set it to High.

6. Check the box for **Mode** and set it to NFM.

7. Click **OK**. *Boom.* You just updated 20 channels in 2 seconds. No more menu diving.

SECTION 4: POWER USER TUNING (FIXING THE RECEIVER)

This is the secret sauce. This is the difference between a $25 toy and a usable tool.

The Baofeng's "Squelch" menu (SQL 0-9) is logically broken by default. Squelch is supposed to be a gate: Level 1 should be a flimsy gate (lets weak signals in), and Level 9 should be a bank vault (only strong signals get in). On factory Baofengs, Level 1 and Level 9 are often mathematically almost identical. This means you can't filter out interference without going completely deaf.

We can fix this in the **Service Settings**.

The Squelch Threshold Fix

1. In CHIRP, on the left sidebar, click **Settings**.

2. Select **Service Settings**. (If you don't see this, ensure you have "Developer Functions" enabled in View, though most modern builds show it by default).

3. You will see values for VHF Squelch and UHF Squelch. They look like weird numbers (e.g., 12, 15, 20...).

The Logic: These numbers represent the raw signal voltage required to open the squelch gate. We need to spread them out so 1 is sensitive and 9 is hard.

Recommended "Tactical Dad" Values (Enter these):

SQL Level	VHF Value	UHF Value	What it feels like
0	0	0	Open (Constant Static)
1	**20**	**20**	Super Sensitive (Hear grass growing)
2	**25**	**25**	Sensitive (Good for rural areas)
3	**30**	**30**	**Standard Daily Use**
4	**40**	**40**	City Driving (Filters light noise)
5	**50**	**50**	Noisy Environment
6	**70**	**70**	High Interference (Near PC/Server)
7	**90**	**90**	Strong Signals Only
8	**110**	**110**	Very Strong
9	**125**	**125**	Next-to-Tower Only

Note: These are safe baselines. If you find Level 3 is still letting in too much static, bump the numbers up by 5.

SECTION 5: VERSIONING & SCENARIOS (THE NARRATIVES)

You don't wear a tuxedo to the gym. Why would you carry a "Disney World" channel plan during a "Hurricane"? Programming is not a "do it once" event. You should save different **Images** (.img files) for different missions.

Narrative 1: The "Hurricane Prep" Image (Static Defense)

The Context: A Category 3 storm is 48 hours out. You are staying put. Power will likely fail. Cell towers will likely jam or lose battery backups after 24 hours. **The Goal:** Conservation and Local Intel.

The Setup Strategy:

- **Channel 0 (Call):** 146.520 (National Simplex).
- **Channel 1 (Home):** Your family simplex channel. Crucial tweak: Set Power to **LOW** (1 Watt).
    - *Why?* You are all in the same house or neighborhood. You don't need 5 Watts to talk to the kitchen. Low power doubles your battery life.
- **Scan List:** VERY short. Home, ARES (Emergency) Repeater, and maybe the Fire Department dispatch. Do not scan 50 channels; you want to save battery cycles.
- **Weather:** NOAA channel programmed in memory but **removed** from scan. You don't want the radio stopping on weather constantly. You will manually switch to it at the top of the hour to check the track, then switch back.
- **Squelch:** Set to 1 or 2. You want to hear weak signals from neighbors asking for help, even if it means tolerating some static.

Narrative 2: The "Disney Road Trip" Image (Dynamic Convoy)

The Context: You are driving two cars from Ohio to Florida. You have handhelds in both cars. **The Goal:** Convoy coordination and situational awareness on the highway.

The Setup Strategy:

- **Channel 1 (Convoy):** A GMRS or Ham simplex channel. Power set to **HIGH**.
    - *Why?* You are dealing with 70mph separation distances and metal car bodies blocking signals. You need the punch.
- **Tone Protection:** Use a DCS/CTCSS tone on your Convoy channel! The highway is RF soup. You don't want your wife to hear every random burst of static as you pass a Walmart or a truck stop. Lock the door with a tone.
- **Route Repeaters:** Use RepeaterBook's "Travel Search" feature. Import major repeaters for cities along I-75 (Cincinnati, Knoxville, Atlanta).
    - *Why?* You aren't chatting; you are listening for traffic accidents or road blockages reported by locals *before* you get stuck in the jam.
- **Scan List:** Convoy Channel (Priority) + Highway Patrol (if analog/legal) + MURS 3 (Truckers often use 151.xxx or CB, but MURS is common for road crews).

SECTION 6: WORKBOOK TOOLS

The CHIRP Workflow Card

(Print this out, laminate it with packing tape, and tape it to your laptop lid or radio case)

1. **Driver Check:** Device Manager -> Ports -> No Yellow Triangle? -> Good.
2. **Physical Connection:** Plug USB -> Radio OFF -> Plug K1 (HARD SNAP) -> Radio ON (100% Vol).
3. **Download:** Radio -> Download -> Select Port/Model -> OK.

4. **The Safety Save:** File -> Save As -> Backup_DATE.img.

5. **Edit/Import:**

    - Import from RepeaterBook (Uncheck all -> Select Top 3).
    - Mass Edit (Right-Click -> Properties) for NFM/High Power.
    - Organize Zones (0-19 Local, 20-59 Repeaters).

6. **Squelch Check:** Settings -> Service Settings -> Verify 20-125 spread.

7. **Upload:** Radio -> Upload to Radio.

8. **The Field Test:** Unplug. Turn off. Turn on. Press VFO/MR to go to Channel Mode. Check Channel 1.

The "Programming Backup Plan" Checklist

When the laptop dies, do you have a path?

- **USB Drive in the Go-Bag:** Contains the "Prolific 3.2.0.0" installer .exe and the CHIRP installer .exe. (You can't download these if the internet is down).
- **Paper Repeater Map:** A printed list of the frequencies in your "Travel Image." If the radio resets, you need to type them in via keypad.
- **The "Rescue Cable":** A spare programming cable. They are flimsy. They break. Two is one, one is none.

Chapter 5

RANGE REALITY LAB

ANTENNAS, PLACEMENT, AND PREDICTABLE PERFORMANCE

You bought the radio. You looked at the box. It said, in bold, aggressive font: **"UP TO 36 MILE RANGE."** You stood in your driveway. Your buddy stood at his house three miles away. You pressed the button... Nothing.

You drove closer. Two miles. Nothing. You drove to the edge of his subdivision. One mile. Finally, you heard a scratchy, static-filled voice that sounded like a robot drowning in a blender. You felt cheated. You assumed the radio was broken, or maybe you just bought a cheap piece of junk. You probably went online and saw people arguing about "fake chips" or "bad batches."

Here is the truth: The radio is fine. The box is a lie. And physics is a harsh, unforgiving mistress. This chapter is the "Red Pill" of radio. We are going to strip away the marketing hype and look at the cold, hard math of Radio Frequency (RF). By the end of this chapter, you won't just know *why* you couldn't talk three miles; you will know exactly how to fix it so you can. We are going to turn "magic" into "mechanics."

SECTION 1: WHY RANGE CLAIMS LIE (THE PHYSICS OF LINE OF SIGHT)

How can they legally print "36 Miles" on the box?

The manufacturer's claim assumes a specific, impossible scenario: You are standing on the peak of a mountain. Your friend is standing on the peak of another mountain 36 miles away. There is a perfect vacuum between you. There are no trees, no buildings, no power lines, and no atmosphere to scatter the signal. In that specific, theoretical universe, yes, 5 watts will travel 36 miles.

In the real world, you are standing in a valley (your driveway), surrounded by RF-absorbing sponges (pine trees), inside a Faraday cage (your car), trying to push a signal through six blocks of stucco, copper wiring, and rebar.

The "Line of Sight" Rule

VHF (Very High Frequency) and UHF (Ultra High Frequency), the bands your Baofeng uses, are **Line of Sight** waves. They behave almost exactly like light from a flashlight. If you shine a flashlight at your neighbor's house, but there is a hill in the way, the light hits the dirt. It does not curve over the hill. It does not go through the hill. It stops.

Radio waves are the same. If there is dirt, rock, or a dense cluster of buildings between your antenna and his antenna, the signal dies. It doesn't matter if you have 5 watts or 50 watts; you cannot power your way through a mountain.

The Equation of "Height is Might" This is the only rule that matters. Power (Watts) is overrated. Height is everything.

- **5 Watts at 5 feet off the ground:** Range ~1-2 miles in suburbs. You are fighting bushes, cars, and fences.

- **5 Watts at 50 feet off the ground:** Range ~10-15 miles. You have cleared the "clutter" of the neighborhood.
- **5 Watts on a mountain top:** Range ~50+ miles. You have a direct line of sight to the horizon.

Tactical Dad Lesson: If you can't reach someone, don't shout louder (add power). Climb higher (add height). Standing on the bed of your truck, climbing a deer stand, or walking to the second floor of your house does more for your signal than buying a $200 amplifier.

The "Sponge" Factor (Attenuation)

Radio waves pass through some things and get absorbed by others. This is called **Attenuation**. Think of it like running through water.

- **Glass/Drywall:** Easy. Minimal loss. You can talk through a wood-frame house easily.
- **Leaves/Trees:** Medium to Heavy loss. Pine needles are specifically tuned by nature to be approximately the length of 70cm waves. A dense forest is basically a "radio black hole" for UHF signals. Wet leaves after a rainstorm are even worse.
- **Concrete/Rebar:** Heavy loss. Urban environments are tough because every building is a cage of steel mesh.
- **Earth/Hills:** Total blockage. Game over. There is no "punching through" the earth.

The Urban Canyon Effect (Multipath)

In a city, you might find that you can't talk to someone three blocks away because of buildings, but you *can* talk to someone if you turn 90 degrees and bounce the signal off a glass skyscraper.

This is **Multipath**. Your signal bounces like a billiard ball. Sometimes these bounces help you (reflecting around a corner), and sometimes they hurt you (two signals arriving at different times, canceling each other out). If you are in a city and the signal is garbage, move five feet to the left. You might step out of a "dead spot" and into a reflection sweet spot.

SECTION 2: ANTENNA UPGRADES THAT ACTUALLY HELP

Your radio came with a short, stiff, rubber-coated antenna. We call this the "Rubber Ducky." It is durable, small, and convenient.

It is also an electrical compromise. It is a "dummy load on a stick." To make it 4 inches long, the manufacturer coiled the wire tightly, which creates resistance and heat. A huge chunk of your battery power is wasted just heating up the rubber coating, rather than radiating out as a signal.

Upgrade 1: The "Whip" (Nagoya NA-771 / Signal Stick)

The single best $20 you can spend is replacing the stock antenna with a "tuned" 1/4 wave or 5/8 wave whip. These are usually 15 to 19 inches long.

- **The Difference:** A stock antenna might radiate 30% of your power. A 15-inch whip might radiate 70-80%. That is effectively doubling your output range without draining your battery faster.
- **Recommended Brands:**
  - **Nagoya NA-771:** The classic. It has a rigid base and a whippy top. *Warning:* There are more fake Nagoyas on Amazon than real ones. Buy from a reputable dealer (like BTech).

- **Signal Stuff "Signal Stick":** Made in the USA by ham radio nerds (HamStudy.org). It is made of **Nitinol** (memory alloy). You can tie it in a literal knot, stuff it in your pocket, sit on it, and when you pull it out, it snaps straight instantly. It is lighter and more durable than the Nagoya. **This is the Tactical Dad choice.**

Upgrade 2: The "Tactical Foldable" (The Tape Measure)

You will see antennas that look like folding rulers or flexible steel tape measures (like the Abbree AR-771). They look very "commando."

- **The Truth:** They are electrically decent, but they are physically heavy.
- **The Risk:** The heavy base puts massive leverage stress on the radio's SMA connector. If you drop the radio, or even just wave it around too vigorously, the leverage can snap the connector right off the circuit board.
- **The Verdict:** Use them if you need to fold the radio into a pouch for transport. But when you deploy it, hold the base of the antenna with your thumb to support the weight.

Upgrade 3: The Vehicle Mag Mount (The Game Changer)

This is where the magic happens. If you use a handheld radio inside a car, you are trapped in a metal box (Faraday Cage). The metal body blocks the signal from getting out, and the electronic noise from the engine (alternator whine) interferes with the signal getting in. You will get maybe 0.5 miles of range.

The Fix: A "Mag Mount" (Magnetic Mount) antenna placed on the **center of the roof**.

- **Why it works:** It's not just that the antenna is outside. The steel roof of your car acts as a **Ground Plane**. It becomes the "other half" of the antenna system, reflecting the signal upward and outward like a mirror.
- **The Result:** A $25 mag mount antenna (like a Nagoya UT-72) connected to your $25 Baofeng will outperform a $500 handheld sitting in the cup holder. **Range often jumps from 0.5 miles to 5-10 miles instantly.**

The Secret Weapon: The "Tiger Tail" (Counterpoise)

Every antenna needs two halves: the radiating element (the stick) and the ground (the other side). On a car, the roof is the ground. On a handheld, **YOU** are the ground. The radio uses your hand and arm to complete the circuit. This is inefficient.

- **The Hack:** A "Tiger Tail" or "Rat Tail" is simply a piece of wire crimped to a ring terminal that slips over the antenna connector and hangs down freely.
- **Length Matters:**
    - **Ham Radio (2m):** ~19 inches.
    - **GMRS / 70cm:** ~6.5 inches.
- **The Effect:** It provides a proper electrical ground. It creates a "Dipole" antenna.
- **The Value:** It costs about 50 cents to make and can boost your signal by 30-40% in marginal conditions.

SECTION 3: PLACEMENT RULES (BODY LOSS & GEOMETRY)

Where you hold the radio matters more than the radio itself. You can have the best antenna in the world, but if you bury it in your armpit, it won't work.

The "Meat Sack" Problem (Body Blocking)

You are mostly a bag of salt water. Salt water is excellent at absorbing radio waves.

- **The Scenario:** You clip the radio to your belt. You wear a headset. The antenna is pressed tight against your kidney.
- **The Physics:** Your body acts as a massive shield. You are absorbing 60-80% of the signal radiating to your rear and sides. You are "shadowing" the signal. You can hear people in front of you, but you are deaf to people behind you.
- **The Fix:** Get the antenna away from your body.
    - *Good:* Holding it in your hand, arm extended slightly, radio vertical.
    - *Better:* Radio in a chest pouch or high on a backpack strap where the antenna clears the shoulder.
    - *Best:* External antenna up a tree or on a mast.

Polarization (Vertical vs. Horizontal)

Baofeng antennas are "Vertical." They transmit waves that ripple up and down. They want to stand straight up.

- **The Mistake:** You hold the radio sideways "gangster style" or lay it flat on a picnic table while eating lunch.
- **The Penalty:** "Cross-Polarization Loss." If your antenna is horizontal and the repeater antenna (or your buddy's antenna) is vertical, you lose about **20dB** of signal.
    - *Translation:* You are receiving 1/100th of the signal strength. That is the difference between "Loud and Clear" and "Total Silence."
- **The Rule:** Keep the antenna vertical. Even if you are lying prone, tilt the radio so the antenna points to the sky.

SECTION 4: COAX AND CONNECTORS (AVOIDING THE VAMPIRES)

If you decide to connect your radio to an external antenna (like on your roof or in your attic), you need a cable. This cable is called "Coax" (Coaxial Cable).

The Gender Confusion (SMA-Female vs. Male)

Baofeng defied industry standards just to mess with us.

- **Most Handhelds (Yaesu/Icom):** The radio has a Pin (Male). The antenna has a Hole (Female).
- **Baofeng:** The radio has a **Pin** (SMA-Male). The antenna has a **Hole** (SMA-Female).

The Trap: You buy an antenna meant for a different brand (like Wouxun or Anytone) and try to screw it on. It fits physically, the threads match, but **neither side has a pin.** Or worse, **both sides have pins. The Consequence:** You transmit into an open circuit. The energy reflects back into the radio and fries the final transistor. Your radio is now a brick. **The Fix:** Always verify the center pin. Look before you screw. One side must have a pin; the other must have a hole.

Adapters are Signal Vampires

Every time you convert from SMA to BNC to UHF to N-Type, you lose signal. Each adapter "eats" a fraction of a decibel.

- *Bad:* Baofeng -> SMA-to-BNC adapter -> BNC-to-UHF adapter -> Cable.
- *Good:* Baofeng -> Cable with SMA connector pre-installed.

- *Rule:* Use the fewest adapters possible.

Coax Loss (The Garden Hose Analogy)

Think of Coax like a garden hose. Signal is water pressure. Friction kills pressure.

- **RG-58 (The Thin Stuff):** This is like a drinking straw. It's cheap and flexible.
  - *The Math:* If you run 50 feet of RG-58 to a UHF antenna, you lose half your power just in the cable.
  - *Use Case:* Fine for short mobile runs (10-15 ft) in a car. Terrible for base stations.
- **RG-8x (The Medium Stuff):** A standard garden hose. Good for 50ft runs. A great balance of flexibility and performance.
- **LMR-400 (The Thick Stuff):** A fire hose. It is thick, stiff, and expensive. But it loses almost no signal.
- **Tactical Dad Rule:** For a car mag mount, the thin cable attached to it is fine. Route it through the **top corner** of the door frame where the weather stripping is thickest/softest. Never pinch coax in a car door jamb or window, if you crush the foam inside, the cable is ruined.

SECTION 5: RUN A RANGE TEST LIKE A GROWN-UP

Stop asking "Can you hear me?" and start collecting data. You need to map your "Circle of Influence."

Knowing you have a "2-mile hard deck" is better than *hoping* you have a 10-mile range.

The "Tactical Dad" Range Protocol

You need two people. One at "Base" (Home) and one "Rover" (Mobile).

1. The Setup:

- **Base Station:** Wife/Kid/Buddy at home. Radio set to **High Power**. Antenna vertical (standing up on the counter, or preferably upstairs near a window).
- **Rover:** You in the car (or on a bike). Radio set to **High Power**.

2. The Route: Drive in a cardinal cross (North, South, East, West). Pick landmarks at 0.5 mile, 1 mile, 2 miles, and 5 miles.

3. The Check-In Script: Don't just say "Hello." Use the **RST** format (Readability, Signal).

- **Readability:** 1 (Unreadable) to 5 (Perfect).
- **Signal:** 1 (Static noise) to 9 (Strong/Quiet).
- *Rover:* "Base, this is Rover. Reaching Checkpoint Alpha. 1 Mile North. How copy?"
- *Base:* "Rover, this is Base. I copy you 5 by 9. Loud and clear. How me?"
- *Rover:* "Base, I copy 5 by 9. Proceeding to Bravo."

4. The Failure Point (The "Fade"): Eventually, you will hit a spot where Base says: *"Rover, I copy you... 3 by... static... broken."* **Stop.** Do not give up yet.

- **Audit the Environment:** Are you in a valley? Are there power lines?

- **The Car Pivot:** Turn the car 90 degrees. The metal frame of the car might be blocking the path back home.
- **The Jailbreak:** Get out of the car. Stand up. Hold the radio high. Did the signal improve? (Usually yes, dramatically).

5. Logging the Data: Go home and open Google Maps (or a paper map). Drop pins where you lost signal. Connect the dots. **This is your Real World Range.** It doesn't matter that the box said 36 miles. Your map says 2.4 miles to the North and 0.8 miles to the South because of that hill. *This map is the most valuable survival tool you own. It tells you exactly where you need to drive to send a message.*

ACTION STEPS & WORKBOOK

Range Test Log

Print this and take it in the car.

- **Date:** ____________
- **Weather:** ___________________
  - (Rain kills UHF range! Test in bad weather to know your "worst case.")
- **Base Antenna:** _____________
- **Rover Antenna:** ____________

Distance	Direction	Location/Landmark	Readability (1-5)	Notes (Static? Choppy? Had to get out of car?)
0.5 Mile	North	*Example*: Gas Station	5	Perfect audio.
1.0 Mile	North			
2.0 Miles	North			
3.0 Miles	North			
Limit	North	*Example*	1	Lost signal at the creek bridge.
1.0 Mile	South			
2.0 Miles	South			
Limit	South			
1.0 Mile	East			
Limit	East			

The "Placement Rules" Cheat Sheet

Tape this to your radio storage bin.

1. **Height is Might:** 1 foot of height = 1 mile of range (roughly). Get high.
2. **Vertical Always:** Never transmit with the radio laying flat.
3. **Escape the Cage:** Inside a car = 90% signal loss. Get an external antenna or get out.
4. **Body Block:** Keep the antenna away from your ribs. High on the shoulder is best. The radio needs to "see" the horizon.
5. **The "Baofeng Twist":** Ensure your adapter has a HOLE (SMA-Female) to match the radio's PIN (SMA-Male). If you see two holes, you are killing your radio.

Chapter 6

AUDIO DISCIPLINE & ON-AIR CONFIDENCE

You have the radio. You programmed the frequencies. You climbed a hill to get a signal. You press the button to speak.

"Uhh, hey honey, it's me. Are you there? Uh... 10-4? Over?"

You sound terrified. You sound unsure. And worse, because you held the microphone wrong, you sound like you are speaking from inside a running blender. This is the most overlooked skill in radio. We spend hours obsessing over antennas, watts, and batteries, but we spend zero minutes thinking about **cadence** and **clarity**. We treat the radio like a cell phone, assuming the technology will fix our bad habits. It won't.

Here is the brutal reality: If your signal is strong (5 bars) but your audio is garbage, **you have failed.** If your audio is crisp but your message is rambling and confusing, **you have failed.** In a high-stress situation, when the power is out, the storm is raging, or you are separated from your group, your voice is the only thing your family has to hold onto. If you sound panicked, high-pitched, and breathless, they will panic. If you sound calm, clear, and professional, they will calm down. Your voice is a tool for command and control.

This chapter teaches you how to sound like a "Level 4 Operator" even if you are just asking if we need milk. We are going to cure "Mic Fright" and replace it with muscle memory.

SECTION 1: HOW TO SOUND CLEAR ON A CHEAP HT

The microphone on a $25 Baofeng is an electret condenser element about the size of a grain of rice. It is not a studio microphone. It has zero pop-filtering, terrible wind rejection, and very poor dynamic range. If you scream into it, it doesn't get louder; it just turns your voice into a square-wave of distortion. You have to work *with* the hardware, not against it.

1. The "Baofeng Delay" (The First Syllable Killer)

This is the single most common mistake new users make, and it makes you sound like an amateur instantly. Modern radios have "Battery Save" features. To extend life, the receiver rapidly turns itself on and off (sleep/wake) hundreds of times a second. When a signal comes in, it takes about 500 milliseconds (half a second) for the chip to say, "Oh, that's a signal," wake up the audio amplifier, and open the speaker.

Furthermore, repeaters are slow. When you press PTT (Push-to-Talk), your radio has to send a CTCSS tone to the repeater. The repeater has to "hear" the tone, verify it, and then turn on its own transmitter. This mechanical handshake takes time.

The Mistake: You press the button and speak instantly.

- *You say:* "Base, this is Rover."

- *They hear:* "...ver."
- *Result:* Confusion. "Say again? Who is this?"

The Fix: The "Press... Breathe... Speak" Rhythm.

1. **Press** the PTT button firmly.
2. **Take a short breath** (roughly 1 second). This feels like an eternity, but it is necessary.
3. **THEN** speak.

Tactical Dad Tip: Use this breath to calm your nerves. That 1-second pause guarantees the line is open, the repeater is active, and your family hears the *first* letter of your name.

2. The "Ice Cream Cone" Grip

Where you hold the radio changes the audio physics.

- **The "Pizza Slice" (Bad):** Holding the radio flat in front of your face like you are eating a slice of pizza. You are speaking *over* the mic, not *into* it. Your voice will sound distant, hollow, and echoey.
- **The "Rock Star" (Bad):** Eating the mic. Lips touching the plastic. You are blasting air directly into the diaphragm.
    - *The Physics:* Fast-moving air from "P" and "B" sounds (Plosives) hits the diaphragm like a hammer. This causes "clipping" (distortion). You will sound like muffled garbage or a blown-out speaker.
- **The "Ice Cream Cone" (Good):** Hold the radio vertical, about 2-3 inches away from your mouth, slightly to the side (45-degree angle).
    - *The Technique:* Speak **across** the mic hole, not directly **into** it. The sound waves enter the mic, but the burst of air from your mouth passes harmlessly by. This creates clear, distortion-free audio and helps cut down on wind noise when you are outdoors.

3. Volume and Cadence (The Adrenaline Dump)

When humans are stressed, two things happen biologically:

1. Our vocal cords tighten (Voice gets higher).
2. We speak faster (trying to dump information).

Radio circuits are designed for normal conversation volume in a lower register. If you shout, the audio "clips" (hits the ceiling of what the radio can process) and turns into jagged static. A whisper is equally bad because it gets buried in the noise floor.

- **Speak Slower:** Under stress, force yourself to talk at 75% of your normal speed. Enunciate every syllable.
- **Speak Monotone:** This sounds weird, but a calm, flat tone cuts through static better than an excited, high-pitched one. Think "Airline Pilot" encountering turbulence, not "YouTube Streamer" playing a video game.
    - *Bad:* "Ohmygod there's a tree down huge tree right across the road!"
    - *Good:* "Break. Tree down. Blocking road. Large size."

SECTION 2: THE "ANTI-CB" MANIFESTO (PLAIN ENGLISH ONLY)

Somewhere along the way, pop culture convinced us that using radios meant talking like a trucker from a 1970s movie. *"Breaker breaker one-nine, what's your 10-20? We got a Smokey taking pictures."*

Stop it. Remove this from your brain immediately. Unless you are a trucker or a police officer in a specific jurisdiction, **do not use 10-codes.**

Why 10-Codes are Dangerous

Codes are not universal. They are a "local dialect."

- "10-50" means "Traffic Accident" in one county.
- It means "Officer Down" in the next county.
- It means "Traffic Stop" in a third.

In a regional disaster (hurricane, wildfire), you might be talking to volunteers, GMRS users, Ham operators, and National Guard members from three different states. If you use a code, you risk being misunderstood. Confusion kills. Even the Federal Government (NIMS/ICS) has mandated "Plain English" for all inter-agency disaster communications.

The "Tactical Dad" Standard: Plain English

Say what you mean. Use standard words. Be boringly obvious.

- Instead of "What's your 10-20?", say **"What is your location?"**
- Instead of "10-4," say **"Received"** or **"Copy."**
- Instead of "10-9", say **"Say again."**

The "Pro-Words" (Professional Words)

While we avoid codes, we *do* use a specific set of standardized words (Pro-Words). These are not codes; they are **procedure**. They function like punctuation marks in a sentence, telling the other person exactly what is happening.

The Word	What It Means	When To Use It
"COPY"	I heard and understood your message.	The universal acknowledgment.
"WILCO"	"Will Comply." I heard you AND I will do it.	When given an instruction. (Don't say "Roger Wilco," it's redundant.)
"STANDBY"	I am busy. Wait. I will call you back.	When you need a minute to think, drive, or write something down.
"SAY AGAIN"	I didn't hear you. Repeat the message.	*Never say "Repeat."* (In military contexts, "Repeat" is a specific command to artillery to fire the same mission again. Don't call in artillery on your house.)
"AFFIRMATIVE"	Yes.	"Yes" is a short, soft word that gets lost in static. "Affirmative" is distinct.
"NEGATIVE"	No.	"No" is too short. "Negative" cuts through noise.
"BREAK"	I am separating two different messages.	"I need milk. BREAK. I also need gas." Use this instead of unkeying the mic.
"CORRECTION"	I made a mistake. Here is the right info.	"Turn Left... CORRECTION... Turn Right."
"OUT"	I am ending the conversation.	"Copy. Out." (Do not say "Over and Out." That means "I'm done talking, go ahead, but I'm hanging up.")

SECTION 3: THE LANGUAGE OF CONFIDENCE (NATO PHONETIC ALPHABET)

This is the mark of a pro. This is the difference between "Did you say B or D?" and instant clarity.

In a static-filled environment, the human ear struggles to distinguish rhyming letters. This is the **"Kill Zone" of the alphabet:**

- **B, C, D, E, G, P, T, V, Z.** They all end in "EEEE."
- **M and N.** They sound identical. **F and S.** They both sound like hiss.

To fix this, NATO developed a set of words where **no two words sound alike.** Even if you only hear half the word "November," you know it isn't "Mike."

The Deep Dive: How to Learn It

Do not just print the chart and stick it in your wallet. You have to wire this into your brain so you can speak it without thinking. When you are spelling your street name to emergency services over a dying battery, you don't want to be looking for a cheat sheet.

The Chart (Tactical Dad Reference)

Number Discipline (The Forgotten Skill)

Just like letters, numbers get confusing. "Nine" sounds like "Five" in static. "Nine" also sounds like "Nein" (No) in German, which caused issues in NATO.

- **9** is spoken as **"NINER"** (To distinguish from Five/None).
- **5** is spoken as **"FIFE"** (To distinguish from Fire).
- **4** is spoken as **"FOWER"** (To distinguish from "For" or noise).
- **3** is spoken as **"TREE"** (To keep it sharp).
- **0** is spoken as **"ZERO"** (Never say "Oh").

The "License Plate" Drill

This is how you learn this in 3 days without flashcards. While driving to work, look at the license plate of the car ahead of you. Translate it instantly.

- Plate: K7P M29
- You say (out loud): *"Kilo Seven Papa... Mike Two Niner."*
- Plate: DVG 304
- You say: *"Delta Victor Golf... Tree Zero Fower."*

Do this for 10 cars a day. It forces your brain to recall the words randomly and rapidly. By Friday, you will be fluent.

SECTION 4: THE "C.L.E.A.R." REPORTING FORMAT

When you get on the radio, you shouldn't ramble. Rambling ties up the frequency, drains the battery, and annoys everyone listening. *"Uhhh, hey, so I'm at the store, and it's kinda crazy, lots of people, um, I think they are out of water, but I saw some Gatorade, do you want that?"*

No. That transmission took 15 seconds to say nothing. We need a standard format. The military uses "SALUTE" reports. We are going to use a civilian adaptation designed for families.

The C.L.E.A.R. Report

- **C - CALLSIGN:** Who are you? (Identify yourself first).
- **L - LOCATION:** Where are you? (Be specific).
- **E - EVENT:** What is happening? (The headline).
- **A - ACTION:** What are you doing about it? (Your status).
- **R - REQUEST:** What do you need? (Resources).

Example 1: The Grocery Run (Low Stress)

"Base, this is Dad. ***Location:*** *Kroger on Main.* ***Event:*** *Store is crowded, out of water.* ***Action:*** *I am buying Gatorade instead.* ***Request:*** *Do we need bread? Over."*

Example 2: The Car Breakdown (High Stress)

"Base, this is Unit 1. ***Location:*** *I-75 Mile Marker 102.* ***Event:*** *Flat tire, safe on shoulder.* ***Action:*** *Changing tire now.* ***Request:*** *Stand by for check-in in 15 minutes. Over."*

Why this works: Even if your wife misses the first part because of static, the structure tells her what is coming. She knows the second thing you say is *Where* and the last thing you say is *What you need*. It forces you to think before you speak. It stops the panic-rambling.

SECTION 5: ETIQUETTE & "DON'T BE THAT GUY"

Radio communities are small. If you act like a goofball, you will be blocked, ignored, or shamed.

The "Don't Be That Guy" List

1. **The "Roger Beeper":**
   - *The Crime:* Having that annoying "BEEP" sound at the end of every transmission. (Menu 39 ROGER on UV-5R).
   - *Why it's bad:* NASA used Quindar tones because they were talking to the moon. You are talking to the suburbs. On a cheap radio, this beep is often 2x louder than your voice, blasting the eardrums of anyone wearing a headset. It marks you as a newbie immediately.
   - *Fix:* Turn it OFF.
2. **The "Baofeng Siren" (The Orange Button of Shame):**
   - *The Crime:* Accidentally bumping the orange CALL button on the side and sending a loud, wailing siren alarm over the airwaves.
   - *Why it's bad:* It sounds like an emergency, but it's usually just you sitting on your radio. It deafens headphone users.
   - *Fix:* Be very careful with the orange button. On some models, you can disable the "Alarm" function in CHIRP (Set "Alarm Mode" to "Site" instead of "Tone").

3. **The "Kerchunker":**
    - *The Crime:* Keying the mic repeatedly to hear the repeater tail ("Ker-chunk"), but never identifying yourself.
    - *Why it's bad:* It triggers the repeater for everyone listening. It's like ringing a doorbell and running away.
    - *Fix:* If you key up, say your callsign. "Testing, this is 'Callsign'."
4. **The "Dead Key-er":**
    - *The Crime:* Sitting on the PTT button while you think of what to say.
    - *Why it's bad:* You are blocking everyone else from using the channel. In radio, only one person can talk at a time. If you are transmitting "Uhhhh...", nobody else can call for help.
    - *Fix:* **Think** *then* **Push**. Not Push *then* Think.
5. **The "Interrupting Cow":**
    - *The Crime:* Jumping in the second someone stops talking.
    - *Why it's bad:* In emergency nets, we leave a "gap" between transmissions to let urgent traffic break in.
    - *Fix:* When you hear someone finish ("...clear"), count "One Mississippi" before you press your button. This gap allows a third party with an emergency to shout "Break!"
6. **The "Heavy Breather" (Accidental VOX):**
    - *The Crime:* Turning on VOX (Voice Operated Switch) by mistake and broadcasting every heavy breath or gust of wind.
    - *Why it's bad:* It keeps the channel open constantly, drains your battery, and annoys everyone.
    - *Fix:* Ensure VOX is set to **OFF** (Menu 4) unless you are specifically setting it up for a controlled task.

WORKBOOK TOOLS

The "Message Discipline" Pocket Card

Print, laminate, and zip-tie to your radio or backpack.

BEFORE YOU PRESS PTT:

1. **LISTEN:** Is the channel clear?
2. **THINK:** What is my C.L.E.A.R. message?
3. **WAIT:** Press PTT, wait 1 second (breath), then speak.

THE C.L.E.A.R. FORMAT:

- Callsign (Me to You)
- Location (Where)
- Event (What)

- **A**ction (Doing)
- **R**equest (Need)

PRO-WORDS:

- **COPY:** Understood.
- **WILCO:** Will do it.
- **SAY AGAIN:** Repeat.
- **CORRECTION:** I messed up.
- **BREAK:** New topic / Interruption.

5 Practice Scenarios (The Drill)

Do these with your partner. One person in the kitchen, one in the yard. Grade each other on Clarity, Phonetics, and Format.

Scenario 1: The Storm Spotter *Context:* You see a tree down blocking the neighborhood entrance. *Goal:* Report it using C.L.E.A.R. and phonetic spelling of the street name. *Drill:* "Base, this is Unit 1. Location: Entrance to **OAK** subdivision. That is **OSCAR-ALPHA-KILO**. Event: Large tree blocking road. Action: Assessing if I can move it. Request: Call public works. Over."

Scenario 2: The Lost Kid *Context:* You are at the county fair. You can't find your son, Jason. *Goal:* Communicate urgency without panic. Use description. *Drill:* "Base, this is Dad. Location: Ferris Wheel. Event: Separated from Jason. Action: Starting grid search north. Request: You stay at the gate. Description: Red Shirt, Blue Hat. Over."

Scenario 3: The Bad Connection *Context:* The signal is scratchy. You need to verify an address (15 Maple St). *Goal:* Use "Say Again" and Number Discipline. *Drill:* "Base, signal is broken. Say again address. Over." -> "I say again: **ONE-FIFE** Maple. **MIKE-ALPHA-PAPA-LIMA-ECHO**. Over."

Scenario 4: The Convoy *Context:* Driving two cars. You need to stop for gas. *Goal:* Brevity. Don't clog the channel. *Drill:* "Unit 2, this is Unit 1. Fuel low. Exiting next ramp. Wilco? Over." -> "Unit 1, this is Unit 2. Copy. Exiting next ramp. Out."

Scenario 5: The Radio Check *Context:* Testing a new antenna. *Goal:* Correct format and patience. *Drill:* "Radio Check, Radio Check. This is 'Name' on 'Channel Name'. How copy? Over."

Chapter 7

POWER, CHARGING, AND ENDURANCE

THE PART EVERYONE REGRETS IGNORING

YOU HAVE THE BEST radio. You have a tuned antenna. You have a perfect channel plan. You have practiced your phonetic alphabet until you dream in "Alpha Bravo."

Then, 12 hours into the blackout, you hear the dreaded *beep-beep*. The screen flashes. The radio dies.

You reach for your charger, but the power is out, so the wall outlet is useless. You run to your car, but you realize your charging cradle requires a 10V wall wart, not a USB plug. You stare at a pile of AA batteries in your junk drawer, but you have no way to put them into the radio.

Your communications plan just failed. Not because of a lack of skill, but because of a lack of electrons.

Most people treat batteries as an afterthought. They use the stock battery, charge it once a month, and assume it will be there when the world ends. This is a fatal error. In a real emergency, **power is the fuel of information.** If you run out of fuel, you are deaf and mute.

This chapter is about hardening your power supply. We are going to move away from proprietary, clunky chargers and embrace a standardized, redundant system that keeps you on the air for weeks, not hours.

SECTION 1: BATTERY TRUTH (THE PHYSICS OF THE BRICK)

The standard battery that comes with a UV-5R (the BL-5) is a Lithium-Ion pack rated at roughly 1800mAh (milliamp-hours). Note that I said "rated at." Cheap Chinese batteries often lie.

That "3800mAh" extended battery you saw on eBay for $12? It's probably just an 1800mAh cell with a bigger plastic shell filled with hot glue and lead weights. Unless you buy from a reputable brand (like BTech or Baofeng Tech), assume the label is optimistic fiction.

The "Duty Cycle" Reality

To understand endurance, you have to understand consumption. Your radio burns power at three drastically different rates. Understanding this math is the difference between your radio lasting 6 hours or 3 days.

1. **Standby (The "Listening" Mode):**
    - *Draw:* ~75 milliamps.

  - *Life:* If you just leave the radio on and nobody talks, a healthy battery lasts about **24 hours**. This is your baseline.

2. **Receive (The "Talking" Mode):**
   - *Draw:* ~300-400 milliamps (depending on volume).
   - *Life:* If you are listening to a busy repeater conversation non-stop, or monitoring a weather alert that is broadcasting continuously, the battery lasts about **4-6 hours**. The audio amplifier consumes significant power just to push sound out of the speaker.

3. **Transmit (The "Flamethrower" Mode):**
   - *Draw:* ~1500 milliamps (1.5 Amps) on High Power (5W).
   - *Life:* If you held the PTT button down continuously, the battery would die in **less than 1 hour** (and the radio would likely melt or go into thermal shutdown). Transmitting is expensive.

The "5-5-90" Rule: Engineers design radios assuming a 5-5-90 duty cycle:

- 5% of the time Transmitting (Brief reports).
- 5% of the time Receiving (Listening to replies).
- 90% of the time on Standby (Silence).
- *Result:* This gives you a "shift life" of about 10-12 hours. If you talk more, you die faster.

The Vampire Settings (What Kills You)

If your battery is dying in 6 hours, you are probably guilty of these crimes. These are the "Vampires" that suck the life out of your pack without you noticing:

- **Scanning:** This prevents the radio from going into "Power Save" sleep mode. Normally, the radio "pulses" the receiver to check for activity. When scanning, the CPU is running full throttle, checking 3 channels every second. Scanning burns power 3x faster than sitting on one channel.
- **Backlight:** If you have the backlight set to "Always On" (or a long timeout), you are wasting power. Set it to 2 seconds. You don't need a flashlight on your screen just to listen.
- **Dual Watch (TDR):** Checking two frequencies prevents deep sleep. The radio has to wake up twice as often to check Priority A and Priority B.
- **High Power Simplex:** You don't need 5 Watts to talk to your wife in the backyard. Switch to Low Power (1 Watt) by tapping the # key. This cuts your transmit consumption by 60%, significantly extending your "Flamethrower" time.

SECTION 2: THE MANDATORY UPGRADE (THE USB-C REVOLUTION)

Stop living in 2010. The standard UV-5R charging cradle is a logistical nightmare. It requires 110V AC power (a wall outlet) and a specific "wall wart" transformer. If the power grid goes down, that cradle becomes a paperweight unless you have a generator running. Even if you have a generator, do you really want to use a precious outlet for a single radio charger?

The Solution: The USB-C Extended Battery.

Baofeng (and BTech) now sells batteries that have a **direct USB-C charging port** built into the battery casing itself.

Why This Changes Everything

1. **One Cable to Rule Them All:** You already charge your phone, your tablet, and your flashlight with USB-C. Now your radio joins that ecosystem. You don't need a special "radio charger" taking up space in your bag. You reduce your cable clutter by 50%.

2. **Charge Anywhere:** You can charge the battery from your car's USB port, your laptop, a solar panel, or, most importantly, a portable power bank. The source doesn't matter, as long as it has a USB port.

3. **Ditch the Cradle:** You can leave the bulky plastic cradle at home. This saves weight and bulk in your Go-Bag. No more worrying about bending the pins on the cradle or losing the AC adapter.

Action Step: Go to Amazon or BTech right now. Buy a **BL-5L (3800mAh) USB-C Battery**.

- *Note:* Ensure it specifically says "USB-C Charging" or "Direct Charge." Some older extended batteries look identical (big bricks) but still require the desktop cradle. Read the description carefully.

- *Cost:* ~$20 - $25.

- *Value:* Infinite. This is the single most important hardware upgrade after the antenna. It modernizes a 2012 radio into a 2025 tool.

SECTION 3: CHARGING STRATEGY (HOME, CAR, BAG)

You need a layered approach to power, just like you have a layered approach to frequencies. Relying on one method is a recipe for silence.

Layer 1: Home Base (Grid Up)

Keep one radio in the desktop cradle, plugged into the wall. This is your "Grab and Go" unit.

- *The Trap:* Do not leave it in the cradle 24/7/365. Lithium batteries hate being at 100% voltage constantly. It degrades their chemistry, leading to "puffy" batteries or reduced capacity.

- *The Fix:* Use a simple outlet timer (like for Christmas lights) to run the charger for 2 hours a day. Or, simply create a routine where you rotate the radio out of the charger once a week to use it, then put it back.

Layer 2: The Vehicle (Mobile)

Your car is a massive generator on wheels. It is the ultimate survival battery.

- *The Old Way:* Buying a "Battery Eliminator" (fake battery that plugs into cigarette lighter). These are okay, but they tether you to the car. You can only talk while sitting in the driver's seat.

- *The New Way:* A USB-C cable plugged into your dashboard USB port, charging your USB-C battery. You can charge *while* you drive to the scene, then unhook and walk away with a full tank of electrons.

Layer 3: The Go-Bag (Grid Down)

This is where the USB-C strategy shines. Instead of carrying a specialized radio charger, you carry a **standard Anker-style Power Bank** (10,000mAh or 20,000mAh).

The Math of Survival:

- Radio Battery = ~1,800mAh (Standard) or ~3,800mAh (Extended).

- Power Bank = 20,000mAh.
- *Loss:* Voltage conversion efficiency is about 80% (some energy is lost as heat).
- *Result:* A single 20,000mAh brick can recharge your standard radio **8 to 10 times.**
- *Endurance:* If one charge lasts 2 days (conservative use), **one brick gives you 20 days of communications.** That is nearly three weeks of autonomy in a package the size of a smartphone.

SECTION 4: THE "AA" OPTION (THE BREAK-GLASS BACKUP)

Lithium batteries are great, until they are dead. When the Anker brick is empty, the car is out of gas, and the sun isn't shining, what is left? **The AA Battery.**

You can scavenge AA batteries from TV remotes, children's toys, abandoned wall clocks, and junk drawers in almost any house in America. They are the currency of the apocalypse.

The AA Battery Case (BL-5 AA)

You can buy a plastic clamshell case (about $8) that snaps onto the back of the UV-5R and takes 6x AA batteries (or 6x AAA depending on the model).

The Pros:

- **Scavenge Capability:** You are no longer dependent on the grid. As long as you can find AAs, you can talk.
- **Instant "Recharge":** No waiting for a USB cable. Swap the cells and you are back on air in 30 seconds.

The Cons (Why this is a BACKUP only):

1. **Size:** It makes the radio physically huge (about 2 inches longer). It won't fit in standard pouches.
2. **Voltage Sag:** AA Alkaline batteries are not designed for high-current draw. When you press PTT on High Power, the batteries panic and the voltage drops instantly. You might get "Low Voltage" warnings even if the batteries are half full.
3. **Spark Gap:** The fit and finish on these cases is often cheap. If you drop the radio, the case pops open and batteries fly everywhere.

The Verdict: Buy one. Put 6 lithium (Energizer Ultimate) AA batteries in it (they have 10-year shelf life and handle cold weather better). Tape it shut with electrical tape to prevent "explosive disassembly" if dropped. Put it in the bottom of your bag. **Do not use it unless you have to.** It is your "Last Stand" power source.

SECTION 5: SOLAR VS. STORED (DON'T GET FOOLED)

You will see "Solar Chargers" marketed to preppers. They show a guy hiking with a tiny solar panel clipped to his backpack, looking rugged. **This is mostly fantasy.**

The Solar Reality

Small, backpack-sized solar panels produce very little current, especially if you are moving. Shadows from trees, changing angles, and overcast skies kill the output.

- To charge a 20,000mAh brick with a small "backpack" panel might take **3 to 4 days of perfect, stationary sunlight.**

- Directly charging a radio from a solar panel is often impossible because the fluctuation in voltage (as clouds pass) confuses the radio's charging circuit, causing it to stop charging entirely.

The Stored Power Reality

A 20,000mAh USB battery brick costs $30, fits in your pocket, and works in a pitch-black basement during a hurricane. It works at night. It works in the rain.

Tactical Dad Rule: Focus on **Storage** first, **Generation** second.

1. Buy two high-quality USB Power Banks.
2. Keep them charged.
3. Only mess with solar if you are planning for a "Long Term (30+ day) Grid Down" scenario. For a 3-day or 7-day event, stored power wins every time. It is reliable math vs. variable weather.

SECTION 6: ENVIRONMENTAL HARDENING (HEAT & COLD)

Batteries are chemical reactions. Temperature controls the speed of the reaction.

The Cold (The Lithium Killer)

If you leave your radio in your car in Minnesota in January (-10°F), the battery will read "Dead" almost instantly. The chemical reaction slows down so much it can't provide the current needed to transmit.

- *The Fix:* **Body Heat.** Keep the radio inside your jacket, close to your chest. The warmth of your body keeps the chemistry active. Use a remote speaker mic (routed out of your zipper) to talk. Do not leave radios in unheated cars overnight in winter, bring them inside.

The Heat (The Capacity Killer)

If you leave your radio on the dashboard in Arizona in July (140°F), you are permanently cooking the chemistry.

- *The Result:* The battery won't explode, but it will lose capacity permanently. A battery that used to hold a charge for 24 hours will now only last 12.
- *The Fix:* Store radios under the seat, in a cooler, or in a light-colored bag. Never in direct sun.

SECTION 7: STORAGE & MAINTENANCE (AVOIDING DRAWER DEATH)

The worst thing you can do to a lithium battery is drain it to 0% and leave it in a drawer for a year. The voltage will naturally drift down over time. If it drops below the "safe" threshold (usually around 2.5V per cell), the battery's internal protection circuit will trip and permanently lock it out. It will never charge again. It is a brick.

The "Quarterly Rotate" Plan

You don't need to be obsessive. You just need a rhythm. Tie this to another household chore so you remember.

- **When:** The first day of every season (Start of Spring, Summer, Fall, Winter). Same day you check smoke detector batteries.
- **The Drill:**

    a. Turn on every radio. Check voltage (Hold 0 key).

b. If below 7.4V, charge it up.

c. Check the Go-Bag power banks. Press the button to check the lights. Top them off if needed.

d. Inspect the AA alkaline batteries for leaks (white crusty powder). Alkaline leaks can destroy the electronics in your backup case.

ACTION STEPS & WORKBOOK

The Power Plan Worksheet

Kit Location	Primary Power	Secondary Power	Backup Power
Home Base	Desktop Cradle (AC)	USB-C Battery + Wall Plug	Case of AA Batteries
Car (Mobile)	USB-C Cable (Dash)	Spare BL-5 Battery (Glovebox)	12V Battery Eliminator
Go-Bag	USB-C Battery (On Radio)	20,000mAh Anker Brick	AA Clamshell Case

The "72-Hour Comms Readiness" Checklist

Can you survive a 3-day weekend without power?

- [ ] **Primary Radio:** Has a USB-C battery installed and fully charged.
- [ ] **Charging Cable:** A USB-A to USB-C cable is IN THE BAG (not borrowed from your phone charger).
- [ ] **Power Bank:** One 10,000mAh+ brick, fully charged (4 lights).
- [ ] **The Backup:** One AA Case loaded with fresh Lithium AA batteries (taped shut to prevent rattling).
- [ ] **Vehicle Check:** Tested the car charger, does it actually work with the engine off? (Some cars kill 12V ports when the key is removed, meaning you can't charge while parked).

Chapter 8

"SECURITY" THE HONEST WAY

PRIVACY, OPSEC, AND WHAT TONES REALLY DO

IF YOU HAVE WATCHED any spy movies, you probably think you can "encrypt" your communications. You imagine pressing a button and having your voice turn into scrambled robot noise that only your team can understand. You picture a specialized headset and a secret frequency that no one else knows exists.

Let me burst that bubble immediately: **On a Baofeng UV-5R, privacy does not exist.**

The airwaves are public property. They are a shared space, like a public park. Anyone with a $25 radio can hear you. Anyone with a $300 police scanner can record you. Anyone with a $20 software-defined radio (SDR) dongle on their laptop can visualize your signal on a map and log every word you say to a hard drive.

If you are transmitting, you are broadcasting.

Security on analog radio does not come from technology. It comes from **behavior**. It comes from what you say, how you say it, and when you choose to stay silent. This is called **OPSEC** (Operational Security). It is the art of denying the enemy information about your intentions and capabilities.

In this chapter, we are going to learn how to hide in plain sight. We are going to stop pretending we are spies and start acting like disciplined operators.

SECTION 1: THE GREAT TONE MYTH (CTCSS/DCS IS NOT ENCRYPTION)

This is the most dangerous misconception in the hobby. It gets people into trouble because it creates a false sense of security.

You go into the menu. You set a "Privacy Tone" (CTCSS) of 103.5 Hz. You test it with your wife. She can hear you, but when you turn her tone off, she can't hear you (because her squelch is closed). You think: *"Great! We are on a private line! Nobody else can hear us!"*

The Truth: CTCSS (Continuous Tone-Coded Squelch System) is not a wall. It is not encryption. It is a filter.

- **What you think it does:** Scrambles your voice so others can't hear.
- **What it actually does:** Tells *your* radio to ignore everyone else. It puts a filter on *your* speaker, not on *their* ears.

The "Earplug" Analogy: Imagine you are in a crowded restaurant.

- **Encryption** would be you and your wife speaking a secret language (Elvish). Nobody understands you, even if they listen closely.
- **CTCSS** is you and your wife putting in specialized earplugs that block all sound *except* your specific voices. You can't hear the other diners. You feel like you are in a private bubble. **But the other diners can still hear you screaming**

across the table.

The 'Monitor' Check: Because your 'earplugs' (Tones) make you deaf to other traffic, you might accidentally talk over someone else who is already using the channel. Always press the **MONI** (Monitor) button on the side of the Baofeng for 2 seconds before you talk. This temporarily pulls out the earplugs (opens squelch) so you can check if the channel is actually clear.

If a bad actor has a scanner, he isn't using tones. He is listening to "Carrier Squelch" (Open). He hears everything, the truckers, the kids playing with walkie-talkies, and you discussing where you hide your spare key.

Tactical Dad Rule: Never say anything on a "Privacy Tone" channel that you wouldn't feel comfortable yelling across a crowded Walmart parking lot. If you wouldn't say it to a stranger, don't say it on the radio.

SECTION 2: THE "SCRAMBLER" TRAP (VOICE INVERSION)

Deep in the Baofeng menu (usually Menu 10 or similar on unlocked models), there is a setting often labeled SCR or SCRAMBLER. When you turn it on, your voice sounds like a duck quacking underwater or a warped vinyl record. It sounds cool. It sounds "tactical."

The Legal Reality: Using voice scrambling on Amateur Radio (Ham) frequencies or GMRS frequencies is **illegal**. The FCC rules explicitly state that messages must not be obscured to hide their meaning. The amateur bands are for public, open communication. If you use it, you are breaking federal law.

The Tactical Reality: Even if you don't care about the law, it is technically weak security. "Voice Inversion" is 1950s technology. It simply flips the high frequencies to low and low to high.

- **The Vulnerability:** Any determined listener can unscramble it with a simple computer program (like Audacity) or even another Baofeng by guessing the inversion frequency. It is not military-grade AES-256 encryption. It is a screen door, not a bank vault.

The "SHTF" Exception: In a true "Without Rule of Law" scenario (zombies, total societal collapse), FCC rules obviously don't apply. In that extreme, life-or-death instance, the scrambler might offer a tiny layer of obscurity against casual listeners who are just scanning for clear speech. But do not rely on it. A sophisticated listener will identify the sound of a scrambler immediately, and it might actually draw *more* attention to you because it sounds like you have something to hide. It is a paper shield.

SECTION 3: OPSEC (OPERATIONAL SECURITY)

If we can't use technology to hide, we must use discipline. We must assume someone is always listening.

1. Channel Discipline (The "Hopping" Plan)

Don't camp on one frequency forever. If you talk on Channel 1 every day at 5:00 PM, you establish a "Pattern of Life." Anyone monitoring will know exactly where to find you.

The Fix: The PACE Plan. In the military, we use a PACE plan for communications: **P**rimary, **A**lternate, **C**ontingency, **E**mergency.

- **Primary:** "Channel 1." We use this for routine checks.
- **Alternate:** "Channel 5." We switch here if Channel 1 has interference or too many strangers.
- **Contingency:** "Channel 7." We switch here if we suspect we are being monitored.
- **Emergency:** "Channel 0 (National Call)." We go here if everything else fails.

Implementation: If you feel uneasy, simply say: "Switch to Blue" (where Blue = Channel 5 in your notebook). Don't say "Switch to 146.580."

2. What NOT to Say (The Red List)

Data Aggregation is real. One piece of info isn't dangerous. Ten pieces are deadly. A listener might hear "Mike" on Monday, "Maple Street" on Tuesday, and "Generator" on Wednesday. By Friday, he knows Mike on Maple Street has a generator.

The Red List (NEVER SAY):

- **Specific Street Addresses:** "I'm at 123 Maple Drive." Instead say: "I'm at Base."
- **Gate Codes or Lock Combinations:** Never.
- **Detailed Inventory:** "I have 50 gallons of gas and a generator." You just put a target on your house.
- **Dates of Absence:** "We are leaving for vacation tomorrow." You just invited burglars.
- **Full Names of Children:** Use nicknames or tactical callsigns.

3. The "Innocuous" Code System

You don't need to sound like a SEAL Team. In fact, sounding like a soldier attracts attention. If you are shouting "Alpha Bravo Tango, perimeter breach!", every listener with a scanner is going to stop and listen to you. You sound interesting. You sound like you have supplies.

You want to sound like a boring suburbanite. You want to be the "Gray Man" of the airwaves.

Instead of using "Tactical Codes" (e.g., "Condition Red"), use **"Innocuous Codes"**, phrases that sound normal but have hidden meaning.

Family Code Worksheet:

The Message	The "Tactical" Way (BAD)	The "Gray Man" Way (GOOD)
"Come home NOW. Danger."	"Initiate Alpha Protocol! Rally Point!"	"Can you pick up a pepperoni pizza?"
"I am being followed."	"Hostile bogie at 6 o'clock!"	"My check engine light just came on."
"Switch to backup channel."	"Go to Channel 4! Secure Comms!"	"Call Uncle Mike."
"Do not come home. It's unsafe."	"Perimeter breached! Abort!"	"The dog is sick, don't let him out."
"I have the supplies."	"Asset secured. In transit."	"I got the milk."

Why this works: A scanner listener hearing "Can you pick up a pizza?" will ignore it. They are looking for excitement. Bore them to death. Hidden in plain sight is better than a lock.

SECTION 4: LOW-PROFILE COMMS HABITS

The best way to stay secure is to be brief. The longer you talk, the easier you are to find.

1. The "Burst" Transmission and Direction Finding

Radio signals can be tracked. This is called "Fox Hunting" or Direction Finding (DF). A person with a directional antenna (Yagi) can swing it around and find exactly where a signal is coming from.

- If you talk for 5 minutes, they can triangulate your exact house.
- If you talk for 5 seconds, they barely have time to grab their radio.

The Rule: Keep transmissions under 10 seconds.

- **Bad:** A 5-minute rant about the traffic, the weather, and your feelings.
- **Good:** "Traffic stopped at Exit 5. Detouring. Out." (5 seconds).

2. Predictability is Death

Do not set a rigid schedule like "We will talk every hour on the hour." This allows a listener to set an alarm, wake up, and record you. Use a **"Window" system**.

- "I will check in between :15 and :20 past the hour." This forces a listener to monitor you constantly, fighting static and boredom. Most will give up.

3. Power Discipline (RF Footprint)

Your radio puts out a signal 'bubble.'

- **High Power (5W):** Big bubble (5-10 miles). Everyone in the county can hear you.
- **Low Power (1W):** Small bubble (1-2 miles). Only your immediate neighbors can hear you.

The Rule: Use the lowest power necessary to make the connection. If your team is 1 mile away, switch to Low Power (tap the # key). There is no tactical advantage to broadcasting your conversation to the next county.

BREAKOUT: THE DIGITAL (DMR) ELEPHANT

You might be asking: *"Why don't we just buy digital radios? I heard they have encryption."*

The Tech: Digital Mobile Radio (DMR) converts your voice into 1s and 0s (data). It sounds like modem noise to an analog radio. It can indeed use AES-256 encryption, which is military-grade. Nobody can listen to that without the key.

Why we aren't doing it in this book:

1. **Complexity:** Programming a UV-5R is like programming a VCR. Programming a DMR radio (like an Anytone 878) is like programming a Linux server. It requires "Codeplugs," "Color Codes," "Time Slots," "Talk Groups," and "Encryption Keys." The learning curve is a vertical wall. Most people give up.
2. **Incompatibility:** If you have a fancy encrypted radio, you can **only** talk to other people with that exact same radio and encryption key. You cannot talk to your neighbor's Baofeng. You cannot talk to the local ARES volunteers. You cannot shout for help on a standard channel. You isolate yourself in a digital fortress.
3. **Cost:** A good DMR radio is $100-$300. A Baofeng is $25. For a family of four, equipping everyone with DMR is a $1,000 investment versus $100.
4. **Failure Modes:** Digital signals are "all or nothing." Analog signals degrade gracefully, you can still understand a scratchy analog voice. A weak digital signal just disappears.

Verdict: Stick to analog for now. Master the basics. Learn to walk before you try to fly a helicopter.

WORKBOOK TOOLS

The OPSEC Audit: *Sit down with your family and fill this out.*

Our "Green List" (Okay to say):

- First names only (or callsigns).
- General landmarks ("The grocery store," "The high school").
- Weather reports / Road conditions.
- "I am safe" / "I am moving."

Our "Red List" (NEVER say):

- "We have plenty of food/ammo/gas." (Invites looting).
- "The alarm system is broken." (Invites burglary).
- "We are leaving the house empty for three days." (Invites squatting).
- "I am alone."

The "Pizza Code" Card

Create 3 codes that only your family knows. Memorize them. Do not write the meanings on the card, only the phrases.

1. **The "Get Out" Phrase:** (Means: Leave immediately, meet at rally point. Do not pack, just go).
   - *Phrase:* ________________ (e.g., "Grandma is asking for you.")
2. **The "Duress" Phrase:** (Means: I am being forced to talk, but I am not safe. Someone is with me).
   - *Phrase:* ________________ (e.g., "I'll be home in 100 minutes" - knowing you are only 10 mins away).
3. **The "Silence" Phrase:** (Means: Turn off the radio immediately, stop transmitting, listen only. We are being hunted).
   - *Phrase:* __________________ (e.g., "Goodnight Moon.")

Chapter 9

THE FAMILY/TEAM COMMS PLAN

FROM GEAR TO CAPABILITY

YOU HAVE SPENT WEEKS learning about antennas, batteries, and repeaters. You are now a competent radio operator. You can hit a repeater 20 miles away and you know the difference between a UV-5R and a UV-82. Congratulations. You are halfway there.

If the grid goes down and you are the only person in your family who knows how to turn on a radio, **you have failed.**

A radio is a team sport. It requires someone to talk *to*. If your wife thinks the radio is just "Dad's weird hobby" and your kids are afraid to touch it because you yelled at them once for pressing the wrong button, you do not have a communications plan. You have a toy collection.

In a crisis, hardware failure is annoying, but **human failure is fatal.** If your family freezes because they don't know the channel, or if they talk over each other because they lack discipline, the expensive radios become paperweights.

This chapter is the hardest part of the book because it involves people, not physics. You can't program people with a cable. We are going to turn your family into a cohesive, capable team.

SECTION 1: SELLING THE PLAN (WITHOUT SOUNDING CRAZY)

You cannot force your family to care about radio. If you start talking about "EMPs," "Civil War," and "Societal Collapse," their eyes will glaze over. They will label you as paranoid, and they will tune you out. You need to sell **utility**, not fear. You need to show them how this tool makes *their* life easier right now, not just in the apocalypse.

The "Convenience" Pitch (The Gateway Drug): Don't say: *"We need this for when the Russians invade."* Say: *"Next time we are at the County Fair or Disney World, you know how the cell service always crashes because there are 50,000 people posting selfies? We can use these to find each other so we don't have to wait in line for food alone. It's basically a VIP line for talking."*

The "Safety" Pitch (For the Spouse): *"If a storm knocks out the cell towers like it did last year, I want you to have a way to call me from the house if you get hurt or if the car breaks down. It's just a backup phone. I hope we never use it, but I'll sleep better knowing you have it."*

The "Cool Factor" Pitch (For the Kids): *"It's a secret channel. We can talk without Mom hearing us from the backyard."* (Trust me, this works. Kids love exclusivity).

- **For Teenagers (The Texting Rebuttal):** They will say, 'I'll just text you.' You reply: 'Texts get stuck when towers are overloaded. Radios don't. Do you want to be the only one stuck at the concert with no ride home because your text didn't go through?' Appeal to their independence.

SECTION 2: ROLES & THE "BUS FACTOR" (WHO IS IN CHARGE?)

In most families, Dad is the "Comms Guy." He programs the radios, charges the batteries, and initiates the calls. He is the single point of failure. **The Problem:** If Dad gets hurt, is driving, is stuck at work, or is handling a physical threat, the system collapses. You are the bottleneck.

The Roles

You need to assign roles based on personality and capability, not just age.

1. **Net Control (The Hub):** Usually you. You direct traffic. You decide which channel the family uses. You prioritize messages. "Unit 2, stand by. Unit 3, go ahead."
    - *Responsibility:* Maintains the master battery supply, monitors the repeater, and makes the final call on movement.
2. **The XO (Executive Officer / Second-in-Command):** Usually the spouse or oldest capable child. *This is the critical role.*
    - **The Protocol:** The XO monitors the radio when Net Control is busy. If Net Control does not check in within 15 minutes of a scheduled time, the XO assumes command. They make the decision: "Okay, Dad isn't answering. We are switching to Plan B. Everyone meet at the house."
    - *Why this matters:* It gives your family permission to act without you.
3. **The Scouts (Rovers):** The kids or mobile units. Their job is simple: Listen and Report.
    - *Responsibility:* Keep the radio ON. Do not touch the channel knob. Answer when called. Report what they see ("I see a road block").

The "Incapacitated Operator" Drill (Morbid but Necessary)

Ask your family this question tonight at dinner: *"If I am unconscious on the floor and the power is out, can you turn on this radio and call for help?"*

If the answer is "I don't know which button is On," or "I don't know what channel the police are on," you have work to do.

- **The Fix:** Put a laminated "Cheat Sheet" on the back of the designated "Kitchen Radio" (the one that stays in the charger).

    a. Turn Top Knob clockwise (ON).

    b. Check Screen says "Channel 1".

    c. Press and Hold the LARGE Side Button (PTT).

    d. Speak: "Emergency. Help."

SECTION 3: YOUR COMMS PLAYBOOK (THE 3 MODES)

You don't wear a raincoat on a sunny day. You shouldn't act like you're in a warzone when you're just parking a trailer. You need different behaviors for different threat levels.

Mode 1: Normal Day (Green)

- **Status:** Grid is up. Cell phones work. The world is normal.

- **Radio Use:** Recreational only. Hiking, camping, convoy driving to the beach.
- **Goal:** Practice without stress. Use the radios to coordinate backing up the RV or finding a parking spot at the mall.
- **The Benefit:** This builds muscle memory. Your family gets used to holding the button, waiting a second, and speaking clearly. If you only bring the radios out during a disaster, they will fumble.

Mode 2: Elevated Alert (Yellow)

- **Status:** Severe Weather Watch, Local Power Outage, Civil Unrest nearby, or "The Cell Network is acting weird."
- **Radio Use:** Monitoring and Readiness.
- **Action Steps:**
  - **Net Control:** Turns on the "Home Base" radio (the one with the good antenna) to scan local repeaters, NOAA weather, and emergency frequencies.
  - **The Team:** Ensures all handhelds are placed in their chargers or have fresh batteries inserted.
  - **The Check-In:** "Everyone take a radio. Keep it on Channel 1. Check in once an hour on the hour. Keep it quiet otherwise."

Mode 3: Emergency Mode (Red)

- **Status:** Cell towers dead. Disaster declared. Family separated. Immediate threat.
- **Radio Use:** Primary Command & Control.
- **Action Steps:**
  - **Deploy the Big Gun:** Connect the Mag Mount antenna (on a cookie sheet or roof) to the main radio for maximum range.
  - **Initiate "Pizza Code":** (See Chapter 8). Use code words to verify safety.
  - **Net Control:** Stays on the radio 24/7 (or runs shifts with the XO).
  - **Silence Rule:** Unless you have "Traffic" (important info), stay off the air to save batteries and keep the frequency clear. No "Just checking in" unless scheduled.

SECTION 4: GAMIFIED DRILLS (MAKE IT FUN)

If you look at your kids and say "Time for a readiness drill," they will groan and hide. If you say "Let's play Radio Hide-and-Seek," they will fight over who gets the radio. Gamification bypasses the boredom and plants the skills deep in their brains.

Game 1: "Radio Hide-and-Seek" (Signal Logic)

- **The Setup:** Dad is "Base" on the front porch with a radio. Kid is "Rover."
- **The Rules:** The Kid hides somewhere in the neighborhood (within safe boundaries).
- **The Gameplay:** The Kid keys up. *"Base, this is Rover. Can you hear me?"*
  - Dad gives a signal report: *"Loud and clear"* or *"Scratchy, you sound like a robot."*

    - Kid gives a clue: *"I see a red mailbox."*

- **The Physics Lesson:** The kid learns quickly that hiding in a concrete culvert kills the signal, but climbing the playground slide makes it crystal clear. They learn **Line of Sight** intuitively through gameplay, not a lecture.

Game 2: "The Scavenger Hunt" (Pro-Words & Listening)

- **The Setup:** Mom and Dad stay home. Kids go to the park or woods.
- **The Rules:** Dad gives instructions over the radio one at a time. The kids cannot move to the next target until they confirm via radio.
- **The Gameplay:**
    - Dad: *"Rover, target one is a pinecone. Over."*
    - Kid: *"Wilco. Finding pinecone."* ... (Pause) ... *"Base, I have the pinecone."*
    - Dad: *"Copy. Target two: A smooth rock. Over."*
- **The Lesson:** This teaches **Pro-Words** (Copy, Wilco) and listening skills. If they don't listen effectively, they don't find the treasure. It teaches them to wait for the transmission to end before acting.

Game 3: "The Zombie Run" (Stress & Silence)

- **The Setup:** A "Grid Down" simulation during a camping trip.
- **The Scenario:** "Zombies" (Dad) are patrolling the woods. Mom and Kids must navigate to a "Safe Zone" (the car) using only the radio to coordinate their movement without getting caught.
- **The Rules:**
    - They must split up (Flanking).
    - They must whisper (Volume discipline).
    - They must use code words ("Zombies are at the fire pit" = Dad is at the campfire).
    - If Dad hears them talking on his radio, he "catches" them (mimics interception).
- **The Lesson:** This teaches **Radio Discipline** (whispering, brevity, codes) and OPSEC under mild adrenaline. It teaches them that the radio is a tool for evasion, not just chatting.

SECTION 5: AFTER-ACTION REVIEW (AAR)

After every trip, game, or actual storm, ask three questions at dinner. This is not a lecture; it is a debrief. Do not skip this.

1. **What broke?** (Did a battery die unexpectedly? Did a belt clip snap? Did the cheap headset hurt your ears?)
2. **What was confusing?** (Did we mix up Channel 1 and Channel 2? Did you not understand what "Wilco" meant?)
3. **What do we fix?** (Buy a spare battery. Laminated the cheat sheet. Practice the phonetic alphabet more).

The Golden Rule of the AAR: **Blame the process, not the person.** If your wife couldn't figure out how to unlock the keypad, don't say "You should have studied." Say "The process failed. We need to disable the keypad lock or make a better checklist."

WORKBOOK TOOLS

The Family Comms Card

Fill this out, laminate it using clear packing tape, and zip-tie it inside every backpack and go-bag.

PRIMARY CHANNEL: 1 (Home - Simplex) **BACKUP CHANNEL:** 5 (Blue - Alt Simplex) **EMERGENCY (Calling for Help):** Channel 0 (National Call)

IF LOST:

1. **STOP.** Stop moving. Sit down.
2. Turn Radio **ON** (Clockwise click).
3. Turn Volume **UP** (Halfway).
4. Hold PTT button, count "1 Mississippi", then say: *"This is [Name]. I am lost. I am at [Landmark]. Over."*
5. Let go and **LISTEN**. Wait 30 seconds before trying again.

DAD'S PHONE: 555-0199 **MOM'S PHONE:** 555-0188

The "Net Control" Handoff Checklist

For the Spouse/XO. Tape this to the "Base Station" radio.

- [ ] **Trigger:** Dad is unresponsive for 15 minutes OR Dad is injured.
- [ ] **Take Command:** Pick up the Base Radio.
- [ ] **The Call:** "All stations, this is [Name]. I am now Net Control. Report in."
- [ ] **Roll Call:** Wait for each child/unit to respond.
- [ ] **Write Down:** Write down the location of all units.
- [ ] **Decide:** Do we stay put (Shelter in Place) or move to the Rally Point?
- [ ] **Broadcast:** Give the order clearly. "All units, return to base immediately."

Chapter 10

TROUBLESHOOTING LIKE A FIELD TECH

THE DIAGNOSTIC PLAYBOOK

THE RADIO WORKED YESTERDAY. Today, you press the button, and nothing happens. Or worse, you hear people talking, but they can't hear you.

Most people panic. They assume the radio is "broken." They start mashing random buttons, hoping to stumble upon a fix. They perform a "Factory Reset," wiping out all their carefully programmed channels, only to find the problem is still there. Finally, they throw the radio in the "Junk Drawer of Shame" and assume they aren't smart enough for this hobby.

Stop.

A radio system is a chain. Audio enters the mic > converts to electricity > modulates into RF > travels up the coax > radiates off the antenna > flies through the air.

If the system fails, it is because **one specific link** in that chain is broken. Your job is not to be a wizard; your job is to find the broken link. This is called **Diagnostics**. It is the difference between a user who complains and an operator who fixes.

In this chapter, we are going to turn you into a Field Technician. We will replace guessing with a structured workflow that isolates the problem in seconds.

SECTION 1: THE 60-SECOND TRIAGE (THE "OODA LOOP")

Before you open a menu, connect a laptop, or start tearing things apart, look at the physical radio. 90% of failures are mechanical, not digital. Run this **OODA Loop** (Observe, Orient, Decide, Act) before you panic.

The "Tactical Dad" Triage Loop:

1. **Is it ON?**
   - *The Check:* Turn the volume knob clockwise. Did you feel the physical "click" past the detent? Did the screen light up?
   - *The Trap:* Sometimes the knob gets bumped to "just barely off" inside a backpack. Or, the knob is turned on, but the volume is set so low it *appears* off. Turn it up to 50%.
2. **Is it FED?**
   - *The Check:* Hold the 0 key for two seconds. The screen will display the voltage (e.g., 8.2V).

   - *The Standard:*
      - **8.0V - 8.4V:** Full Charge. Green light.
      - **7.4V - 7.9V:** Functional. Yellow light.
      - **Below 7.4V:** The Danger Zone. The radio might turn on and receive audio, but the moment you press PTT (which demands high current), the voltage will sag, and the processor will reboot or the transmission will cut out. It is effectively dead.

3. **Is it CONNECTED?**
   - *The Antenna Check:* Grab the antenna near the base. Twist it clockwise. Was it loose? A loose antenna can burn out your transmitter because the energy has nowhere to go.
   - *The Mic Check (The K1 Gap):* If using a hand mic, push the plug in hard. The "K1" connector is notorious for wiggling loose by 1mm. This creates a "Zombie State" where the speaker pins connect (you hear static), but the mic audio pin disconnects (you transmit silence). Push it until it snaps. *Field Fix:* If it keeps popping out, use your pocket knife to carefully shave 1mm of plastic off the base of the plug. This allows it to seat deeper into the radio body.
4. **Is it UNLOCKED?**
   - *The Check:* Look at the screen. Is there a tiny (Key icon) or L visible?
   - *The Symptom:* You press buttons to change the channel, but nothing happens. You panic, thinking the keypad is dead.
   - *The Fix:* Press and hold the # (Hash/Key) button for 2-3 seconds until the icon disappears. This feature exists to stop your hip from changing channels, but it often confuses new users.

If those four pass, the hardware is alive. Now we hunt for the software glitch.

SECTION 2: SYMPTOM FLOWCHARTS (LOGIC TREES)

Don't guess. Follow the path. By asking binary (Yes/No) questions, we eliminate variables.

Symptom 1: "I Can't Hear Anything" (The Deaf Radio)

Scenario: You are standing next to your buddy. His radio is chattering away with traffic. Your radio is sitting there in silence.

- **Step 1: Check the Volume.**
   - Turn the knob all the way up. Sometimes "Low" is actually "Silent" on cheap potentiometers. (Yes > Next).
- **Step 2: Check the Squelch (The Gate).**
   - Press and hold the MONI (Monitor) button on the side (usually the small button below PTT). This bypasses all programming and forces the speaker open.
   - *Result A (Silence):* If you hold the button and hear **absolute silence**, your speaker wire is broken or your audio amplifier is dead. The radio is physically broken. Swap it.
   - *Result B (Static):* If you hear a loud **SSSSHHHHH**, your speaker works. The radio is receiving. The problem is a setting that is keeping the "Gate" closed. > Next.

- **Step 3: Check the Tone Squelch (R-CTCS).**
  - Look closely at the left side of the screen. Do you see a tiny "CT" or "DCS" icon?
  - *The Problem:* You have a "Receive Tone" set. Your radio is politely waiting for a specific password (e.g., Tone 103.5) before it unmutes. The people talking aren't sending that password, so your radio ignores them.
  - *The Fix:* Go to Menu 11 (R-CTCS) and Menu 10 (R-DCS). Turn them both to **OFF**. You want to hear everything.
- **Step 4: Check the Frequency.**
  - Look at the small numbers (e.g., 146.520). Are you on **exactly** the same frequency as your buddy? A difference of 0.005 MHz is enough to make you deaf or make voices sound like "Duck Talk."
- **Step 5: Check the Active Band (The Arrow).**
  - Look for the tiny black arrow (►) on the left side of the screen. Is it pointing to the Top Row (A) or Bottom Row (B)?
  - *The Trap:* If you are programmed on Top (A), but the arrow is on Bottom (B) and Dual Watch is OFF, you will not hear anything on A.
  - *The Fix:* Press the EXIT/AB button to move the arrow to the correct line.

Symptom 2: "I Can Hear, But Can't Transmit" (The Mute Radio)

Scenario: You press PTT to talk. You shout your message. Nobody answers. They continue talking as if you don't exist.

- **Step 1: Check the Indicator Light.**
  - Press the PTT button. Does the little LED light on the front turn **RED**?
  - *NO:* Your PTT button is physically broken or the battery is too dead to fire the transmitter circuit.
  - *YES:* The radio is generating RF power. The signal is leaving the building. > Next.
- **Step 2: Check the Offset (Repeater vs. Simplex).**
  - Are you trying to talk "Simplex" (direct) to a friend?
  - Look at the top of the screen. Do you see a tiny + or - icon?
  - *The Problem:* You are listening on 146.520, but when you press the button, the radio jumps to 146.580 (the offset). Your friend is still listening on .520. You are shouting into the void on a different frequency.
  - *The Fix:* Go to Menu 25 (SFT-D) and turn it **OFF**.
- **Step 3: Check the Mic (The Mechanical Failure).**
  - Are you using a headset or hand mic?
  - *The Test:* **Unplug it.** Throw it on the seat. Pick up the radio and talk directly into the built-in mic.
  - *The Result:* If they can hear you now, your headset wire is broken internally. This is the #1 cause of transmit failure in cheap gear. The wires inside those $10 headsets are as thin as hair and break if you look at them wrong.

Symptom 3: "I Can Hit Simplex, But Not the Repeater"

Scenario: You can talk to your wife in the driveway perfectly. But when you try to hit the "County Repeater" 10 miles away, you get nothing.

- **Step 1: The "Kerchunk" Test.**
    - Tap the PTT button for 1 second and release. Listen immediately.
    - *The Sound:* Do you hear a burst of static or a "beep" return immediately after you let go? That is the "Repeater Tail." It means the repeater heard you and kept its transmitter open for a second.
    - *NO (Silence):* You are not opening the repeater. The door is locked. > Next.
- **Step 2: Check the Encode Tone (T-CTCS).**
    - Almost all repeaters require a "Tone Key" (CTCSS) to open. If you don't send the key, they ignore you.
    - Go to Menu 13 (T-CTCS). Is it set to **OFF**?
    - *The Fix:* Look up the repeater on RepeaterBook. Find the "Tone" (e.g., 103.5). Enter that value into Menu 13.
- **Step 3: Check the Offset Direction.**
    - Are you shifting the right way?
    - *The Problem:* The repeater listens on a frequency *lower* than its output (-), but you are transmitting *higher* (+). You are knocking on the back door instead of the front door.
    - *The Fix:* Check the database. Toggle Menu 25 (SFT-D) between + and -. Try again.
- **Step 4: The 'Reverse' Check.**
    - While on the repeater channel, tap the * SCAN button once (do not hold). An R will appear on the screen.
    - *What it does:* It forces your radio to listen to the *Input* frequency (what you normally talk on).
    - *The Test:* If you hear other people talking clearly while in Reverse mode, but you can't hit the repeater, the repeater might be down (or you are too close to the other person). If you hear nothing, no one is talking.

SECTION 3: THE "DESENSE" MYSTERY (WHY CITIES KILL RADIOS)

This is a specific, maddening problem that plagues the Baofeng UV-5R because of its low cost. It confuses new users more than anything else.

The Symptom: You are standing in the middle of a city. You have a great antenna. You should be hearing everything. Instead, you hear **nothing**. The radio is totally silent. Occasionally, it might "burp" a blast of static, then go silent again.

The Cause: "Front End Overload" (Desensitization). The receiver chip in a Baofeng (RDA1846) is like a very sensitive ear. It listens to everything at once. If you are standing near a powerful Pager Tower, FM Radio Station, or Cell Tower, that strong signal screams into the Baofeng's ear. The radio reacts by "putting its fingers in its ears" (Automatic Gain Control). It turns its sensitivity down to zero to protect itself. **Result:** It can't hear the strong tower (because it's on a different frequency), but now it *also* can't hear your friend whispering on the ham band. It is deafened by the noise floor.

The Diagnostic Trick: "The Antenna Twist"

1. Tune to the frequency you want to hear (e.g., NOAA weather or a busy repeater).

2. Unscrew your antenna.

3. Slowly lift the antenna off the connector so the center pin *barely* touches (or disconnects entirely, just hovering near it).

4. **Result:** If you suddenly hear the voice clearly for a split second, your radio was overloaded.

    ◦ *Why this works:* By disconnecting the antenna, you acted as an "Attenuator", you blocked the massive noise source (the local tower), allowing the radio to unplug its ears. Because the radio has surprisingly good sensitivity, it can still hear your friend through the tiny gap, but the noise source is too weak to trigger the overload.

The Fix: You cannot fix the chip. You have to move away from the interference source (the radio tower on the building next door) or use a better radio (superheterodyne) for city environments.

SECTION 4: YOUR SPARES KIT (MODULAR SWAPS)

In the field, you don't solder. You don't rewrite code. You **Swap**. If a part fails, you replace the unit and move on. Troubleshooting in the rain is miserable; swapping parts is fast.

The "Tactical Dad" Field Repair Kit: *(Keep this in a quart-sized Ziploc bag in your Go-Bag)*

1. **Spare Battery (USB-C):** The #1 failure point is lack of power. A dead battery looks exactly like a broken radio.

2. **Spare Antenna (The "Rubber Duck"):** If you trip and snap your fancy 15-inch whip antenna, or the connector bends, you need a backup. Keep the original stubby antenna in the bag. It's better than nothing.

3. **Spare Mic/Earpiece:** The wires on cheap headsets break internally constantly. Carry a cheap spare. If your main mic dies, you swap it in 10 seconds.

4. **The "Rescue Cable":** A USB programming cable.

    ◦ *Why:* Sometimes, if a Baofeng battery dies completely while scanning, the radio's CPU glitches. It might reset to Chinese language or lose all memories. The cable lets you re-upload your "Known Good" image from your laptop to restore the brain.

5. **The "Cheat Sheet":** A printed list of your frequencies. If the radio dies and you have to use a friend's radio, you need to know the numbers to type in.

The Monthly Maintenance Drill:

- **Inspect:** Look at the gold contacts on the back of the battery. Are they shiny? If they are green, dull, or blackened, clean them with a **pencil eraser**. The rubber acts as a gentle abrasive to remove oxidation without scratching the gold plating.

- **Tighten:** Check the two tiny screws on the belt clip. Check the SMA antenna connector. Vibration loosens everything. Finger-tighten them.

- **Charge:** Top off the spare battery. Lithium batteries lose a tiny bit of charge every month.

BREAKOUT: THE PRO-TOOLS (FOR THE NERDS)

You don't *need* these to be effective, but they remove the mystery and let you see the invisible.

1. The SWR Meter (Surecom SW-33)

- *What it does:* It measures "Standing Wave Ratio." It tells you if your antenna is actually radiating power into the air, or reflecting it back down the wire into the radio (which creates heat).
- *Why get it:* If you build your own antennas or use mag mounts on a car, this proves they are safe. A high SWR (over 3:1) means your antenna is broken or untuned and will burn out your transmitter.

2. The Signal Stick (SDR Dongle)

- *What it does:* A $25 USB stick that plugs into a laptop. It shows a "waterfall" visual of radio signals on the screen.
- *Why get it:* You can visually *see* your radio's signal. You can see if it is "splattering" (transmitting messy noise on adjacent channels), which is a common failure mode for dying Baofengs. It also lets you see if you are transmitting off-frequency.

3. The Multimeter

- *What it does:* Checks voltage and continuity.
- *The Hack:* Use the "Continuity" (Beep) mode to test headset cables. Touch one probe to the plug and one to the earpiece wire. Wiggle the wire. If the beeping stops, the wire is broken inside the insulation. This finds "intermittent" failures that drive you crazy.

ACTION STEPS & WORKBOOK

Printable Field Flowchart

Print this small (index card size) and laminate it. Tape it to your radio pouch.

START: Radio won't TX/RX.

1. **Check Power:**
    - Hold 0 key. Is Voltage > 7.4V?
    - **NO:** Swap Battery immediately.
    - **YES:** Go to Step 2.
2. **Check Physical:**
    - Is Antenna tight? Is Mic plug fully seated?
    - **NO:** Tighten/Push.
    - **YES:** Go to Step 3.
3. **Check Audio:**
    - Press MONI button. Do you hear static?

- **NO:** Audio Amp dead. Swap Radio.
- **YES:** Go to Step 4.

4. **Check Settings (The "Deaf" Fix):**
    - Are Tone Squelch icons (CT/DCS) visible on screen?
    - **YES:** Turn Menu 10 & 11 to OFF.
    - **NO:** Go to Step 5.
5. **Check Transmission (The "Mute" Fix):**
    - Press PTT. Does RED light turn on?
    - **NO:** Check PTT button / Battery.
    - **YES:** Check Offset.
        - Simplex? Ensure + / - icons are GONE (Menu 25 OFF).
        - Repeater? Ensure + / - matches RepeaterBook.
6. **Still Dead?**
    - Perform **Antenna Twist Test** (Check for City Overload).
    - Unplug Hand Mic (Check for bad wire).
    - **Last Resort:** Menu 40 (RESET ALL). *Warning: Deletes everything.*

Monthly Maintenance Checklist

- **Physical Check:** Antenna connector straight? Belt clip screws tight? Volume knob smooth?
- **Battery Check:** Voltage check on primary radio + spare battery in bag.
- **Contact Cleaning:** Use pencil eraser on battery contacts.
- **Audio Check:** Call a friend/family member. Ask: "How is my audio? Am I scratchy? Am I quiet?"
- **Program Check:** Did you accidentally delete a channel? Scroll through your zones to verify memory map.

Chapter 11

THE UPGRADE PATH

WHEN TO KEEP THE BAOFENG, WHEN TO STEP UP

YOU HAVE MASTERED THE Baofeng. You have a working channel plan, you know how to troubleshoot a dead mic, and you have made contact with a repeater 20 miles away. You feel competent. You feel like an operator.

But you have also noticed the cracks.

You noticed that when you drive near the hospital or a cell tower, your radio goes deaf right when you need it most. You noticed that the scanning speed is painfully slow, you miss entire conversations while the radio chugs through your list of 20 channels. You dropped it on the driveway, and the battery latch cracked, requiring a tactical application of duct tape to keep the power on.

You are ready for the next level.

The temptation is to go online and buy the "Baofeng Pro Max Ultra V2" with 10 Watts, a camouflaged case, and a gold-plated keypad. **Don't do it.** Buying a slightly more expensive Baofeng is like putting a spoiler and racing stripes on a golf cart. It might look cool, but underneath, it is still a golf cart. It has the same engine, the same flaws, and the same limitations.

This chapter is your roadmap out of the "Budget Tier." We are going to look at upgrades that actually increase your **capability**, not just your credit card bill. We will focus on tools that solve specific problems.

SECTION 1: THE CHEAP UPGRADES (MAXIMIZING THE $25 RADIO)

Before you drop $200 on a new radio, make sure you have wrung every ounce of performance out of the one you have. Often, a $20 accessory fixes the problem you thought required a new radio.

The $20 Upgrade: The Signal Stick Antenna

- *The Gain:* Better range, less durability anxiety.
- *The Why:* As discussed in Chapter 5, the stock antenna is an electrical compromise. It is often untuned, meaning a portion of your battery power is converted into heat inside the antenna coil rather than radio waves in the air.
- *The Pick:* The **Super-Elastic Signal Stick**. It is made of Nitinol (memory alloy). You can tie it in a knot, stuff it in a cargo pocket, or bend it in half in your go-bag. When you pull it out, it snaps straight instantly. Unlike the stock antenna, it won't kink. Unlike the heavy "tactical folding" tape-measure antennas, it is lightweight and won't snap the SMA connector off your radio circuit board.

The $25 Upgrade: The Extended Battery (USB-C)

- *The Gain:* Endurance and charging flexibility.

- *The Why:* The stock charging cradle is a logistical choke point. It requires 110V AC wall power. If the grid is down, or if you are in your truck, that cradle is a paperweight unless you have an inverter.
- *The Fix:* A 3800mAh extended battery with a **direct USB-C port**. This changes the radio from a "toy" to a "tool." You can charge it in your truck, from a solar panel, or from the same Anker power bank you use for your phone. It effectively gives you infinite field endurance without carrying a dedicated charger brick.

The $15 Upgrade: A Better Mic

- *The Gain:* Audio clarity and command presence.
- *The Problem:* The cheap "earbud" style earpieces included with the radio break internally if you look at them wrong. The stock internal mic picks up wind noise easily.
- *The Pick:* The **CommMountain** or **BTech** heavy-duty speaker mic. It has a larger speaker diaphragm (louder, richer audio), a clickier PTT button (tactile feedback even with gloves), and a thicker, coiled cable that resists fraying. Clipping this to your collar keeps the radio protected on your belt while keeping the audio near your ear, essential for "Gray Man" operations where you don't want to be holding a radio in your hand.

SECTION 2: WHEN YOU HAVE OUTGROWN THE BAOFENG

How do you know it's time to spend real money? When the **hardware** limits your **mission**. If you find yourself saying, "I can't hear them because of the interference," or "I'm afraid to take this in the rain," it is time to upgrade.

The "Superheterodyne" Factor (The Big Reason)

This is the technical term you need to know. It is the single biggest difference between a $25 radio and a $150 radio.

- **The Baofeng (Direct Conversion Receiver):** This uses a "System on a Chip" (RDA1846). Imagine you are trying to listen to a friend whisper in a crowded stadium. A Direct Conversion chip hears *everything*, the whisper, the shouting match next door, the band playing, and the traffic outside, all at once. It tries to filter it digitally, but if the overall noise is too loud, the chip gets overwhelmed ("Desense"). It effectively goes deaf to protect itself.
- **The Upgrade (Superheterodyne Receiver):** This is like wearing industrial noise-canceling headphones that physically block out every sound frequency *except* the exact pitch of your friend's whisper. It uses physical filters (crystals and coils) to reject interference *before* it hits the brain of the radio.

You need a Superheterodyne radio if:

1. **You live in a dense urban environment (RF Soup):** If you are near cell towers, hospitals, or paging transmitters, a Baofeng is useless. It will be silent even when people are talking. A Superhet will hear clearly.
2. **You work Public Service events:** If you volunteer for marathons, bike races, or parades where 50 people are transmitting nearby on different frequencies, a Baofeng will lock up.
3. **Life-Safety Reliability:** If you are relying on this radio to call for a medevac or fire support, you don't want it to go deaf just because you parked next to a radio tower.

Durability and Water

The UV-5R is "splash resistant" at best. If you drop it in a creek, it dies. If you drop it on concrete, the battery latch often shears off, ejecting the battery and resetting the radio. **You need a better radio if:**

1. **Water:** You kayak, fish, or hunt in the rain.
2. **Impact:** You are clumsy (or your kids are).

3. **Environment:** You want a radio that can survive being tossed into the bed of a truck with firewood.

SECTION 3: THE GOLD STANDARD HANDHELDS (BUY ONCE, CRY ONCE)

If you are going to spend $150+, buy a radio that will last 20 years. Do not buy midway "upgrades" that are just rebranded Baofengs with a different plastic shell.

The Tank: Yaesu FT-60R (~$150)

This is the AK-47 of ham radio. It was released in 2004 and is still sold today because it is perfect.

- **Pros:** True Superheterodyne receiver (bulletproof in cities). Incredibly durable (people run them over with trucks and they still work). It includes an AA battery shell, so you can power it with Energizers bought from a gas station in an emergency. Huge aftermarket support.
- **Cons:** Heavy. Old-school NiMH battery (not Lithium), so it discharges faster on the shelf if left for months. No USB charging out of the box (requires a specific cable). The screen is small and orange.
- **Verdict:** The ultimate survival radio. It just works. If the world ends, this is the radio you want in your hand.

The Modern Workhorse: Wouxun KG-UV9D Mate (~$180)

- **Pros:** Superheterodyne. Large color screen. Holds a huge battery. **Cross-Band Repeat:** This is a killer feature. You can leave this radio in your car (connected to a mag mount antenna on the roof) and use a small Baofeng to talk to it. It will receive your weak signal and "repeat" it out to the world with 10 watts of power. It turns your car into a mobile repeater station.
- **Cons:** The menu system is complex. It has a learning curve compared to the simplicity of the Yaesu.
- **Verdict:** The best "modern" analog handheld for power users who want features, performance, and the ability to extend their range via cross-band repeat.

The GMRS Specialist: Wouxun KG-935G (~$170)

- **Pros:** IP66 waterproof (can take a hose spray or heavy rain). Specifically tuned for GMRS frequencies. Incredibly loud speaker (cuts through road noise or wind). Simple, locked-down interface that is hard to mess up.
- **Cons:** Locked to GMRS frequencies (legal requirement). You cannot transmit on Ham frequencies.
- **Verdict:** If your family uses GMRS exclusively (no Ham licenses), this is the Rolls Royce. It feels like a Motorola police radio in your hand.

SECTION 4: THE "MOBILE" LEAP (THE SMARTER SECOND PURCHASE)

Most people buy a Baofeng, then immediately buy a better handheld. This is often a strategic mistake. **Your second radio should be a Mobile Radio (Truck Radio).**

A handheld is a tactical compromise (low power, poor antenna location). A mobile radio is a strategic asset.

Why a Mobile?

- **Power:** Handheld = 5 Watts. Mobile = 50 Watts. You are **10x louder**. You can hit repeaters 50-70 miles away that your handheld doesn't even know exist. You can punch through trees and terrain that stop a Baofeng cold.
- **Antenna:** It uses a dedicated antenna on the roof, solving the "Faraday Cage" issue permanently. Height + Power =

Range.

- **Power Source:** It runs off your car battery. You have weeks of runtime (if the engine runs) compared to hours on a battery pack.

The Recommendations

- **The Budget Beast: Midland MXT400 (GMRS) / Retevis RT95 (Ham).** Simple, 25-50 watts, no screen clutter. Rugged and cheap (~$150). Great for mounting under a seat or in a glovebox.
- **The Gold Standard: Icom IC-2730A (Ham).** A dual-band workhorse with a **detachable faceplate**. You can mount the heavy radio body under the seat and put the small screen/knobs on your dashboard using a suction cup or vent mount. It scans insanely fast and allows you to monitor two frequencies simultaneously (Dual Receive). (~$300).

The Tactical Dad Strategy: Put the 50-Watt mobile in the "Family Escape Vehicle" (the SUV or Truck). If you have to evacuate, that radio is your lifeline to the outside world when cell towers fail. Keep the Baofengs for "last mile" scouting when you have to leave the vehicle.

SECTION 5: HOW TO UPGRADE WITHOUT WASTING MONEY

Don't end up with a drawer full of incompatible junk. The "hidden cost" of upgrading is often the accessories.

1. Connector Standardization (The Gender Trap) Radio manufacturers do not agree on connectors.

- **Baofeng:** Uses **Kenwood K1** audio plugs (2-pin) and **SMA-Female** antennas (Radio has a pin).
- **Yaesu/Icom:** Use **Single-pin** or specialized audio plugs and **SMA-Male** antennas (Radio has a hole).
- **The Trap:** Your collection of Baofeng mics and antennas *will not fit* a Yaesu. You cannot just swap them.
- **The Fix:** Factor the cost of a new mic and antenna into your purchase. Or, buy **SMA-Male to SMA-Female adapters** (use sparingly, as they weaken the signal). When you pick a "System" (e.g., Yaesu), stick with it so your accessories transmit across your fleet.

2. The "Buy Once" Rule It is better to have **one** $150 radio that works 100% of the time than **six** $25 radios that work 50% of the time. Start with one quality upgrade for yourself (Net Control). Keep the Baofengs as loaners/backups for the kids, neighbors, or cache supplies. Do not buy a "slightly better" Chinese radio hoping it will be "good enough." Make the jump to the Big Three (Yaesu, Icom, Kenwood) or high-end Wouxun and never look back.

SECTION 6: THE LONG GAME (COMMUNITY & LICENSING)

Gear is fun, but skills save lives. A 50-watt radio is useless if you don't know who to talk to.

Get Your General License

The Technician license (Level 1) gets you local VHF/UHF. This is "Line of Sight." The **General license** (Level 2) gets you HF (High Frequency). This allows you to talk **around the world** by bouncing signals off the ionosphere.

- *Why:* In a regional disaster (hurricane wiping out the whole state), VHF can't reach help because the local repeaters are down or out of power. HF can reach a guy in Ohio or Texas who can call the Coast Guard for you. It is the ultimate grid-down backup.

Join a Local Net (Without Being Weird)

Find a local "repeater net" (a weekly on-air meeting). Look on RepeaterBook or local club websites for schedules.

- **Don't:** Barge in asking about "SHTF scenarios," "Zombie defense," or "what gun is best." You will be ignored and labeled as "that guy."
- **Do:** Listen first. Check in when asked. Say "This is [Callsign], checking in, no traffic."
- **The Value:** These guys know where the dead spots are in your county. They know which repeaters have battery backup and which ones die in an hour. They are your local intelligence network. "Elmers" (mentors) love helping new guys who are humble, polite, and willing to learn.

WORKBOOK TOOLS

The Upgrade Decision Matrix

Check the boxes that apply to you to find your next purchase.

- [ ] **The Urbanite:** I live in a dense city center or near a radio tower. My Baofeng often goes silent. (**Need: Superheterodyne Handheld like Yaesu FT-60R**).
- [ ] **The Commuter:** I need to communicate from inside a vehicle to a base > 5 miles away. I spend a lot of time in my truck. (**Need: 50W Mobile Radio**).
- [ ] **The Outdoorsman:** My radio will definitely get wet, dropped in mud, or abused. (**Need: IP67 Waterproof Radio like Wouxun KG-935G**).
- [ ] **The Park Dad:** I just want to talk to my kids in the park or at the mall. (**Stick with Baofeng. Spend money on better antennas**).

The "Next 30 Days" Practice Plan

- **Week 1 (Programming):** Program your radio with the "Chapter 2" Channel Plan using CHIRP. Ensure you have a "Simplex" home channel and 3 local repeaters programmed with correct offsets and tones.
- **Week 2 (Range Lab):** Conduct the "Range Reality Test" (Chapter 5). Map your coverage from your house to the grocery store, the school, and the highway on-ramp.
- **Week 3 (Family Drill):** Teach your family the "Pizza Code" (Chapter 8) and run the "Hide and Seek" drill (Chapter 9). Get them comfortable pressing the button and speaking clearly.
- **Week 4 (The Live Test):** Check into a local net. Listen for the start time, wait for the "Visitor Check-In," and put your callsign out there. Break the ice.

CONCLUSION

YOU ARE NOT A BAOFENG OWNER; YOU ARE AN OPERATOR

You have reached the end of the manual. By now, your radio should be programmed, your batteries charged, and your family briefed on the "Pizza Code." You have the hardware. You have the frequency lists. You have the cheat sheets.

But owning a hammer doesn't make you a carpenter. And owning a radio doesn't make you a communicator.

The difference isn't the gear. It's the mindset. It is the shift from "having" a thing to "being" a capability. This conclusion is not just a summary; it is a final calibration of your expectations.

THE TALE OF TWO DADS

Let me tell you a story about two men who live on the same street in the suburbs of Atlanta. Both men consider themselves "prepared." Both men bought the exact same Baofeng UV-5R off Amazon three years ago.

Dad A: The Gear Collector Dad A feels good about his purchase. He put the radio in his "SHTF Box" in the back of the closet, right next to the freeze-dried food and the survival knife. He left it in the original packaging because he wanted it to be "fresh" for the apocalypse. He downloaded a frequency list from the internet but never actually programmed it because the software looked hard and he promised himself he'd "get to it next weekend." He assumes that when the power goes out, he will turn it on, and it will just work like a cell phone.

Fast forward three years. A severe ice storm hits North Georgia. Power lines snap. The cell towers lose battery backup after four hours. The house is dark, the temperature is dropping, and his wife is stuck somewhere on the highway. Dad A tears open the box by candlelight. He tries to turn the radio on. Nothing. The battery is dead from three years of self-discharge. He frantically digs for the AA battery pack and spills the alkaline batteries on the floor, cursing as he realizes one has leaked white crust inside the case, but manages to scrape it clean and get it to light up. He hears static. He presses the PTT button to call his wife. Nothing happens. He didn't set the offset. He didn't check the frequency.

He shouts into a dead microphone while his kids look at him, scared and cold. His wife asks, "Does it work?" and he has to admit, "I don't know." He owns a radio, but he has zero capability.

Dad B: The Operator Dad B isn't a Navy SEAL. He's an accountant. But every Sunday night while taking out the trash, he turns his radio on for 10 minutes. He programmed his local repeater and his simplex home channel using CHIRP. He taught his 10-year-old son how to do a radio check and taped a laminated "Cheat Sheet" to the back of the battery.

When the same ice storm hits, Dad B doesn't panic. He knows the cell towers are failing because he heard the chatter on the local ARES repeater 20 minutes ago. He initiates 'Yellow Alert' status and calls his wife on the "Home" channel before the phones die (she knew to turn her radio on because of the protocol). "Traffic is gridlocked on Hwy 9. Stay put at the store. I am monitoring Channel 1." His wife doesn't panic because she has practiced this. She stays warm in the store. Dad B monitors the weather alerts on his radio and knows exactly when the road crew clears the main street. He guides her home safely using side roads he confirmed were open via radio reports.

Dad B didn't have better gear. He had **proficiency**.

You have spent this entire book becoming Dad B. Do not let that skill atrophy.

THE WEEKLY RHYTHM (15 MINUTES TO STAY SHARP)

Radio skills are perishable. If you don't use it, you lose it. Batteries die. Frequencies change. Settings get bumped in your bag. The "Set it and forget it" mentality is a trap that leads to failure.

You need a rigid schedule. If you wait for "when I have time," it will never happen.

The "Sunday Night Protocol" (1900 Hours) *Tie this to a habit you already have, like taking out the trash or prepping coffee for Monday.*

1. **Power Up (2 Minutes):** Turn on every radio in your fleet. Check the voltage (Hold 0 key).
    - *The Standard:* If it's below 7.6V, put it on the charger immediately. If it's fully charged (8.2V+), rotate it off the charger. Lithium batteries hate sitting at 100% on a charger for months; it kills their capacity.
2. **The Physical Audit (3 Minutes):**
    - Check the belt clip screws (they vibrate loose over time).
    - Check the antenna connector (is it tight?).
    - Check the volume knob (is it smooth, or did it get bent?).
    - *Why:* Finding a broken belt clip now is annoying. Finding it while running to a shelter is a disaster.
3. **The "Live" Check (5 Minutes):**
    - **Receive Test:** Tune to the local NOAA weather station. Verify you can hear clear audio. This proves your speaker works and your antenna is receiving.
    - **Transmit Test:** Tune to your primary Repeater. Press PTT for 1 second ("Kerchunk"). Listen for the tail (the return static/beep). This proves your transmit circuit, tone, and offset are still correct.
4. **The Family Drill (5 Minutes):**
    - Hand a radio to your kid/spouse in the other room.
    - **You:** "Base to Unit 1. Radio Check."
    - **Them:** "Unit 1. Loud and Clear."
    - *Note:* This isn't just testing the radio; it's testing **them**. It keeps them comfortable with the PTT button so they aren't afraid of it in an emergency. It normalizes the sound of your voice over RF.
5. **Stow (3 Minutes):** Turn them off. Return them to the "Ready Rack" or Go-Bag. Verify the spare battery is present and the "Cheat Sheet" is still attached.

Total Time: 15 Minutes. **Result:** 100% confidence that your lifeline works.

THE FINAL EXAM: EARN YOUR "OPERATOR" STATUS

You have read the book. Now prove you can do the work. Complete these three scenarios. Once you check these boxes, you are no longer a "newbie." You are an asset to your family.

Scenario 1: The "Blackout" Simulation (Simplex Mastery)

The Setup: Turn off the main breaker to your house (or just pretend). Unplug your Wi-Fi router. Put your cell phones in a drawer on "Airplane Mode." **The Mission:**

1. Give a radio to a family member and send them to the end of the block (or a nearby park).
2. Establish contact on your **Simplex Home Channel** (e.g., Channel 1).
3. Have them read a random sentence from a book.
4. You must write it down perfectly. If you miss a word, ask for a repeat using pro-words ("Say Again").
5. **The Twist:** Switch to Low Power (1 Watt) by tapping the # key. Can you still hear them? If not, move to the second floor or stand on a chair. **The Lesson:** You will learn that "Height is Might." You will see exactly how walls and terrain eat your signal, and how moving five feet can change "static" to "clear." You learn that 1 Watt from the second floor beats 5 Watts from the basement.

Scenario 2: The "Listening Post" (Scanner Mastery)

The Setup: A Saturday afternoon. **The Mission:**

1. Go to RepeaterBook and find the frequency for your local Fire Dispatch or EMS (if analog) or a busy Ham repeater.
2. Program it into your radio using the keypad (Field Rescue Method), no computer allowed.
3. Listen for 30 minutes.
4. **The Goal:** Write down three specific pieces of information you heard (e.g., "Fire truck dispatched to 123 Main St," "Operator KB4XYZ is testing a new antenna," or "Accident at Mile Marker 42").
5. **The Lesson:** This proves you can extract **intelligence** from the airwaves. In a crisis, information is as valuable as water. Knowing where the fire is allows you to drive the other way.

Scenario 3: The "Grid Down" Message (Repeater Mastery)

The Setup: Drive 5 miles away from your home. Park your car. **The Mission:**

1. Access your local repeater.
2. Wait for a break in traffic. Listen first!
3. Key up and say: "This is 'Your Callsign,' testing a new mobile setup. Can I get a signal report?"
4. **The Goal:** Get a stranger to respond with "Loud and clear."
5. **The Lesson:** This proves your signal can leave your immediate neighborhood. You have touched the outside world. You are not alone. It also breaks the psychological barrier of "Mic Fright." Speaking to a stranger on the air is the final hurdle to becoming an Operator.

FINAL CALL TO ACTION

You bought this radio because you felt a responsibility. You looked at the fragile state of the world, the storms, the outages, the uncertainty, and you decided to do something about it. You decided to be the one who stays connected when the lights go out.

That is a noble instinct. Protect it.

Don't let this radio become just another piece of plastic in a drawer collecting dust. Keep it charged. Keep it programmed. Keep your family practiced.

When the phones die, and the internet goes dark, and the rest of the world is deaf and blind... **you will be the one with the answers.**

Base, this is Tactical Dad. The net is open. Out.

Jared Johnson (KF0RTU)

Jared@MorseCodePublishing.com

APPENDIX A

THE ONE-PAGE CHEATSHEETS

Access printable versions of all appendix documents here:

CHEATSHEET 1: THE "NO-PANIC" RESET

IF THE RADIO ACTS WEIRD:

1. **Check Battery:** Is voltage > 7.4V? (Hold 0).
2. **Check Antenna:** Is it tight?
3. **Check Tones:** Are "CT" or "DCS" icons on the left? Turn them OFF.
4. **Check Offset:** Is there a + or - at the top when you are trying to talk Simplex? Turn Menu 25 OFF.

THE NUCLEAR OPTION (FACTORY RESET):

- *Warning: Erases all channels.*
- Menu 40 (RESET) -> ALL -> Press MENU.
- Change Language back to English: Menu 14 -> ENG.

CHEATSHEET 2:THE TWO BRAINS (VFO vs. MR)

THE GOLDEN RULE:

- **VFO (Frequency Mode):** Used for **Programming** and **Searching**. (Manual input).
- **MR (Channel Mode):** Used for **Talking** and **Scanning**. (Saved presets).

HOW TO SWITCH:

- Press the orange VFO/MR button on the front.
- **IF LOCKED:** Turn radio OFF. Hold MENU button. Turn ON. (Some models).

VISUAL CHECK:

- **CHANNEL MODE:** You see small numbers (001, 022) on the right side of the screen.
- **FREQUENCY MODE:** No small numbers on the right. Just the big frequency.

TROUBLESHOOTING:

- *"I can't type in a new number!"* > You are in **Channel Mode**. Switch to VFO.
- *"I can't change the channel!"* > You are in **VFO Mode**. Switch to MR.
- *"Keypad won't work!"* > Check for Lock icon. Hold # key for 2 seconds.

CHEATSHEET 3: REPEATER MATH (OFFSETS & TONES)

THE LOGIC: Repeaters listen on one frequency (Input) and talk on another (Output). You must tell your radio how to knock on the door.

THE 4 VARIABLES:

1. **RX Freq:** What you listen to (The big number on screen).
2. **Offset:** How far to shift (Distance).
3. **Direction:** Which way to shift (+ or -).
4. **Tone:** The password key (CTCSS/DCS).

MENU SHORTCUTS (UV-5R):

- **Menu 13 (T-CTCS): TRANSMIT TONE.** (The Key). Set this to the repeater's PL Tone (e.g., 103.5).
- **Menu 11 (R-CTCS): RECEIVE TONE.** (The Filter). **KEEP THIS OFF** unless you have a specific reason. It makes you deaf to traffic that doesn't have the tone.
- **Menu 25 (SFT-D): SHIFT DIRECTION.** (+ / - / OFF).
- **Menu 26 (OFFSET): SHIFT AMOUNT.** (0.600 for 2m / 5.000 for 70cm/GMRS).

THE "REVERSE" TRICK:

- Press the * SCAN button quickly (tap).
- If you see an R on screen, you are listening to the **INPUT**.
- *Use:* If you hear the person clearly in Reverse mode, you are close enough to talk Simplex.

CHEATSHEET 4: SCANNING & SQUELCH

SCANNING RULES:

- **Start Scan:** Hold * SCAN button for 2 seconds.
- **Stop Scan:** Press EXIT or PTT.
- **Skip Channel:** In CHIRP, mark "Skip". On radio, delete channel.

SCAN MODES (MENU 18):

- **TO (Time):** Pauses for 5 seconds, then moves on (even if they are talking). *Bad for convos.*
- **CO (Carrier):** Pauses while signal is present. Resumes when signal drops. *Best for monitoring.*
- **SE (Search):** Stops when it finds a signal and stays there. *Good for finding new stuff.*

SQUELCH (MENU 0):

- **0:** Open (Static).
- **1:** Sensitive (Weak signals). *Standard Setting.*
- **5-9:** Insensitive (Strong signals only). *Use in city/interference.*

FM RADIO:

- Press side CALL button once. (Listen to music).
- *Note:* If a signal comes in on your main channel, the radio effectively pauses the music so you can answer.

APPENDIX B

REPEATER WORKSHEET & TRAVEL LIST

INSTRUCTIONS:

1. Go to RepeaterBook.com or use the app.
2. Fill this data out **in pencil** (repeaters change).
3. Keep this folded inside your radio case or taped to your sun visor.
4. **Crucial:** The "Input Tone" is the key to the door. Without it, the frequency is useless.

PART 1: THE "HOME BASE" LIST (CRITICAL INFRASTRUCTURE)

These are the repeaters covering your daily life: Home, Work, School, and the commute in between.

______________________ ______________________

______________________ ______________________

______________________ ______________________

______________________ ______________________

______________________ ______________________

______________________ ______________________

PART 2: THE TRAVEL / CONVOY LIST

Fill this out before a road trip. Look for repeaters along your interstate route every 30-50 miles.

TRIP DESTINATION: ____________________ **DATE:** ________

TRIP DESTINATION: ____________________ **DATE:** ________

TRIP DESTINATION: ____________________ **DATE:** ________

TRIP DESTINATION: ____________________ **DATE:** ________

PART 3: THE SIMPLEX STANDARD (NO REPEATER NEEDED)

Your pre-agreed frequencies for direct radio-to-radio talk.

________________________ ________________________

________________________ ________________________

________________________ ________________________

________________________ ________________________

QUICK REFERENCE: FREQUENCY BANDS

- **2 Meter Ham:** 144.000 – 148.000 MHz
- **70cm Ham:** 420.000 – 450.000 MHz
- **GMRS:** 462.550 – 462.725 MHz (Channels 15-22)
- **MURS:** 151.820 – 154.600 MHz (Green Dot / Blue Dot)
- **NOAA Weather:** 162.400 – 162.550 MHz (Receive Only)

APPENDIX C

CHIRP WORKFLOW CARD & BACKUP LOG

INSTRUCTIONS: Use the "Backup Log" to track your .img files. Trust me, in six months you won't remember if Backup_Final_v2.img is the good one or the broken one.

PART 1: THE CHIRP WORKFLOW CARD (CUT OUT)

THE GOLDEN RULES OF PROGRAMMING

1. **CABLE CHECK:** Plug USB in first. Check Device Manager. No Yellow Triangle? Good.

2. **PHYSICAL CONNECTION:** Turn Radio OFF. Push K1 plug hard until it SNAPS. Turn Radio ON (100% Volume).

3. **READ FIRST:** Radio $\rightarrow$ Download from Radio. *Always do this first.*

4. **SAFETY SAVE:** File $\rightarrow$ Save As $\rightarrow$ Factory_Backup_[Date].img. *Never touch this file again.*

5. **EDIT:** Import from RepeaterBook, copy/paste from other tabs, or manual entry.

6. **WRITE:** Radio $\rightarrow$ Upload to Radio.

7. **VERIFY:** Unplug radio. Turn OFF/ON. Press VFO/MR to Channel Mode. Check Channel 1.

PART 2: PROGRAMMING BACKUP LOG

Track your file versions here. Don't guess.

RADIO MODEL: ______________ **SERIAL/ID:** ______________

RADIO MODEL: ______________ **SERIAL/ID:** ______________

RADIO MODEL: ______________ **SERIAL/ID:** ______________

RADIO MODEL: ______________ **SERIAL/ID:** ______________

PART 3: THE "RESCUE USB" CHECKLIST

Keep a USB thumb drive in your Go-Bag with these files. If the internet dies, you can still program.

- [ ] **CHIRP Installer:** The latest .exe or .dmg file.
- [ ] **Prolific Driver:** The 3.2.0.0 Windows installer (for the cable).
- [ ] **Factory Image:** The original read from your radio.
- [ ] **Current Config:** Your latest "Known Good" .img file.
- [ ] **PDF Reference:** A copy of this book or the RepeaterBook PDF for your state.

APPENDIX D

RANGE TEST LOG & RESULTS TRACKER

INSTRUCTIONS:

1. Print this sheet.
2. Grab a clipboard and a pen.
3. Conduct the test with a partner (Base Station + Rover).
4. **Goal:** Find the "Hard Deck" (the maximum reliable distance) in every direction.

PART 1: TEST PARAMETERS

Variable changes affect results. Record what you are using today.

DATE: ________________ **TIME:** ______________

WEATHER: ______________________________ (Rain/Fog kills UHF!)

PART 2: THE CARDINAL LOG

Drive North, South, East, and West. Check in at landmarks. Use the ***RST System*** *(Readability 1-5, Signal 1-9).*

- **Readability 5:** Perfect.
- **Readability 3:** Understandable but difficult.
- **Readability 1:** Unreadable.

NORTH ROUTE

SOUTH ROUTE

EAST ROUTE

WEST ROUTE

PART 3: RESULTS TRACKER & ADJUSTMENTS

MY "HARD DECK" (Reliable Radius): Test 1

- **North:** ________ Miles
- **South:** ________ Miles
- **East:** _________ Miles
- **West:** _________ Miles

MY "HARD DECK" (Reliable Radius): Test 2

- **North:** ________ Miles
- **South:** ________ Miles
- **East:** _________ Miles
- **West:** _________ Miles

APPENDIX E

COMMS PLAN, DRILL CARDS & AAR

INSTRUCTIONS:

1. **The Wallet Card:** Fill out, fold, laminate, and put in every family member's wallet or backpack.
2. **The Drill Cards:** Cut these out. Draw one from a hat on a Saturday morning to gamify your practice.
3. **The AAR:** Keep a stack of these in your radio binder. Fill one out after every storm or major drill.

PART 1: THE FAMILY COMMS WALLET CARD

(Cut along the outer line. Fold in the middle to make a double-sided card)

SIGNAL PLAN & CONTACTS

PRIMARY CH: ____________________

BACKUP CH: ____________________

EMERGENCY: Channel 0 (National Call)

MOM: ____________________

DAD: ____________________

OUT OF AREA: ____________________

RALLY POINT: ____________________

CODES & PROTOCOLS

CODE: "GET OUT" (LEAVE NOW)
PHRASE: ____________________

CODE: "DURESS" (SEND HELP)
PHRASE: ____________________

CODE: "SILENCE" (DANGER)
PHRASE: ____________________

IF LOST: STOP MOVING.

1. Turn Radio ON. Vol UP.
2. Hold Button. Count "1".
3. Say: **"I am [Name]. I am at [Place]."**

FOLD HERE

PART 2: THE "TACTICAL DAD" DRILL CARDS

(Cut these out. Laminate them. Use dry-erase markers to fill in specifics)

DRILL 1: RADIO HIDE-AND-SEEK (Signal Mapping)

- **Objective:** Teach kids about "Line of Sight" and signal quality.
- **Roles:** Base (Dad), Rover (Kid).
- **The Mission:** Rover hides within 3 blocks. Must check in every 2 minutes.
- **The Test:**
    - [ ] Rover transmits from a low spot (creek/ditch). Base reports signal quality.
    - [ ] Rover transmits from a high spot (slide/hill). Base reports signal quality.
    - [ ] Rover describes landmark. Base confirms.
- **Success Criteria:** Kid understands that *Height = Clarity*.

DRILL 2: THE SCAVENGER HUNT (Pro-Words)

- **Objective:** Teach "Copy," "Wilco," and listening skills.
- **Roles:** Command (Mom/Dad), Team Alpha (Kids).
- **The Mission:** Find 3 objects. Command gives instructions via radio ONLY.
- **The Test:**
    - [ ] Command: "Target 1 is a pinecone. Over."
    - [ ] Team: "Wilco. Searching." ... "Target secured."
    - [ ] Command: "Copy. Target 2 is a red leaf."
- **Success Criteria:** Kids wait for the transmission to end before moving. No interrupting.

DRILL 3: THE **ZOMBIE RUN (Silent Ops)**

- **Objective:** Teach volume discipline and whispering.
- **Roles:** Zombies (Dad), Survivors (Mom/Kids).
- **The Mission:** Survivors must get from Point A to Point B without the Zombie hearing them on the radio or in person.

- **The Rules:**
  - Use Low Power (1W).
  - Use Headphones/Earpieces if available.
  - Whisper transmissions.
  - Use Code Words for the Zombie location.
- **Success Criteria:** Survivors reach the "Safe Zone" undetected.

PART 3: AFTER ACTION REVIEW (AAR) LOG

Fill this out after *every drill or real-world event. Be honest.*

EVENT: ________________________________
DATE: ____________ **DURATION:** ____________
WEATHER: ______________________

HARDWARE FAILURES:

- [ ] Battery died unexpectedly on Unit: ______
- [ ] Antenna loose/broken on Unit: ______
- [ ] Belt clip failed on Unit: ______

ACTION PLAN: What to do differently next time?

APPENDIX F

MAINTENANCE SCHEDULE (MONTHLY/QUARTERLY)

PART 1: THE MONTHLY "READY CHECK" (10 MINUTES)

Goal: Ensure the gear will turn on TODAY if needed.

- [ ] **Voltage Check:** Turn on every radio. Hold 0.
    - *Standard:* If < 7.6V, charge it. If > 8.0V, it's good.
- [ ] **Power Bank Check:** Press the button on your USB Anker/power bricks. Do they show 3-4 lights? Top off if needed.
- [ ] **Physical Tightness:**
    - Twist the antenna. (They loosen in bags).
    - Wiggle the Belt Clip. (Screws back out over time).
    - Check the Volume Knob. (Is it bent/stiff?).
- [ ] **The "Kerchunk" Test:** Transmit on your local repeater for 1 second. Did you hear the tail? (Verifies the transmitter works).

PART 2: THE QUARTERLY "DEEP CLEAN" (30 MINUTES)

Goal: Prevent corrosion and update software. Do this at the start of every season (Spring, Summer, Fall, Winter).

- [ ] **The AA Inspection (CRITICAL):** Open your backup "Clamshell" cases.
  - *Check:* Are the batteries leaking white powder?
  - *Action:* If leaking, throw them out and clean contacts with vinegar. If good, rotate them (put them in the TV remote) and buy fresh ones for the kit.
- [ ] **Contact Scrub:** Remove the main battery. Look at the gold contacts on the radio and the battery.
  - *Action:* Take a **pencil eraser** and scrub the gold contacts until they are shiny. This removes invisible oxidation that causes power failures.
- [ ] **Cable Audit:** Inspect your hand mic and earpiece cables.
  - *Check:* Look for fraying, especially where the wire enters the plug.
  - *Action:* If insulation is cracked, wrap with electrical tape or replace.
- [ ] **Go-Bag Rotation:** Take the radio out of the bag. Turn it on. Make a test call.
  - *Why:* Electronics hate sitting dormant. Capacitors need to charge.
- [ ] **Digital Backup:** Connect the radio to the laptop. Open CHIRP.
  - *Action:* Download the current config. Save it as Backup_[Season]_[Year].img. This ensures you have a copy of any frequency changes you made on the fly.

PART 3: THE ANNUAL "BATTERY AUDIT"

Do this once a year (e.g., New Year's Day).

- [ ] **Capacity Test:** Leave your radio on "Scanning" mode. Time how long it takes to die.
  - *Standard:* If it dies in under 4 hours, the battery is end-of-life. Replace it.
- [ ] **Replace Disposables:** Replace all Lithium AA batteries in your backup kits, even if they aren't dead. Fresh batteries = Peace of mind.

Before You Put This Book Away...

If this book helped you finally understand your Baofeng radio, program your channels, use GMRS more confidently, understand repeaters, or simply feel less intimidated by the little radio in your hand, please consider leaving a review wherever you purchased it.

It does not have to be long. A sentence or two is more than enough.

You might simply share what helped you most:

Did the book make programming easier?

Did it help you understand privacy tones, repeaters, or channels?

Did it give you more confidence for emergency communication?

Did it help your family build a clearer communication plan?

Did it finally make your Baofeng feel useful instead of confusing?

Your review helps more than you might realize. It helps the next beginner decide whether this is the guide that can get them unstuck. It helps someone who is overwhelmed by menus, tones, offsets, licenses, and conflicting online advice find a clear path forward.

And more importantly, it helps grow the radio community.

Because this hobby does not continue through radios alone. It continues through people who welcome beginners.

So when you hear a nervous, shaky voice on your local repeater saying, "This is my first time on the air," be the person who comes back.

Welcome them.

Give them a fantastic signal report.

Tell them they are doing just fine.

That small moment might be the reason they stay in the hobby, keep learning, and become the kind of operator who helps someone else down the road.

Thank you for reading this book, supporting independent publishing, and helping keep practical radio communication alive.

Stay prepared.
Stay courteous.
Stay on the air.

Build Your Library

The Operator's Upgrade Path

You have mastered the Baofeng. You have the hardware, the channel plan, and the mindset. To truly secure your connection to the outside world, you need to expand your legal privileges, your range, and your hardware options.

Here are three guides from **Morse Code Publishing** designed to take you from "Guy with a Radio" to "Communications Commander."

1. GET LEGAL, GET LOUD

Ham Radio Technician Class Study Guide

You know how to use the radio now. But do you have the credentials to use it openly?

Operating "under the radar" is fine for the apocalypse, but it prevents you from practicing on Tuesday afternoons. To truly test your range, join local repeater nets, and build a network of skilled friends, you need your Technician License.

Don't let the exam scare you. You don't need a degree in electrical engineering. This guide strips away the confusing jargon and focuses on the no-nonsense facts you need to pass.

The Hook: Stop worrying about the FCC and start transmitting with authority.

Best For: The operator who wants to join the local ARES/RACES emergency groups.

2. WHEN LOCAL ISN'T ENOUGH

Ham Radio General Class License Study Guide

Your Baofeng is a "Line of Sight" tool. It dominates the neighborhood. But what happens if a hurricane wipes out the power grid for the entire state? What if the local repeaters are dead?

You need to reach the outside world.

Upgrading to your **General Class License** unlocks the High Frequency (HF) bands. This is the "Big League" of radio—bouncing signals off the ionosphere to talk to Texas, Maine, or Europe without satellites or the internet. This guide takes the intimidation out of the General exam, breaking down the math and theory into plain English.

The Hook: Break the line-of-sight barrier. Secure your link to the outside world.

Best For: The prepared citizen who needs regional and national reach.

Complete Combined Glossary

Combined Glossary for Baofeng and GMRS Radio Beginners

1/4 Wave Antenna: A common mobile antenna type. It is compact, simple, and often needs a metal ground plane, such as a vehicle roof, to work well.

5/8 Wave Antenna: A longer mobile antenna that can provide a stronger, lower-angle signal than a 1/4 wave antenna. Often useful for mobile range, but less compact.

70cm Band: The UHF amateur radio band from 420–450 MHz. It is commonly used for local repeaters, handheld radios, and urban communication. GMRS channels also sit in the nearby UHF range.

2m Band: The VHF amateur radio band from 144–148 MHz. It is popular for local ham repeaters, outdoor use, and longer line-of-sight communication in open areas.

Air Time: The amount of time you spend transmitting. Keeping air time short saves battery, reduces repeater congestion, and is good radio etiquette.

AM — Amplitude Modulation: A mode where the strength, or amplitude, of the radio wave changes to carry audio. It is common in aviation and broadcast radio, but not the normal voice mode for most Baofeng/GMRS use.

Analog: A traditional voice signal that varies continuously, like regular FM voice on most Baofeng radios. This is different from digital voice systems such as DMR.

Antenna Gain: A measure of how an antenna focuses radio energy in a certain direction. More gain does not create more power, but it can direct the signal more effectively.

Battery Eliminator: An adapter that lets a handheld radio run from a vehicle's 12-volt power outlet instead of the radio battery.

Birdie: An internal signal or false signal created by the radio itself. It can sound like interference even when no real outside signal is present.

BNC Connector: A quick-connect antenna connector often found on test equipment, scanners, and some radio gear. Baofeng handhelds usually use SMA connectors instead.

Bubble-Pack Radio: A basic walkie-talkie sold in plastic retail packaging, often for FRS/GMRS use. These are simple to use but usually less flexible than programmable radios.

Call Sign: The unique identifier issued by the FCC to a licensed radio operator. Ham operators must use their call signs when transmitting.

Carrier: A radio signal being transmitted. When someone presses PTT, the radio sends a carrier, even before voice is added to it.

Channel: A saved memory slot in the radio that stores a frequency and, when needed, settings such as tone, offset, and power level.

Channel Mode / Memory Mode (MR): The mode where you scroll through saved channels instead of manually typing frequencies. This is the best mode for daily use once your radio is programmed.

CHIRP: Free, open-source software used to program many Baofeng radios from a computer. It is much faster and easier than programming dozens of channels by keypad.

Coax / Coaxial Cable: The cable that connects a radio to an external antenna. Think of it like a hose that carries radio energy from the radio to the antenna.

Control Operator: The licensed person responsible for making sure a station is operated legally. In amateur radio, the control operator is accountable for transmissions from the station.

Courtesy Tone: A short beep sent by many repeaters after someone unkeys. It lets listeners know the repeater heard the transmission and that the other person has stopped talking.

Cross-Band Repeat: A feature on some radios that receives on one band and retransmits on another. Most basic Baofeng handhelds do not provide true cross-band repeat.

CSQ — Carrier Squelch: A setting where the receiver opens for any signal strong enough to break squelch, with no tone required.

CTCSS — Continuous Tone-Coded Squelch System: A low, sub-audible tone sent along with your voice. Repeaters often require it as a "key" to open the repeater. It is often called a PL tone or privacy tone, but it does not make your conversation private.

DCS — Digital Coded Squelch: A digital tone/code used like CTCSS to open a repeater or control squelch. If a repeater listing shows something like D023N, use the DCS menu, not the CTCSS menu.

Decibel (dB): A logarithmic unit used to compare signal strength, gain, loss, and power. In radio, dB helps describe big changes in manageable numbers.

Desense / Desensitization: When a strong nearby signal overloads the receiver and makes it less able to hear weaker signals. Cheap handheld radios can desense near strong transmitters, towers, or interference sources.

Direct: Another word for simplex communication: radio-to-radio without a repeater.

DPL — Digital Private Line: Motorola's trade name for DCS. Despite the word "private," it does not encrypt or hide your conversation.

DTMF — Dual-Tone Multi-Frequency: The touch-tone sounds used by phone keypads. Some radios can send DTMF tones to control certain repeaters or remote systems.

Dummy Load: A test device that safely absorbs radio energy without radiating a signal. It lets technicians test transmitters without actually broadcasting.

Duplex: Communication using two different frequencies, one for transmitting and one for receiving. Repeaters use duplex operation.

Duty Cycle: The percentage of time a radio can safely transmit without overheating. Transmitting creates heat and drains the battery much faster than listening.

Effective Radiated Power (ERP): The practical power sent out by the antenna after accounting for transmitter power, antenna gain, and system losses.

FCC — Federal Communications Commission: The U.S. agency that regulates radio communication, including ham radio, GMRS, FRS, and commercial radio services.

Feedline: The coax cable running from the radio to the antenna.

Ferrite: A material used to reduce unwanted electrical noise or interference on cables. Ferrite chokes are often clipped around power or data cables.

FM — Frequency Modulation: The voice mode used by most Baofeng, GMRS, FRS, MURS, and local VHF/UHF repeaters. In FM, the frequency shifts slightly to carry your voice.

Frequency: The exact "address" on the radio spectrum where communication happens, measured in hertz. Baofeng users often work with MHz, such as 146.520 or 462.550.

Frequency Mode (VFO): The manual mode where you type in a frequency directly. This is the mode you use to explore, test, and program new channels.

FRS — Family Radio Service: A license-free UHF radio service used by many basic walkie-talkies. FRS has lower power and fewer privileges than GMRS.

Full Quieting: A strong, clean received signal with little or no background hiss. If someone says you are "full quieting," your signal sounds excellent.

Gain: The ability of an antenna to focus radio energy in a useful direction. Gain is about focus, not magic power creation.

GMRS — General Mobile Radio Service: A licensed UHF radio service used for families, groups, off-roaders, farms, neighborhoods, and preparedness. In the U.S., GMRS requires an FCC license but no exam.

Ground Plane: A metal surface under an antenna that helps it radiate properly. A vehicle roof can act as a ground plane for many mobile antennas.

Ham Radio / Amateur Radio: A licensed radio service used for hobby communication, emergency communication, experimentation, and public service. In the U.S., amateur radio requires passing an exam.

Handheld / HT — Handheld Transceiver: A portable two-way radio, such as a Baofeng UV-5R. "HT" is common ham shorthand.

Hardline: Heavy-duty coaxial cable designed for permanent installations with low signal loss.

Harmonics: Unwanted signals transmitted at multiples of the intended frequency. Poorly filtered transmitters can create harmonics that interfere with other services.

High Power: The higher transmit-power setting on a handheld radio. It can improve range but drains the battery faster and creates more heat.

Interference: Unwanted noise or signals that disrupt reception. Interference can come from other radios, electronics, power supplies, LED lights, vehicles, or strong nearby transmitters.

Intermod / Intermodulation: Interference caused when two or more strong signals mix together and create unwanted signals inside a receiver.

K-Correction / K-Index: A measure of geomagnetic activity. It matters more for HF propagation than typical Baofeng/GMRS use, but high solar activity can affect radio conditions.

Kerchunk: Pressing PTT briefly to trigger a repeater without speaking. It is commonly done to test whether a repeater can hear you, but it is considered poor etiquette if done repeatedly or without identifying.

Keying Up: Pressing the PTT button to transmit.

Line of Sight (LOS): The basic rule for VHF/UHF communication: signals travel mostly in straight lines. Hills, buildings, trees, and terrain can block or weaken the signal.

Low Band: Lower VHF frequencies, often below the common 2m ham band. Most Baofeng beginner use focuses on VHF high band and UHF instead.

Low Power: A lower transmit-power setting that saves battery and reduces unnecessary signal range. Useful for close-range simplex communication.

Mag Mount: A magnetic antenna mount, often used on a vehicle roof. It provides a convenient temporary antenna setup and can use the vehicle roof as a ground plane.

Memory Mode (MR): The saved-channel mode where you scroll through programmed channels. This is the safest mode for family plans and daily use.

MHz — Megahertz: One million hertz. Most Baofeng and GMRS frequencies are written in MHz, such as 146.520 MHz or 462.675 MHz.

Mobile Radio: A higher-power radio designed to be installed in a vehicle or at a desk. Mobile radios usually offer more power and better audio than handhelds.

Monitor Button (MONI): A side button that opens the squelch so you can hear raw static or weak signals. It is useful for proving the speaker works and checking marginal signals.

MURS — Multi-Use Radio Service: A license-free VHF radio service in the U.S. with limited channels and low power.

Narrowband / Narrow FM (NFM): A narrower FM setting often using 12.5 kHz channel spacing. Many GMRS/FRS and business-style channels use narrowband settings.

NOAA Weather Radio: A continuous weather broadcast service in the U.S. It is one of the best "known-good" signals for testing whether your radio can receive.

Offset: The difference between a repeater's transmit frequency and receive frequency. Your radio listens on one frequency and automatically shifts up or down when you press PTT.

Over: A word used to show you are done speaking and are waiting for the other person to reply.

PL Tone: Motorola's trade name for CTCSS. Many repeater listings and operators still use "PL tone" to mean the required transmit tone.

Prepper Comms: A practical approach to using radios for emergencies, backup communication, family coordination, and grid-down planning.

Privacy Codes / Privacy Tones: Marketing terms often used for CTCSS or DCS tones. These tones do not create privacy, secrecy, or encryption. They only control when a radio opens its speaker.

PTT — Push-to-Talk: The side button you hold down to transmit. Press to talk, release to listen.

Q-Codes: Short radio abbreviations such as QSL or QTH. They are common in ham radio, but beginners can usually use plain English, especially on local FM repeaters.

Radio Check: A short transmission asking whether others can hear you and how your signal sounds.

Range: The distance your radio can communicate. With VHF/UHF handhelds, range depends heavily on height, terrain, buildings, antenna quality, and line of sight.

Receive (RX): Listening to a signal. RX means your radio is receiving.

Repeater: An automated station, often on a tower, hill, or tall building, that receives a weak signal and retransmits it with more power and better coverage.

Repeater Tail: The short burst, beep, or open-squelch sound after someone releases PTT on a repeater. It confirms the repeater was active and gives others a moment to respond.

Reverse: A function that lets you listen to the repeater input frequency. It can help you tell whether another station is close enough to hear directly.

Roger Beep: A beep sent when PTT is released. Many operators dislike it on ham and GMRS because it can be annoying or unnecessary.

RX: Receive. The radio is listening.

Scan: A mode where the radio cycles through channels or frequencies looking for activity.

Scanner: A receiver designed mainly to monitor many channels. A Baofeng can scan, but it is much slower and less capable than a dedicated scanner.

Scrambler: A voice-obscuring feature found on some radios. Scrambling is generally not legal for normal ham/GMRS use and should not be used as a substitute for privacy.

Simplex: Direct radio-to-radio communication on the same frequency without a repeater. Simplex is excellent for short-range, grid-independent communication.

Skip: Long-distance signal travel caused by atmospheric conditions. Skip is common on HF and some lower bands, but it is not something to rely on for Baofeng/GMRS emergency planning.

SMA Connector: The small threaded antenna connector used on many handheld radios. Baofeng radios commonly use an SMA-style connector, so antenna compatibility matters.

Squelch (SQL): A setting that keeps your speaker quiet until a signal is strong enough to open it. Too low means constant static; too high can make the radio seem deaf.

Squelch Tail: The burst of noise heard when a transmission ends before the squelch closes.

Step Size: The spacing the radio uses when tuning frequencies. If the step size is wrong, the radio may "round" the frequency and refuse to land on the number you typed.

SWR — Standing Wave Ratio: A measure of how well antenna energy is leaving the antenna instead of bouncing back toward the radio. High SWR can reduce performance and potentially damage equipment.

Talkaround: Using simplex to talk directly instead of going through a repeater. Useful when stations are close enough to communicate without the tower.

Time-Out Timer (TOT): A safety setting that stops transmission after a set amount of time. It prevents overheating, stuck microphones, and accidental long transmissions.

Tone: A CTCSS or DCS signal used to open a repeater or control squelch. A tone is not encryption.

Tone Scan: A radio function that searches for the CTCSS or DCS tone being used on a signal.

Transceiver: A radio that can both transmit and receive. A Baofeng UV-5R is a transceiver.

Transmit (TX): Sending a signal. TX means your radio is transmitting.

TSQL — Tone Squelch: A setting that requires a tone on received signals before your speaker opens. Beginners should usually leave receive tone settings off unless they have a specific reason.

UHF — Ultra High Frequency: The radio range from 300 MHz to 3 GHz. GMRS and the 70cm ham band are UHF. UHF often works well in towns, buildings, and vehicles but is still line-of-sight.

Unity Gain: An antenna with no gain compared to the reference antenna, often described as 0 dB gain.

VHF — Very High Frequency: The radio range from 30–300 MHz. The 2m ham band is VHF. VHF often performs well outdoors and across open terrain.

VOX — Voice Operated Exchange: A hands-free setting that transmits automatically when it hears your voice. Beginners should usually keep VOX off to avoid accidental transmissions.

Wall Wart: A plug-in power adapter used to power or charge small electronics.

Wideband / Wide FM: A wider FM setting, often using 25 kHz spacing. Some ham repeaters use wideband, while many shared services use narrowband.

Wouxun: A popular Chinese radio brand known for higher-quality GMRS and ham handhelds compared with many bargain radios.

Yagi Antenna: A directional antenna that focuses signal in one direction, often shaped like a small TV antenna. Useful when you want to reach or listen toward a specific location.